SMART GRID STANDARDS

SMART GRID STANDARDS

SPECIFICATIONS, REQUIREMENTS, AND TECHNOLOGIES

Takuro Sato

Waseda University, Japan

Daniel M. Kammen

University of California, Berkeley, USA

Bin Duan

Xiangtan University, China

Martin Macuha

France Telecom Japan Co. Ltd., Japan

Zhenyu Zhou

North China Electric Power University, China

Jun Wu

Shanghai Jiao Tong University, China

Muhammad Tariq

FAST National University of Computer and Emerging Sciences, Pakistan

Solomon Abebe Asfaw

University of California, Berkeley, USA

WILEY

Library of Congress Cataloging-in-Publication Data

Sato, Takuro.
  Smart grid standards : specifications, requirements, and technologies / Takuro Sato, Daniel M. Kammen, Bin Duan, Martin Macuha, Zhenyu Zhou, Jun Wu, Muhammad Tariq, Solomon Abebe Asfaw.
    pages cm
  Includes bibliographical references and index.
  ISBN 978-1-118-65369-2 (cloth)
  1. Smart power grids–Standards. I. Title.
  TK3105.S25 2014
  621.3102′18–dc23

                                                                          2014004867

Typeset in 11/13pt Times by Laserwords Private Limited, Chennai, India

1   2015

Contents

About the Authors

Takuro Sato received B.Eng. and Ph.D. degrees in electronics engineering from Niigata University, Niigata, Japan, in 1973 and 1994, respectively. He was with Research and Development Laboratories, Oki Electric Industry Co., Ltd., Tokyo, Japan, where he worked on pulse-code modulation transmission systems and mobile telephone systems and contributed standardization activities of mobile data transmission for CCITT SG17 from 1983 to 1997, wideband code division multiple access (W-CDMA) system for the Telecommunications Industry Association (TIA)/T1 from 1990 to 1995 and the 3rd Generation Partnership Project (3GPP) in 1995–1996. He was a Professor in the Department of Information and Electronics Engineering, Niigata Institute of Technology in 1995. He established the venture companies Key Stream to provide large-scale-integrated circuits for Wi-Fi systems in 2000 and WiViCom to provide wireless system design in 2001 as industry–academia joint collaboration activities. He has been the Dean of the Graduate School of Global Information and Telecommunication Studies in Waseda University since 2004. His current research is next-generation mobile communication systems, Smart Grid/energy, and social information infrastructure networks. He has been the chairman of the IEICE ICT-SG (Information Communication and Telecommunication in Smart Grid) since 2010. He is a fellow member of the Institute of Electrical and Electronic Engineers (IEEE).

Daniel Kammen is the Class of 1935 Distinguished Professor of Energy with appointments in the Energy and Resources Group, The Goldman School of Public Policy, and the Department of Nuclear Engineering at the University of California, Berkeley. Kammen directs the Renewable and Appropriate Energy Laboratory (RAEL) and the Transportation Sustainability Research Center (TSRC). During 2010–2011, Kammen served as the first Chief Technical Specialist for Renewable Energy and Energy Efficiency. Kammen is the author of over 300 peer-reviewed papers, 50 government reports, and has testified in front of the US House and Senate more than 40 times. He now serves as a Fellow of the US State Department's Energy and Climate Partnership for the Americas (ECPA). Kammen is a coordinating lead author for the Intergovernmental Panel on Climate Change (IPCC) that shared the 2007 Nobel Peace Prize.

Bin Duan received M.Sc. and Ph.D. degrees from Beijing University of Aeronautics and Astronautics (BUAA), China, in 1992 and Xiangtan University, China, in 2004, respectively. Currently, he is a Professor and Associate Dean of College of Information Engineering, Xiangtan University. He was a Visiting Professor at the University of Virginia, US, in 2012. He is the leader of several projects of the National Natural Science Foundation of China (NSFC) in the field of Smart Grid, and the Vice Leader of a Foundation Project of National 863 Plan of China. He is the Vice Dean of Standardization Research Institute for Campus Card of Education Management Information Center, Ministry of Education, China. His research interests include information security, Smart Grid, and software engineering. He has published two books, obtained five patents, and published numerous papers in top Chinese journals and related SCI/EI international journals as well as conferences. He is a senior member of the Chinese Electrotechnical Society. He is also a member of IEEE.

Martin Macuha received his M.Sc. degree in telecommunications from Slovak University of Technology, Bratislava, Slovakia in 2007, and Ph.D. in wireless communication from Waseda University, Tokyo, Japan in 2011. In 2007, he joined Orange Slovakia and in 2008 he was a research student at Waseda University, Tokyo, Japan. In 2011 he was a research associate with Waseda University and since 2012 he has been with Orange Labs in Tokyo, Japan. His research interests include wireless technologies, Smart Grid communication networks, heterogeneous networks, and distributed systems.

Zhenyu Zhou received his M.Sc. degree and Ph.D. in wireless communication from Waseda University, Tokyo, Japan in 2008 and 2011, respectively. From 2011 to 2012, he was the chief researcher in the Department of Technology, KDDI, Tokyo, Japan. From 2012, he has been the Associate Professor, and the Vice Director of Institute of Smart Networking Technology, in the School of Electrical and Electronic Engineering, North China Electric Power University, Beijing, China. His research interests include wireless communication systems, wireless sensor networks, demand response, and data mining for Smart Grid. He is the author of over 30 peer-reviewed papers, and two patents. He is a member of IEEE and IEICE.

Jun Wu was born in Hunan, China. He received his Ph.D. degree from the Graduate School of Global Information and Telecommunication Studies (GITS), Waseda University, Japan, in September 2011. From December 2011 to 2012, he was a special researcher at the Research Institute for Secure Systems (RISEC), National Institute of Advanced Industrial Science and Technology (AIST), Japan. He is currently an Assistant Professor in Shanghai Jiao Tong University, Shanghai, China. His research interests include sensor network security and sensor network application in Smart Grids. He is a member of IEEE.

Muhammad Tariq received his M.Sc. from Hanyang University, Seoul, South Korea in 2009 under a Pakistan government (HEC) scholarship, and his Ph.D. from Waseda University, Tokyo, Japan in 2012 under a Japanese government (MEXT) scholarship. Currently, he is an Assistant Professor at the Department of Electrical Engineering, FAST-NUCES, Peshawar, Pakistan. He is an approved Ph.D. supervisor of the Higher Education Commission (HEC) of Pakistan in the field of electrical engineering. His research interests include wireless ad hoc and sensor network systems, and wired and wireless systems in smart power grids. He won the student paper award in IEEE VTC 2010, and outstanding presentation award in JSST 2011. He won a Brain Korea 21 (BK21) research grant from the Ministry of IT, South Korea for the session 2008–2009. He is the author/coauthor of 25 research papers, including SCI indexed and peer-reviewed journals, as well as local and international conference proceedings. He is a member of IEEE, IEICE, Japan Society of Simulation Technology (JSST), and Pakistan Engineering Council (PEC).

Solomon Abebe Asfaw received his undergraduate degree in Physics from Bahir Dar University, Bahir Dar, Ethiopia; an M.Sc. degree in Physics from the Norwegian University of Science and Technology, Trondheim, Norway; a second M.Sc. and Ph.D. degree specializing in energy system modeling from Ben-Gurion University of the Negev, Sede Boqer, Israel. He was a recipient of the 2010 Wolf prize for outstanding Ph.D. students in Israeli Universities. He is currently a postdoctoral fellow at the University of California, Berkeley. Solomon's research interests include very high grid penetration of intermittent renewable energy resources (solar and wind) with and without energy storage, the role of storage design and dispatch, long-term planning of power grid, and so on. His findings have been published in peer-reviewed journals, as book chapters, and conference proceedings.

Preface

The adverse effects of climate change as well as the achievements of sustained development require changing the present-day practice of energy production, transmission, distribution, and consumption. Developing smarter electric power grids is believed to be a key path to realizing this goal. Here, we could simply consider a Smart Grid as a hub of heterogeneous technological and policy measures that will make the future power grid more efficient, reliable, and clean. Nowadays, various Standard Developing Organizations (SDOs) and industries are working to develop Smart Grid–related standards and technologies, while governments throughout the world are gradually issuing conducive directives toward modernizing their power grids. This book "Smart Grid Standards: Specifications, Requirements, and Technologies," presents a summary of worldwide progress in creating Smart Grid standards and their future trends, Smart Grid development policies, and key projects initiated by countries around the world, as well as the cooperation and collaboration between national, regional, and international SDOs.

The book is not intended to provide a comprehensive description of the wide range of Smart Grid technologies. Rather, it is a result of collaboration between authors working in various research areas. Their research interests include, inter alia, environment and sustainability, energy technology, electric power, power electronics, and information and communication technologies. Thus, the primary purpose of the book is to bring together various aspects of Smart Grid, such as advances in grid automation, clean energy technologies, and challenges on their interoperability, potential technological, and policy paths to overcome some of the challenges, and advances in the area of smart home and demand response measures. We hope that this book provides a broader picture of the Smart Grid concept. Moreover, this book should especially be a suitable text for a course on the Smart Grid that may be taken by both undergraduate and graduate engineering students. It should also be useful, as a concise reference on Smart Grid, for researchers working in a wide variety of fields in physical, policy, and engineering sciences.

Takuro Sato
Waseda University, Tokyo, Japan

Daniel M. Kammen
University of California, Berkeley, USA

Bin Duan

Xiangtan University, Xiangtan, China

Martin Macuha

Orange Lab, France Telecom Japan Co. Ltd., Tokyo, Japan

Zhenyu Zhou

North China Electric Power University, Beijing, China

Jun Wu

Shanghai Jiao Tong University, Shanghai, China

Muhammad Tariq

FAST National University of Computer and Emerging Sciences, Peshawar, Pakistan

Solomon Abebe Asfaw

University of California, Berkeley, USA

January 15, 2014

Acknowledgments

The Smart Grid incorporates one of the key technological advances required to overcome various challenges related to energy production, transmission, distribution, and consumption. Nowadays various standard development organizations (SDOs) and industries are working to develop Smart Grid–related standards and technologies, while governments throughout the world are gradually issuing conducive directives toward modernizing their power grids. This book "Smart Grid Standards: Specifications, Requirements, and Technologies," presents a summary of worldwide progress in creating Smart Grid standards and their future trends, Smart Grid development policies, and key projects initiated by countries around the world, as well as the cooperation and collaboration between national, regional, and international SDOs.

This book is a result of collaborations between authors working in various research areas. Their research interests include, inter alia, environment, energy, electric power, power electronics, and information and communication technologies. We hope that this book provides a broader look into the concept of Smart Grid and could be useful for students, engineers, researchers, businessmen, and policy makers who are working in the field.

Our sincere thanks go to Mr. Mingxin Hou, the Commissioning Editor at John Wiley & Sons, for his numerous support and assistance from the beginning to the end of the preparation of the manuscript.

We also thank anonymous reviewers from various parts of the world, including the United States, the European Union, Asia, and the Middle East for their valuable comments and suggestions during the reviewing processes.

Special thanks also go to students and volunteers including Dr. Yi Jiang, Dr. Yanwei Li, Dr. Keping Yu, Prof. Song Liu, Mr. Jiran Cai, for their hard work and assistance in preparing the draft of this book. In addition, Solomon Abebe Asfaw is also very grateful for the financial support that he has received from Philomathia Foundation while contributing to this book.

We also thank the International Electrotechnical Commission (IEC) for permission to reproduce information from its International Standards IEC 60870-5-101 ed.2.0 (2006), IEC 60870-6-503 ed.2.0 (2002), IEC 61970-1 ed.1.0 (2005), IEC 61968-1 ed.1.0 (2003), ISO/IEC 15045-1 ed.1.0 (2004), ISO/IEC 18012-1 ed.1.0 (2004), and IEC Smart Grid Standardization Roadmap ed.1.0 (2010).

Gratitude is extended to Ms. Clarissa Lim, Project Editor at John Wiley & Sons, for her professional editorial work and review of the manuscript. Her professionalism and experience have greatly enhanced the quality and value of this book. We would also like to thank Ms. Shelley Chow, Project Editor at John Wiley & Sons, for her help in the permission applications.

Last, but not least, our special gratitude goes to our supporting and considerate family members.

Takuro Sato

1

An Overview of the Smart Grid

1.1 Introduction

The world population has been increased rapidly ever since the industrial revolution in the eighteenth century. Toward building a more affluent society, the development of many industries has led to significant increase in energy demand on a global scale. As a result, the increase in CO_2 (carbon dioxide) emissions has threatened the global environment and traditional fossil energy sources have reached their limits on the Earth. Compared to traditional fossil energy, nuclear energy began to be regarded as clean, safe, and reliable. However, in March 2011, this belief was shaken due to the Fukushima Daiichi Nuclear Power Plant accident in Japan. As a result, many countries and international bodies are now showing overwhelming interest in the use of renewable technologies for the production of clean and safe energy. However, maintaining a balance between the demand and supply of energy, fulfilling the national energy requirements through renewable energy, and abandoning energy production from fossil fuels pose many challenges. For example, energy supply which is based on the natural solar power, wind power, and hydroelectric power is unstable and cannot meet the strict requirements of electric systems. In order to make renewable energy into a stable energy resource, it is necessary to monitor power supply and demand in real time and to obtain a balance between supply and demand by integrating conventional electric grid with up-to-date information and communication technologies.

In December 2009, the United Nations organized the fifteenth session of the Conference of the Parties to the United Nations Framework Convention on Climate Change (UNFCCC) (COP 15) and the fifth session of the Meeting of the Parties to the Kyoto Protocol (COP15/MOP5) in Copenhagen, Denmark. The industrial world reached an agreement with the long-term goal of preventing the global average temperature from rising more than 2 °C, that is, 3.6 °F, above preindustrial levels [1]. The first European Union (EU) Electricity Directive issued was Directive 96/62/EC whose primary objective was to liberalize the electric market in 1996 and this was then extended to the gas

Smart Grid Standards: Specifications, Requirements, and Technologies, First Edition. Takuro Sato,
Daniel M. Kammen, Bin Duan, Martin Macuha, Zhenyu Zhou, Jun Wu, Muhammad Tariq and Solomon Abebe Asfaw.
© 2015 John Wiley & Sons, Ltd. Published 2015 by John Wiley & Sons, Ltd.

market in 1998. The Directive was repealed and replaced by Directive 2003/54/EC as the second EU Electricity Directive in 2003, and was transposed by the EU Cogeneration Directive 2004/08/EC in 2004 [2]. In September 2007, the European Commission issued the third package of legislative proposals, which was then approved as Directive 2009/72/EC by the European Parliament and European Council in July 2009 [3]. Since March 2011, the Gas and Electricity Directives of the third package for an internal EU gas and electricity market have been transposed into a national law to introduce smart meters to the extent of around 80% by 2020 [4]. The European Parliament and European Council also established the Agency for the Cooperation of Energy Regulators (ACER) to promote the internal energy market for both electricity and natural gas in Europe [5].

The United States (US) has conducted a restructuring of electric power networks by issuing the American Recovery and Reinvestment Act (ARRA) in 2009 to enforce upgradation of the obsolete electric power networks. In January 2010, the National Institute of Standards Technology (NIST) of the US issued the NIST Framework and Roadmap for Smart Grid Interoperability Standards Release 1.0. NIST has initiated the Smart Grid Interoperability Panel (SGIP) under the Energy Independence and Security Act of 2007 (EISA 2007) to coordinate standards development for the Smart Grid with various organizations such as the Open Smart Grid User Group (OpenSG User Group), Institute of Electrical and Electronics Engineering (IEEE), Internet Engineering Task Force (IETF), Telecommunications Industry Association (TIA), and ZigBee Alliance.

Japan, an advanced industrial country, built a stable electric power network with nuclear power being considered as the main and stable source of power supply. However, this consideration underwent a total change after the accident at the Fukushima Daiichi nuclear power plant on March 11, 2011. After this accident, interest in promoting the research and development in setting out a policy for renewable energy technologies for the Smart Grid has grown rapidly.

The Smart Gird requires maintaining a balance between energy supply and demand through real-time monitoring enabled by bi-directional ICT. Since the cost of electricity generation varies depending on not only the method of generating electric power but also the consumption time, location, and quantity, the purchased electricity price accordingly changes due to different situations of customers. Therefore, it is important for utilities and customers to exchange information about electricity supply and demand with each other. Furthermore, it is important to ensure not only the reliability of power generation, storage, and transmission systems but also the reliability of the information and communication systems. For example, many users use e-mail, Short Messaging Service (SMS), or Internet web pages through either a Computer or Personal Computer/Laptop or mobile phone for energy monitoring and management at home. However, transmitting critical data of electricity usage and customer privacy information through the Internet cannot meet the strict requirements of electricity systems such as latency and security. In particular, cyber security technologies will play an important role in the Smart Grid for ensuring reliable operations.

The challenges for realizing the next generation Smart Grid lies in the gaps between market needs and existing standards, and the lack of interoperability among standards. In order to bridge the gaps between future requirements and existing standards, various Standards Developing Organizations (SDOs) have been trying to promote a standardization process of the Smart Grid. The critical role of standards for the Smart Grid has already been realized worldwide by governments and industrial organizations, which advocates the development and adoption of standards to ensure that today's investments in the Smart Grid remain valuable in the future; to ensure products from multiple manufacturers to interoperate seamlessly; to catalyze innovations; to support consumer choice; to create economies of scale to reduce costs; and to open global markets for Smart Grid devices and systems. As pointed out by the International Electrotechnical Commission (IEC), which is one of the major international standardization organizations for issuing standards related to the Smart Grid, a higher level of syntactic and semantic interoperability is required for the various products, solutions, technologies, and systems which build up the Smart Grid system [6]. Interoperability is necessary to ensure the smooth exchange and use of information between different systems or components. Two major domains of interoperability are syntactic interoperability and semantic interoperability. Syntactic interoperability ensures the ability of communication and exchange of information between different systems through standardized data formats and protocols, a typical domain where much of the work of IEC and other SDOs has focused on. Semantic interoperability is the next step of syntactic interoperability, which ensures the ability of different systems to interpret the exchanged information through standardized information exchange reference models. Besides SDOs, there are also many technical consortia, forums, and panels, which are actively involved in promoting the standardization process of the Smart Grid. This chapter will provide an overview of the current status of the Smart Grid in both developed and developing countries. The organization of this chapter is as follows: Section 1.2 provides an overview of major Smart Grid-related organizations, including SDOs, regulatory organizations, technical consortia, forums, and panels, and marketing/advocacy organizations; Section 1.3 introduces the development of the Smart Grid in the United States; Section 1.4 introduces the development of the Smart Grid in the European Union; Section 1.5 introduces the development of the Smart Grid in Japan; Section 1.6 introduces the development of the Smart Grid in South Korea; Section 1.7 introduces the development of the Smart Grid in China; and Section 1.8 gives the conclusion.

1.2 An Overview of Smart Grid-Related Organizations

In this subsection, we provide an overview of major Smart Grid-related organizations, including SDOs, regulatory organizations, technical consortia, forums and panels, and marketing/advocacy organizations [7]. In general, SDOs are the organizations that develop, revise, coordinate, and amend technical standards. SDOs not only deal with different types of standards to address applications or sets of applications but also deal with specifications that lead to formal standards which are approved by law. Some of the standards are informal or voluntary as they are adopted by industries but not formally approved by law. Besides SDOs, there are various technical consortia, forums and panels, regulatory organizations, and marketing/advocacy organizations, which are also actively involved in developing or evaluating Smart Grid-related technical specifications and cooperating with SDOs in promoting the standardization process. It is noted that the classification of organizations in Figure 1 is just for illustration purpose as some organizations are active in a broad scope and it would been difficult to classify them into a single category.

1.2.1   SDOs Dealing with the Smart Grid

SDOs are classified according to their roles, positions, and domains of applications. SDOs can be local, regional, or international organizations, and might be governmental, semi-governmental, or non-governmental entities. Governmental SDOs are usually profitable organizations while semi and non-governmental organizations are usually non-profit organizations.

1.2.1.1   International Electrotechnical Commission (IEC)

IEC is among the most well-established and largest SDOs along with the International Organization for Standardization (ISO), and the International Telecommunication Union (ITU). It is a nongovernmental international SDO that prepares and publishes international standards for electrical, electronics, power generation, transmission, distribution, and associated technologies. Standards developed by IEC also cover home appliances and office equipment, semiconductors, fiber optics, batteries, nanotechnology, and renewable energy systems and equipments. The IEC also supervises conformity testing in order to certify whether an equipment, system, or component conforms to its international standard.

IEC issued the IEC Smart Grid Standardization Roadmap in 2010, which outlines the gaps between requirements for the Smart Grid and existing standards. In the roadmap, IEC has specified communication, security, and planning as three general requirements for all Smart Grid aspects. Besides these three general requirements, IEC has also specified 13 specific applications and requirements to cover the main areas and applications of the Smart Grid, which are the following: (i) smart transmission system and transmission level applications, (ii) blackout prevention/EMS (Energy Management System), (iii) advanced distribution management, (iv) distribution automation, (v) smart substation automation-process bus, (vi) Distributed Energy Resources (DERs), (vii) advanced metering for billing and network management,

(viii) demand response/load management, (ix) smart home and building automation, (x) electric storage, (xi) E-mobility, (xii) condition monitoring, and (xiii) renewable energy generation. Other requirements which are necessary for implementing the Smart Grid but are not limited to Smart Grid applications and systems have also been specified, including Electromagnetic Compatibility (EMC), Low Voltage (LV) installation, object identification, product classification, properties, and documentation, and user cases.

The existing IEC core standards are shown in Figure 1.1, and summarized as follows:

- **IEC/TR**, Technical Report, **62357**: framework of power automation standards and description of the Service Oriented Architecture (SOA) concept.
- **IEC 61850**: substation automation and beyond.
- **IEC 61970**: EMS, Common Information Model (CIM), and Generic Interface Definitions (GIDs).
- **IEC 61968**: Distribution Management System (DMS), CIM, and Component Interface Specification (CIS).
- **IEC 62351**: security.

IEC has focused on new areas such as Advanced Metering Infrastructure (AMI), which is specified by IEC 62051–62059 and IEC/TR 61334; DER, which is specified by IEC 61850-7-410: -420, and Electric Vehicles (EVs) which is specified by IEC

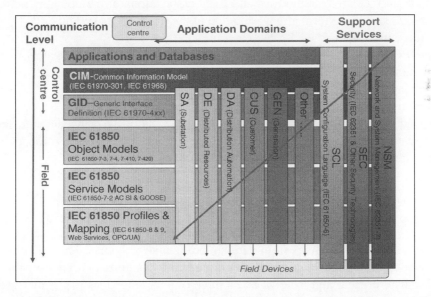

Figure 1.1 IEC 61850 models and the Common Information Model (CIM). Reproduced from IEC Smart Grid Standardization Roadmap ed. 1.0 Copyright © 2010 IEC Geneva, Switzerland. www.iec.ch [6] with permissions from International Electrotechnical Commission (IEC)

61851. These areas are not traditional standardization topics and pose new requirements for IEC standards. In order to promote the standardization of the Smart Grid, IEC has outlined the following general recommendations [6]:

Recommendation G-1

The Smart Grid can have multiple shapes and concepts and the concepts are not unified. Furthermore, legacy systems and existing mature domain communication systems must be incorporated and used. The IEC should avoid standardizing applications and business models and focus in standardizing necessary interfaces and product requirements.

Recommendation G-2

The potential of Smart Grid standardization and in particular the potential of IEC/TR 62357 should be promoted by the IEC. Stakeholders should be informed through numerous ways about the possible applications developed by TC 57.

Recommendation G-3

The IEC should not focus on the harmonization of various technical connection criteria, which are subject to different standards, regulations, and specifications. TC 8 could be responsible for specifying general minimum requirements, but the detailed standardization of these issues is out of the scope of IEC standardization.

Recommendation G-4

The IEC should cooperate closely with stakeholders, organizations, and other important regulatory authorities and trade associations in the domain "markets." The IEC should perform an investigation of the market data systems.

Recommendation G-5

The IEC should cooperate with NIST in the prioritized action fields and offer consultation for local or regional standards, and acknowledge the works of the NIST roadmap.

Recommendation G-6

Production control should be integrated with the enterprise management using the achievement of technology and innovation of management. A new model rather than the original CIM model should be built to solve this problem.

1.2.1.2   International Organization for Standardization (ISO)

The ISO, was founded in 1947, with its headquarters in Geneva, Switzerland. It is composed of representatives from various local standard bodies. The ISO publishes worldwide proprietary, industrial, and commercial standards [8]. The Organization has English, French, and Russian as its three official languages [9]. It manages the specific projects or sends experts to participate in the technical work, subscriptions from local member bodies, and publishes standards once developed and approved.

1.2.1.3 International Telecommunication Union (ITU)

The ITU, which is located in Geneva, Switzerland, was originally founded in May 1865 as the International Telegraph Union. ITU is a specialized body of the United Nations that is responsible for issues that concern ICT [10, 11]. Its responsibilities include promoting international cooperation in assigning satellite orbits, coordinating the shared global use of the radio spectrum, improving telecommunication infrastructure in the developing world, and promoting the development of international standards. The ITU is active in various diverse applications and domains such as wired and wireless communication technologies, 2G/3G mobile communication network services, broadband Internet, aeronautical engineering and maritime navigation, radio astronomy, satellite-based meteorology, triple play services (voice, video, and data), TV broadcasting, and Next Generation Networks (NGN).

1.2.1.4 SAE International

The Society of Automotive Engineers (SAE) International is a global association for engineering professionals and researchers in the aerospace, automotive, and commercial-vehicle industries [18]. SAE International creates and manages more aerospace and ground vehicle standards than any other entity in the world. It has more than 128 000 members globally and the membership is granted to individuals. In the field of EVs, SAE International has published numerous standards to provide references for performance rating of EV batteries, battery system safety, determination of the maximum available power from a rechargeable energy storage system, packaging of EV batteries, communications between EVs and utility grid and Electric Vehicle Service Equipment (EVSE), communications between EVs and customers, interoperability with EVSE, and so on. Many standards issued by SAE International such as SAE J2847, SAE J1772, and SAE J2836 have been identified by the NIST as critical standards for the development of the Smart Grid.

1.2.1.5 Institute of Electrical and Electronics Engineers (IEEE)

The IEEE was founded in 1884 as the American Institute of Electrical Engineers (AIEE). Its headquarters is in New York City. The IEEE is a professional body with more than 400 000 electrical and electronics engineers, among whom around 51% are living in the United States [12, 13]. It has dedicated itself to advancing technological innovation and excellence. Basically, it is incorporated under the Not-for-Profit Corporation Law of New York [14]. By the early 21st century, a total of 38 societies had been formed within the IEEE, and more than 900 active standards had been developed by IEEE members and related stakeholders. In January 2010, the IEEE launched the IEEE Smart Grid Web Portal to bring together a broad array of sources within it, including education, news, and intelligence. The IEEE Smart Grid Web Portal is the first phase of the IEEE Smart Grid, which is an initiative launched by the IEEE to provide expertise and guidance for those involved in the Smart Grid worldwide. The IEEE has developed numerous standards related to the Smart Grid, including IEEE P2030, IEEE 802 series, IEEE SCC 21 1457, IEEE 1159, IEEE 762, and IEEE SCC 31. Some of these standards are covered and introduced in detail in this book.

1.2.1.6 European Committee for Standardization (CEN)

The European Committee for Standardization (CEN) is a nonprofit organization, which was founded in 1961. CEN is a regional organization, which is officially recognized by European Union as a European standard body. The CEN aims to promote the European economy in global trading and the welfare of European citizens and the European environment. Its objectives are to provide an efficient infrastructure to various stakeholders for the development, maintenance, and distribution of coherent sets of standards and specifications. It has 30 national members who are working together to develop Standards for European internal market in various sectors.

1.2.1.7 European Committee for Electrotechnical Standardization (CENELEC)

The European Committee for Electrotechnical Standardization (CENELEC) was founded in 1973. As a European SDO, CENELEC is responsible for European standardization in the area of electrical engineering. Based in Brussels, CENELEC is a nonprofit organization under Belgian law. Its members are the national electrotechnical standardization bodies from most European countries. CENELEC, the European Telecommunications Standards Institute (ETSI), and CEN have formed a Joint Working Group (JWG) on standards for the Smart Grid. Standards coordinated by these agencies are regularly adopted in many countries outside Europe, which also follow European technical standards. Before the CENELEC, the other two European organizations that were responsible for electrotechnical standardization are the CENELCOM (European Committee for the Coordination of Electrotechnical Standards in the Common Market) and CENEL (European Electrical Standards Coordinating Committee) [15].

1.2.1.8 Telecommunications Industry Association (TIA)

The TIA is an association that develops consensus-based industry standards for a wide variety of ICT products. It is accredited by the American National Standards Institute (ANSI). Currently, it represents nearly 400 member companies, with 12 engineering committees under its standard and development department that develop guidelines for satellites, telephone terminal equipment, accessibility, private radio equipment, cellular towers, data terminals, VoIP devices, structured cabling, data centers, mobile communications, multimedia multicast, vehicular telemetric, healthcare ICT, M2M communication, and smart networks [16].

1.2.1.9 Internet Engineering Task Force (IETF)

The Internet Engineering Task Force (IETF) is a large open international community of network operators, vendors, designers, and researchers concerned with the evolution of the Internet architecture and the smooth operation of the Internet. It is open to any interested individual. The IETF has its working groups where its actual technical work is performed. The working groups are organized into several areas according to specific topics such as routing, transport, and security [17].

1.2.1.10 Alliance for Telecommunications Industry Solutions (ATIS)

The Alliance for Telecommunications Industry Solutions (ATIS), which is located in Washington, D.C., has developed numerous standards for the telecommunications industry. ATIS has more than 250 member companies, including broadband providers, commercial mobile radio service providers, competitive local exchange carriers, cable providers, consumer electronics companies, digital rights management companies, equipment manufacturers, Internet service providers, and so on. ATIS is accredited by the ANSI, and is one of the six organizational partners for the 3rd Generation Partnership Project (3GPP), and is a founding partner of oneM2M, which is a common M2M service layer for various hardware and software.

1.2.2 Technical Consortia, Forums, and Panels Dealing with the Smart Grid

1.2.2.1 Wi-Fi Alliance

The Wi-Fi Alliance is a global nonprofit organization founded in 1999 with the goal of driving adoption of high-speed wireless local area networking technologies. The key sponsors of the Wi-Fi Alliance include Cisco, Dell, Apple, Huawei, Broadcom, Intel, Motorola, Samsung Electronics, and Texas Instruments. The Wi-Fi Alliance has developed the Wi-Fi Certified, which is a program for testing products to conform to the IEEE 802.11 standard in terms of interoperability, security, reliability, and so on. Wi-Fi Certified devices will carry a Wi-Fi Certified logo, which ensures interoperability among different manufactures. The Wi-Fi technology is widely used at home around the world and the adoption rate continues to grow. The goals of the Wi-Fi Alliance are to provide a highly effective collaboration forum, to promote the development of the Wi-Fi industry with new technology specifications and programs, and to realize seamless product connectivity through testing and certification.

1.2.2.2   ZigBee Alliance

The ZigBee Alliance was established in 2002 as an open, nonprofit association. Its members include companies, universities, and government agencies worldwide. The ZigBee Alliance has focused on developing green, low-power, and short-range wireless networking standards for monitoring, control, and sensor applications. The two specifications which have been developed by the ZigBee Alliance are the ZigBee Specification, which includes the ZigBee PRO and ZigBee feature sets, and the Zig-Bee RF4CE Specification, which was designed for simple, two-way device-to-device control applications. Furthermore, the ZigBee Alliance has also developed numerous leading standards for building automation, health care, home automation, input device, light link, network devices, remote control, retail services, smart energy, and telecommunication services. Similar to the Wi-Fi Certified program, the ZigBee Certified program certifies ZigBee products from different manufactures to ensure interoperability and quality.

1.2.2.3   WiMAX Forum

The Worldwide Interoperability for Microwave Access (WiMAX) Forum is a worldwide consortium established in 2001 to promote and accelerate the introduction of WiMAX-based services into the marketplace. WiMAX is based on the IEEE 802.16-2004, and IEEE 802.16e-2005 standards. IEEE 802.16-2004 is also called the fixed WiMAX and was developed as a wireless backhaul technology, while IEEE 802.16e-2005 is called the mobile WiMAX and was developed as a replacement for cellular phone technologies such as Global System for Mobile communication (GSM) and Code Division Multiple Access (CDMA). Similar to the Wi-Fi Alliance and the ZigBee Alliance, the WiMAX Forum has established the WiMAX Forum Certification Program to certify the interoperability of IEEE 802.16e products though conformance and interoperability tests. Devices or products which pass the tests can carry the WiMAX Forum Certified logo.

1.2.2.4   UCA International Users Group

The UCA International Users Group (UCAIug) is a nonprofit corporation established to promote the integration and interoperability of electric/gas/water utility systems through the use of international standards-based technology. The UCAIug consists of both utility users and supplier companies, and it provides a forum in which various stakeholders in the utility industry can work in collaboration to deploy standards for real-time applications. The UCAIug works closely with various SDOs as a User Group of many international standards such as IEC 61850, the CIM, advanced metering, and demand response via the Open Automated Demand Response (OpenADR).

1.2.2.5   National Electrical Manufactures Association (NEMA)

The National Electrical Manufactures Association (NEMA) was founded in 1926 to provide a forum for the standardization of electrical equipment. It consists of more than 400 member companies which manufacture electrical equipment used in power generation, transmission, distribution, factory automation, control, and medical systems. The annual sale of NEMA-scoped products has exceeded $120 billion [19]. NEMA has focused on the development of standards and providing solutions for emerging technical, regulatory, and economic issues. It has published over 600 standards and technical papers in building systems, electronics, industrial automations, insulating materials, lighting systems, medical imaging, power equipment, security imaging, wires, and cables. Furthermore, it has helped launch the Electrical Safety Foundation International (ESFI) to increase electrical safety awareness and promote the safe use of electrical equipment.

1.2.2.6   Organization for the Advancement of Structured Information Standards (OASIS)

The Organization for the Advancement of Structured Information Standards (OASIS) is a nonprofit consortium founded in 1993 to promote the development, convergence, and adoption of open standards for security, cloud computing, SOA, web services, Smart Grid, emergency management, and other areas [20]. It consists of more than 5000 members in 10 member sections: OASIS AMQP (Advanced Message Queuing Protocol), OASIS CGM (Computer Graphics Metafile), OASIS eGov, OASIS Emergency, OASIS IDtrust, OASIS LegalXML, OASIS Open CSA (Composite Services Architecture), and OASIS Web Service Interoperability (WS-I). Each member section is formed to meet the needs of a specific group and maintains its own identity as a distinct organization. The advantage is that each member section can focus on its own interest while having access to OASIS infrastructures, resources, and expertise.

1.2.2.7   HomePlug Power line Alliance

The HomePlug Power line Alliance is a trade association organization that promotes the adoption and implementation of cost-effective and standard-based home power line networks and products. It has developed several home power line technologies, including HomePlug AV/AV2 for broadband networks applications such as HDTV (High-definition Television) and VOIP (Voice Over Internet Protocol). It has also developed HomePlug Green PHY (Physical Layer) for low-cost and low-power applications such as demand response, load control, and home and building automation. Furthermore, it has cooperated with the IEEE 1901.2 working group to develop a low-frequency, narrowband Power Line Communication (PLC) certification and marketing program, named Netricity PLC, which can be used for narrowband low-frequency communications such as Smart Grid to meter applications.

1.2.2.8   HomeGrid Forum (HGF)

The HomeGrid Forum (HGF) is an industry alliance, which has been formed to promote the development and adoption of the International Telecommunication Union-Telecommunication Standardization Sector, Gigabit Home Networking (ITU-T G.hn) standard. ITU-T G.hn is the first standard to define a single standard for all major wired communication media including power lines, phone lines, and coaxial cables. In order to ensure compliance and interoperability, HGF has launched certification programs based on plugfests, compliance, and interoperability testing technology. Products that pass the HGF test will carry a HomeGrid logo.

1.2.2.9   GridWise Architecture Council (GWAC)

The GridWise Architecture Council (GWAC) was formed by the US Department of Energy (DoE) to help identify areas for standardization which ensures interoperability among different electric system components. Its members represent the many constituencies of the electricity supply chain and users. GWAC has made efforts to promote the move from control-based interactions toward transaction-oriented interactions, which requires significant information exchange between electric system devices and electricity consumers. The term "transactive energy" means that the decisions of how to consume the energy are made on the basis of economic or market constructs while considering grid reliability constraints. Examples of transactive energy applications are the GridWise Olympic Peninsula Project [21], TeMIX [22], and the Pacific Northwest Smart Grid Demonstration Project [23]. GWAC has also cooperated with NIST to form the Domain Expert Working Groups (DEWGs) to assist NIST with addressing Smart Grid interoperability issues.

1.2.3   Other Political, Market, and Trade Organizations, Forums, and Alliances

Besides various SDOs and technical consortia, forums, and panels, there are also many other political, market, trade, and regulatory organizations, forums, and alliances, which are actively involved in promoting the development of the Smart Grid. In this section, we briefly introduce some of these organizations, forums, and alliances.

1.2.3.1   International Energy Agency (IEA)

The International Energy Agency (IEA) was founded in response to the 1973 oil crisis with the role to coordinate response to major disruptions in oil supply by releasing emergency oil stocks [25]. Its current work is to ensure reliable, affordable, and clean energy. The four core areas of IEA's works are energy security, economic development, environmental awareness, and engagement worldwide. The IEA consists of 28 member countries including Australia, Canada, the Czech Republic, Denmark Finland, Germany, Hungary, Italy, Japan, Luxembourg, New Zealand,

Norway, the Slovak Republic, Spain, Turkey, United Kingdom, United States, and so on. Besides member countries, the IEA has also developed close relationships with nonmember countries such as China, India, Brazil, Russia, and Thailand.

1.2.3.2   Clean Energy Ministerial (CEM)

The Clean Energy Ministerial (CEM) is a high-level global forum for clean energy technologies first hosted by the then US Secretary of Energy Steven Chu at the UN Framework Convention on Climate Change conference of parties in December 2009 [26]. Since then, it has attracted worldwide participation from governments, taking into account 80% of global greenhouse gas emissions and 90% of global clean energy investment. The three core focus areas of the CEM are energy efficiency, clean energy supply, and clean energy access. It has developed 13 action-driven, transformative clean energy initiatives [27]: Electric Vehicles Initiative (EVI), Global Superior Energy Performance Partnership (GSEP) Initiative, Super-Efficient Equipment and Appliance Development (SEAD) Initiative, Bioenergy Working Group Initiative, Carbon Capture, Use, and Storage Action Group (CCUS) Initiative, Multilateral Solar and Wind Working Group Initiative, Sustainable Development of Hydropower Initiative, twenty-first century Power Partnership Initiative, Clean Energy Education and Empowerment (C3E) Women's Initiative, Clean Energy Solutions Center Initiative, Global Lighting and Energy Access Partnership (Global LEAP) Initiative, Global Sustainable Cities Network (GSCN) Initiative, and International Smart Grid Action Network (ISGAN) Initiative. Among these 13 initiatives, ISGAN enables multilateral collaborations among governments to improve the understanding of Smart Grid technologies, practices, policies, and so on. Currently, ISGAN has involved 22 member countries: Australia, Austria, Belgium, Canada, China, Finland, France, Germany, India, Ireland, Italy, Japan, Korea, Mexico, Norway, the Netherlands, Russia, Spain, Sweden, Switzerland, the United Kingdom, and the United States.

1.2.3.3   Demand Response and Smart Grid Coalition (DRSG)

Demand Response and Smart Grid Coalition (DRSG) is a trade association that consists of various companies in the areas of demand response, smart meters, and the smart grid technologies. DRSG consists of executive-level members and associate-level members, who provide demand response and Smart Grid technologies and services. The member companies are working together to promote the development and adoption of demand response and Smart Grid technologies.

1.2.3.4   China Electricity Council (CEC)

The China Electricity Council (CEC), which was founded in December 1988, is a joint organization of China's power enterprises and institutions. The CEC consists of 1188 members, among which the State Grid Corporation of China (SGCC) is the president member. The CEC functions as a bridge between government and its member companies and institutions by forwarding their requests to the government and protecting the legal rights of its members.

1.2.3.5 Global Smart Grid Federation

The Global Smart Grid Federation was founded in April 2010 to facilitate collaboration among various stakeholders to promote development and research of Smart Grid technologies. It consists of the following members:

- GridWise Alliance (USA): a forum found in 2003 in which various stakeholders of the energy supply chain work cooperatively to promote the transformation of the industrial-age electric grid into the information age.
- India Smart Grid Forum (ISGF) (India): a Public Private Partnership (PPP) initiated by the Ministry of Power, Government of India, to promote the development of Smart Grid technologies in India.
- Japan Smart Community Alliance (JSCA) (Japan): an alliance established by the New Energy and Industrial Technology Development Organization (NEDO) to promote collaboration of various stakeholders from electric power, gas, automobile, information and communications, electric machinery, construction and trading industries, public sector, academia, and so on.
- Smart Grid Flanders (Belgium), Smart Grid Canada (Canada), Danish Intelligent Energy Alliance (Denmark), EDSO, European Distribution System Operator, for Smart Grids (EU), Smart Grid Great Britain (UK), Norwegian Smartgrid Centre (Norway), Smart Grid Ireland (Ireland), Israel Smart Energy Association (Israel), Industrial Technology Research Institute (ITRI), Korea Smart Grid Association (Korea), Smart Grid Australia (Australia).

1.2.3.6 National Institute of Science and Technology and Smart Grid Interoperability Panel

NIST was found in 1901 as a nonregulatory federal agency and is one of the nation's oldest physical science laboratories. Currently, NIST is part of the US Department of Commerce, with a mission to promote US innovation and industrial competitiveness by advancing measurement science, standards, and technologies through the NIST laboratories, the Hollings Manufacturing Extension Partnership, the Baldrige Performance Excellence Program, and the Technology Innovation Program. The SGIP was initiated by NIST in 2009 to promote the development of a SmartGrid framework for interoperability standards. Therefore, SGIP was called the "best vehicle for developing Smart Grid interoperability standards" by the Federal Energy Regulatory Commission (FERC) [24]. SGIP has assisted NIST with its fulfillment of the EISA 2007. SGIP has focused on seven core domains of Smart Grid: operations, markets, service providers, bulk generation, transmission, distribution, and customer. These seven core domains are represented by 22 stakeholder categories that are working closely and cooperating with each other to promote the development and adoption of interoperability standards for the Smart Grid.

1.3   Status of the United States (US)

In the United States, FERC is responsible for regulating the interstate transmission of electricity, natural gas, and oil. However, the regulation of retail electricity and natural gas sales to consumers is outside of the FERC's responsibility. FECS issued the FERC Energy Policy Act in 1992 to take a step forward toward electric deregulation, and issued the FERC Orders 888 and 889 in 1996, which led to the creation of the Open Access Same-Time Information System (OASIS, formerly Real-Time Information Networks). The effects of partial deregulation in retail markets was promoted as a means of increasing competition. However, in the 2000–2001 California electricity crisis, only the wholesale prices were deregulated but retail prices were still regulated by the government. As a result, major utilities such as PG&E and SoCal Edition experienced financial deficit that is caused by paying more to the electricity wholesalers than the revenue charged from the customers. Investment in new power plants and distribution infrastructures dropped sharply. Aging, outdated electrical equipment and overloaded distribution lines were becoming a major problem in the United States. In the 2003 Northeast blackout, a software bug in the alarm system caused the second most widespread blackout in history, which affected an estimated 10 million people in Ontario and 45 million people in eight other states [28]. Therefore, in order to upgrade the aging electric infrastructures and power grid, and to stimulate the economy, the Obama government has regarded the Smart Grid as one of the core components of the economy stimulus package. The development of the Smart Grid in the United States can be divided into three domains as shown in Figure 1.2: strategy development and planning, policy and law enforcement, and government and company pilot projects.

1.3.1   Strategy Development and Planning

In the strategy development and planning domain, the DoE has issued the DoE National Transmission Grid Study in 2002, in which the following six topics are reviewed [29]: transmission system operation and interconnection, reliability management and oversight, alternative business models for transmission ownership and operation, transmission planning and the need for new capacity, transmission siting and permitting, and advanced transmission technologies. The study has specified 51 recommendations to ensure a robust and reliable transmission grid for the twenty-first century. On 2 and 3 April 2003, 65 senior executives representing various stakeholders held a meeting to discuss the future of North America's electric system. In July 2003, DoE issued the "Grid 2030 – A National Vision for Electricity's Second 100 Years," to help implement

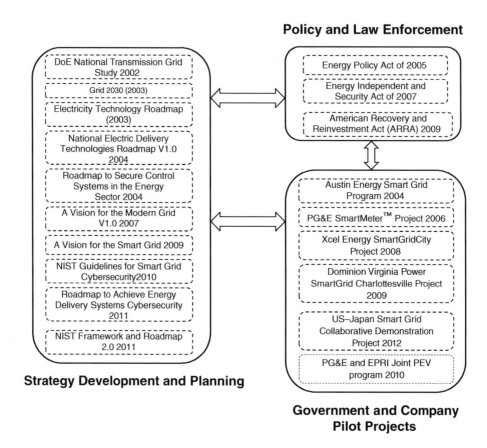

Figure 1.2 The three domains: strategy development and planning, policy and law enforcement, and government and company pilot projects

the 51 recommendations and to modernize the electricity delivery system in the United States [30]. Also in 2003, the Electric Power Research Institute (EPRI) issued the Electricity Technology Roadmap – Meeting the Critical Challenges of the twenty-first century. The roadmap has specified five destinations for realizing a sustainable global energy economy by 2050 [31]: (i) strengthening the power delivery infrastructure, (ii) enabling the digital society, (iii) boosting the economic productivity and prosperity, (iv) resolving the energy/environment conflict, and (v) managing the global sustainability challenge.

In January 2004, the DOE issued the National Electric Delivery Technologies Roadmap Version 1.0, which suggests paths to build America's future electric delivery system and outlines challenges faced by various stakeholders. In January 2006, the DOE issued the Roadmap to Secure Control Systems in the Energy Sector, which has outlined a strategic framework to achieve the vision that by 2015 US energy sector would be able to survive an intentional cyber assault without large impacts on critical applications [32]. In March 2007, the National Energy Technology Laboratory (NETL) issued A Vision for the Modern Grid V1.0, which defined seven characteristics for the Modern Grid [33]: (i) self-heals, (ii) motivates and includes the customer, (iii) resists attack, (iv) meets twenty-first century power quality requirements, (v) accommodates all generation and storage options, (vi) enables markets, and (vii) optimizes assets and operates efficiently. It is noted that in June, 2009, NETL issued A Vision for the Smart Grid, in which the word "Modern Grid" has been changed to "Smart Grid" compared to the document issued in 2007 [34].

In the United States, NIST is the central institute to promote the standardization of the Smart Grid. The EISA 2007 has directed NIST to coordinate the development of specifications, protocols, models, and standards to achieve interoperability. NIST will submit standards to FERC, which will adopt these standards through the rule-making process. There are various agencies and institutions working cooperatively with NIST in developing standards related to the Smart Grid. NIST has released the final version of NIST Framework and Roadmap for Smart Grid Interoperability Standards 2.0 on February 28, 2012, which has specified the following 15 Priority Action Plans (PAPs) [35]:

- PAP00 Meter Upgradeability Standard
- PAP01 Role of Internet Protocol (IP) in the Smart Grid
- PAP02 Wireless Communications for the Smart Grid
- PAP03 Common Price Communication Model
- PAP04 Common Schedule Communication Mechanism
- PAP05 Standard Meter Data Profiles
- PAP06 Common Semantic Model for Meter Data Tables
- PAP07 Energy Storage Interconnection Guidelines
- PAP08 CIM for Distribution Grid Management
- PAP09 Standard DR and DER Signals
- PAP10 Standard Energy Usage Information
- PAP11 Common Object Models for Electric Transportation
- PAP12 Mapping IEEE 1815 (DNP3) to IEC 61850 Objects
- PAP13 Harmonization of IEEE C37.118 with IEC 61850 and Precision Time Synchronization
- PAP14 Transmission and Distribution Power Systems Model Mapping
- PAP15 Harmonization of Power Line Carrier Standards for Appliance Communications in the Home
- PAP16 Wind Plant Communications
- PAP17 Facility Smart Grid Information Standard
- PAP 18 Smart Energy Profile (SEP) 1.X to 2.0 Transition and Coexistence

More details about the NIST Framework and Roadmap for Smart Grid Interoperability Standards can be found in Chapter 8 and references therein. NIST has positioned the Internet as the core network for the realization of smart grid. In Internet, there already exists a set of protocols which define how information is packaged, transmitted, and shared across the Internet users. Similar to that concept, a working group of NIST has been developing core set of protocols that can be used to build a new network for the Smart Grid. A core set which consists of more than 150 individual Request for the Comments (RFCs) protocols has been developed for communications and cybersecurity. In the field of energy, NIST has set up six subject areas: alternative energy, electric power metrology, energy conservation, storage and transport, fossil fuels, and sustainability. Within each subject area, numerous programs and projects have been launched to promote the development and research of Smart Grid standards.

Since the study of the Smart Grid in the United States has mainly focused on the integration of ICT technologies, there are many information and networking companies such as IBM, ACCENTURE, Oracle, SAP, and CISCO participating in the development of the Smart Grid. In order to operate safely and reliably, it is necessary for the Smart Grid to adopt practical countermeasures against cyber threats. Therefore, NIST has formed the Smart Grid Interoperability Panel–Cyber Security Working Group (SGIP-CSWG). SGIP-CSWG has worked since June 2009 to develop a cyber security strategy for the Smart Grid that addresses prevention, detection, response, and recovery issues. The cybersecurity strategy should ensure interoperability of solutions across different domains and components of the Smart Grid. As a result, the SGIP-CSWG has issued the National Institute of Standards Technology Interagency Report (NISTIR) 7628 Guidelines for Smart Grid Cybersecurity in 2010. It is a report for individuals and organizations, including vendors, manufacturers, utilities, system operators, researchers, and so on, who will be addressing cybersecurity for Smart Grid systems. The guideline is comprised of three volumes: Volume 1-Smart Grid Cybersecurity Strategy, Architecture, and High-Level Requirements; Volume 2-Privacy and the Smart Grid; and Volume 3-Supportive Analyses and References. One hundred and eighty-nine high-level security requirements and 137 interfaces have been identified in the guideline. Multiple levels of security implementations are recommended to countermeasure the diverse and evolving cyber security threats.

In 2011, the DoE issued the Roadmap to Achieve Energy Delivery Systems Cybersecurity, which was developed as a replacement to the 2006 Roadmap to Secure Control Systems in the Energy Sector. In this report, a strategic framework has been developed to ensure that energy delivery systems are capable of surviving cyber incidents without large impacts on critical functions.

1.3.2   Policy and Law Enforcement

In order to guarantee energy production in the United States, the US Congress passed the Energy Policy Act of 2005 (EPAct 2005) on 29 July 2005. The EPAct 2005 has specified energy management requirements in the following areas: metering and reporting, energy-efficient product procurement, energy savings performance contract, building performance standards, renewable energy requirement, and alternative fuel use [36].

Later, the United States enforced the EISA 2007, which was originally named the Clean Energy Act of 2007. One of the major goals of the EISA 2007 is to move the United States toward greater energy independence and security. Under EISA 2007, NIST has been given the key role for coordinating development of a framework for interoperable Smart Grid standards. NIST has launched a three-phase plan to promote the development and adoption of Smart Grid interoperability standards. When the Obama administration was born, he put the Smart Grid at the center of a Green New Deal, an economic stimulus package through which he plans to create three million jobs in energy, education, healthcare, and infrastructure. On 13 February 2009, President Obama signed the ARRA into law, under which there is a total of $4.5 billion US energy grant for developing Smart Grid technologies [37]. On 27 April 2009, Obama announced the launch of the Advanced Research Projects Agency-Energy (ARPA-E), which was established to fund energy technology projects.

1.3.3 Government and Company Pilot Projects

Both government agencies and energy companies have also launched numerous Smart Grid projects. In 2004, the Austin Energy Corporation launched the Austin Energy Smart Grid program, with plans to offer real-time meter information by phone or Internet, management of smart appliances through the Web, and remote turn-on and turn-off service [38]. In 2006, the Pacific Gas and Electric (PG&E) initiated the SmartMeter™ project, in which more than nine million gas and electric meters were installed by 31 July 2012 [39]. With the SmartMeter system, residential electric usage data are recorded every hour, while commercial electric usage data are recorded every 15 min. The gas usage data are recorded once a time per day. Both the electric and gas usage recorders are transmitted using wireless communication networks. The SmartMeter system is expected to provide services to all customers.

In 2008, the Xcel Energy Corporation launched the "SmartGridCity project" in Boulder, Colorado to promote the development and adoption of Smart Grid technologies such as the SmartMeter and Home Automation Network (HAN). The SmartGridCity consists of four main components: Smart Grid infrastructure, smart meters, My Account website, and in-home smart devices. A user can log into the My Account website to track electricity use on daily, hourly, or 15-min intervals.

In October 2009, Dominion Virginia Power launched the $20 million SmartGrid Charlottesville project. In this project, more than 46 500 smart meters were installed. This project aims to test battery storage systems, develop automatic reporting of outages, allow for quicker restoration of service, increase customer convenience through remote turn-on and turn-off services, enable remote meter readings, and develop a demonstration program for light-emitting diode (LED) street lights [40].

In 2010, the PG&E and EPRI initiated an innovative Plug-in Electric Vehicle (PEV) pilot program, aimed to test and validate the security, scalability, and functionality of smart charging technology integrated with the Smart Grid [41]. Two leading companies in Smart Grid and PEV charging infrastructure, namely, Silver Spring Networks and ClipperCreek, Inc., joined the pilot program to promote the integration of PEV charging stations with the electricity grid.

In 2012, the US and Japan Smart Grid Collaborative Demonstration Project was conducted for verifying microgrid demonstration in Los Alamos, smart house in Los Alamos, microgrid demonstration in commercial areas of Albuquerque, and collective research on the overall project. Companies including Shimizu Corporation, Toshiba Corporation, Sharp Corporation, Meidensha Corporation, Tokyo Gas Co., Ltd, Mitsubishi Heavy Industries, Ltd, Fuji Electric Co., Ltd, Furukawa Electric Co., Ltd, Furukawa Battery Co., Ltd, Public Service Company of New Mexico, Sandia National Laboratories (SNL), the University of New Mexico, and regional US utility firms are currently working on it. In this project, a microgrid which is comprised of a 50 kW photovoltaic (PV) power generation system, a 240 kW gas-engine generator, an 80 kW fuel cell system, and a 90 kW battery storage system have been deployed to provide power on the demand side [42].

1.4 Status of the European Union (EU)

1.4.1 Activities of the European Union

Toward the establishment of common rules for the internal electricity market in Europe, the European Union issued the First EU Electricity Directive in December 1996. In the First Directive, the European Union obligates the liberalization of the retail market, accounting separation of power transmission and distribution businesses of vertical-type electric power companies, and obligates the approval of constructing new electric power generation infrastructures for the purpose of opening the market for new companies within the EU power market.

The European Union issued the Second EU Electricity Directive in 2004. The primary purpose of the Second Directive is to ensure the energy supply based on principles of consumer protection. The second purpose is to introduce competition in the energy markets. The third purpose is to introduce competition in the supply market. The fourth purpose is to guarantee the transmission and distribution network connection. The fifth purpose is to separate electric power generation, electricity sales business, and power transmission and distribution business from each other. The final purpose is to issue an annual report on the electricity and gas market.

In 2006, the European Union issued the European Smart Grids Technology Platform-Vision and Strategy for Europe's Electricity Networks of the Future, with the aim to create a joint vision for the European electricity networks of 2020 and beyond. The Vision should ensure that the future electricity networks of Europe have the following four features: flexibility, accessibility, reliability, and economy of operation. The key elements of the Vision include a toolbox of proven technical solutions, a harmonized regulatory and commercial framework, shared technical standards and protocols, information and communication systems, and interfaces for both new and old designs [43]. In 2007, the European Union issued the Strategic Research Agenda (SRA) for Europe's Electricity Networks of the Future, which defines and promotes research themes and projects to address the key elements of the Vision. The research agenda has identified 5 research areas, and 19 research tasks, which are necessary to realize the Vision.

In 2009, the Gas and Electricity Directives of the Third Package for an Internal EU Gas and Electricity Market was adopted and it was introduced into force by the European Commission Directorate General for Energy and Transport, and was transposed into national law by Member States in 2011. The purpose was to create a genuine internal energy market in which European consumers could choose gas and electricity supplying services from different companies at reasonable prices, and to enable all energy suppliers including small companies be able to access the market. The European Union established a task force in November 2009 for this purpose. The third legislative package obligates Ownership Unbending (OU), Independent System Operator (ISO), and Independent Transmission Operator (ITO) for the purpose of separating electric power generation and transmission systems. The ACER was established in 2010 as a successor to the European Regulators' Group for Electricity and Gas (ERGEG) to regulate the third legislative package. In addition, the third legislative package gives priority to investment in new renewable energy generation infrastructures, and to strengthen the protection of consumers.

Toward the end of 2009, the European Commission set up the Smart Grid Task Force (SGTF) to provide policy and regulatory directions for the deployment of the Smart Grid. In March 2010, the European standardization organizations CEN and CEN-ELEC held an informal meeting to discuss the standardization problems of Smart Grid technology in Europe. This meeting set up a working group to actively promote the development and research work of the Smart Grid standards. In 2011, the CEN/CENELEC/ETSI Joint Working Group (JWG) issued the final report of the CEN/CENELEC/ETSI Joint Working Group on Standards for Smart Grid, which outlines the standardization requirements for implementing the European vision of the Smart Grid [44]. The report provides an overview of current standards and activities and identifies the necessary steps to be taken to realize the vision.

In March 2012, the European Union issued the Smart Grid SRA 2035 SRA Update of the Smart Grid SRA 2007 for needs by the year 2035 [45]. The SRA 2035 describes the research topics and priorities that will be necessary for further development of the electricity system from 2020 to 2035 and beyond. The European Union promotes low-carbon related technology, bio energy, carbon capture and storage, power grid, hydrogen and fuel cells, nuclear power, smart city, solar power, and wind power based on the European Strategic Energy Technologies (SETs) Plan.

In June 2012, the European Union founded the European Network of Transmission System Operators for Electricity (ENTSO-E) to represent all electric Transmission

System Operators (TSOs) in the European Union. In ENTSO-E, the TSOs cooperate with each other regionally or on the European scale, and activities were organized through three committees: the System Development Committee, the System Operations Committee, and the Market Committee. The climate and energy policy of the European Union sets the following targets for 2020: cutting greenhouse gases by at least 20% of 1990 levels and increasing use of renewable energy sources (wind, solar, biomass, etc.) to 20% of total energy consumption. In order to increase energy efficiency by 20%. The ENTSO-E issued the scenarios A and B (corresponding the conservative case and best estimate case respectively). In July, 2012, the ENTSO-E published the final Ten-Year Network Development Plan (TYNDP) package. The TYNDP package comprises of six regional investment plans (Rglps): Baltic Sea, Continental South East, Continental Central East, Continental South West, Continental Central South, and North Sea [46]. The TYNDP package has also specified the Scenario Outlook and Adequacy Forecast (SO&AF) 2012–2030. In SO&AF 2012–2030, four visions for 2030 have been specified: Slow Progress, Money Rules, Green Transitions, and Green Revolution. These four visions are different enough from each other to capture realistic future pathways and challenges. There will be more than 100 transmission projects of pan-European significance, which total about 52 300 km of Extra High Voltage Routes, and require a total investment of more than €100 billion [46]. These projects will result in a 170 million tons reduction of CO_2, of which 150 million tons CO_2 is due to the deployment of renewable energy resources, and the other 20 million tons CO_2 is due to the effect of market integration.

The European Union has actively promoted the use of wind power and the use of solar energy as renewable energy, and has established some priority electricity corridors. In order to integrate the offshore wind capacities in the Northern Seas, the European Union has established the North Seas Countries' Offshore Grid Initiative (NSCOGI). Further, in order to integrate new renewable generation of Western Europe with other consumption centers, the European Union has launched the South Western Electricity Interconnections to increase interconnections between Member States in Western Europe. Other priority electricity corridors include the North–South electricity interconnections in central Eastern and South Eastern Europe and the Baltic Energy Market Interconnection Plan. The European Union has initiated the Framework Programme (FP) to provide funding for research, technological development, and demonstration. In the field of DER, several projects have been launched, including the FP5 (1998–2002) DISPOWER project, MICROGRIDS project, the FP6 (2002–2006) EU-DEEP project, IRED project, FENIX project, and the FP7 (2007–2013) iGREEN-Grid project, and ADDRESS project. More details about DER can be found in Chapter 4 and references therein.

1.4.2 Activities of EU Member Countries

EU member countries have also been actively involved in the research and development of the Smart Grid. Germany amended the Energy Industry Act in 1998 and started liberalization of the electricity market based on the first EU Electricity Directive. The restructuring of the electric power industry has occurred consequent to this. Following the Japanese Fukushima Daiichi nuclear power plant accident in March 2011, Germany determined to withdraw from nuclear power in June 2011, and to close all nuclear power plants in Germany by 2022. Companies such as E.ON which invested heavily in nuclear power plants had to cut jobs and switch to other energy generation technologies such as renewables. At present in Germany, there are four electric power generation companies: E.ON, RWE, EnBW, and Vattenfall Europe. The electricity transmission companies are TenneT TSO GmbH, 50 Hz Transmission, Amprion, and TransnetBW. In the distribution business, there are more than 800 distribution companies in the country following the liberalization of the electricity retail market.

Germany turned aggressive in developing solar energy and began the Feed-in Tariff (FiT) in 2000. Using FiT, each energy technology is given a different price from other energy technologies, on the basis of the particular cost of generation of that energy. Solar PV is offered a higher price than cheap wind power, and the tariff rates also depend on the size and location of the PV systems. PV manufacturers such as Q-Cells continued to grow but have suffered recently because of competition from cheap solar products from China. Owing to a net loss of more than €8 million, Q-Cells was acquired by Hanwha Korea as Hanwha Q. Cells, and continue to manufacture solar cells and panels. Solar PV installations in Germany have increased dramatically in the past, and are expected to continue to grow in the future. Germany has founded a consortium with 12 founders, including nine Germany companies: Deutsche Bank, Munich Re, HSH Nordbank, Siemens, Solar Millennium, Schott Solar, E.ON, RWE, and M+W Zander. The consortium will build a giant solar plant in the Moroccan desert, the first of many renewable energy power stations which would together cover 15% of Europe's electricity demand by 2050 [47].

The German Federal Ministry for the Environment, Nature Conservation, and Nuclear Safety (BMU) and Federal Ministry of Economics and Technology (BMWi) jointly launched the "E-Energy" demonstration project plan in 2008 [48]. The E-Energy project includes six E-Energy demonstration projects, and seven electricity mobility projects. More than €140 million of funding have been provided by both government and private companies. One example is the Cuxhaven demonstration project. The Cuxhaven demonstration project (e-Telligence) is carried out in Cuxhaven, which has a high percentage of renewable energies, for example, wind power, and a high percentage of electric storage systems, for example, cold stores and indoor swimming pools. A complex control system using Smart Grid technologies has been developed to balance the fluctuations of renewable energies, and to establish a renewable energy market.

The Cuxhaven demonstration project uses the modern ICT to improve the energy supply system and realize the integration of renewable energies, including wind power, solar power, and biomass. Consumers, DER systems, and electric storage systems, are integrated by establishing Virtual Power Plants (VPPs) to improve energy use efficiency. Smart meters have been installed for more than 2000 families, and on-line visualization of electricity usage has been provided. In order to ensure the interoperability among different devices, a distributed information processing platform was established on the basis of the IEC 61850 and IEC 61970/IEC 61968 standards. The communications of the electricity market information and the control and status information are based on the CIM and the IEC 61850, respectively.

France amended the municipal law in 1999 according to the First EU Electricity Directive, for the purpose of enacting electricity deregulation in 2000. The EDF (Électricité de France), a monopoly company in the electricity market, used to produce 22% of the EU's electricity through nuclear power plants. The dependence on nuclear power of France has jumped to as high as 85%, which is the highest in the world. Besides EDF, other power generation companies are GDF Sues (Gaz de France) and SNET (Société nationale d'électricité et de thermique), whose market shares are weaker compared to EDF. In the transmission and distribution business, both the transmission company RTE (Réseaud 'de Transport électricité) and the distribution company ERDF (Électricite Réseau Distribution France) are 100% subsidiary companies of EDF. Therefore, the separation of electrical power generation from power transmission makes very slow progress.

In order to reform the French electricity market, the Nouvelle Organization du Marché de l'Électricité (NOME) law was passed in 2010 to require EDF to sell its nuclear outputs to competitors at a rate specified by the CRE (Commission de régulation de l 'énergie/Energy Regulatory Commission). Under the NOME law, EDF has to sell 100 TWh per year to other electricity suppliers, which ensures that other electricity suppliers can offer competitive prices to customers [49].

In the United Kingdom (UK), after the liberalization of the electric market by enforcing the new Electricity Act 1989, the Distributed Generation Co-ordination Group (DGCG) was established to conduct a purchase obligation of renewable energy by introducing Renewable Obligation Certificates (ROCs) in 2002. A ROC is a green certificated issued by the authority to energy suppliers. ROC can be traded with other suppliers and each supplier needs a sufficient number of ROCs to meet its obligation. When the ROCs are not sufficient, the supplier has to pay an equivalent amount as fee, which can be distributed back to other green renewable energy suppliers.

In 2003, the Department of Trade and Industry (DTI) issued the Energy White Paper to define a goal with 60% reduction of GHG (greenhouse gas) emissions by 2050. In 2005, the Electricity Networks Strategy Group (ENSG) was launched as the organization's activities to DGCG continue [50]. The UK regulator Office of the Gas and Electricity Markets (Ofgem) and the government agency, that is, the Department of Energy and Climate Change (DECC), have initiated the DECC/Ofgem Smart Grid Forum, which provides a platform for various electricity network companies to cooperate closely to address significant challenges faced by the implementation of the Smart Grid. The ENSG, which is jointly chaired by both DECC and Ofgem, published the A Smart Grid Vision in November 2009, and A Smart Grid Routemap in

February 2010. In July 2010, DECC, Ofgem, and the Gas and Electricity Markets Authority (GEMA) jointly published a prospectus to propose the installation of electricity and gas smart metering in Great Britain. It is expected that from 2012 to 2019, more than 50 million smart meters will be installed in 30 million homes and smaller businesses in Britain [51].

In Denmark, the EDISON demonstration project has been launched to work out how the Smart Grid would integrate electric power system to meet the needs of most distributed wind power integration and the development of new energy vehicles. The Danish grid company Energinet, has invested in this project. IBM and Siemens are also involved in the construction of this project. More details of the Edison Project can be found in Chapter 4.

In Italy, Enel has already deployed a large number of smart meters and the electricity consumption data are transmitted to the utilities by GSM/GPRS (General Packet Radio Service), PSTN (Public Switched Telephone Network), KNX, and so on. Spain launched numerous Smart Grid projects such as the STAR project (Spanish acronym for Remote Grid Management and Automation System), Substation to Grid (S2G) wireless substation monitoring project, Movele project, and the Smart Community project. Spain has also cooperated with Japan to launch the Japan–Spain Innovation Program (JSIP) in 2010.

1.5 Status of Japan

The overall utilization efficiency and power supply reliability of the Japanese power system are in the leading position globally. In April 2009, the Japanese Prime Minister Taro Aso gave a speech, "Japan's Future Development Strategy and Growth Initiative toward Doubling the Size of Asia's Economy." One of the future development strategies is to enable Japan to become the leading country in the field of low-carbon emission, which includes becoming the number one solar power nation in the world, and the first nation to popularize eco-cars [52]. In July 2009, the Federation of Electric Power Companies (FEPCs) of Japan started to discuss how to develop the Japanese version of the the Smart Grid and incorporate solar power generation. In November 2009, the Japanese Ministry of Economy, Trade, and Industry (METI), launched the conference on next generation energy and social system, to discuss how to build a low-carbon society.

In April 2010, METI selected Yokohama City, Toyota City, Kansai Science City, and Kitakyushu City as areas for demonstration projects [53]. The Yokohama project in Kanagawa Prefecture was promoted by Yokohama, Accenture, Nissan, Toshiba, Meidensha, Panasonic, Tokyo Electric Power Company, and Tokyo Gas, to deploy large-scale renewable energy (27 000 kW PV system), to introduce smart home and building technologies for 4000 households, to deploy 2000 next-generation vehicles, and to reduce CO_2 emissions by 30% by 2025 compared to the level of 2004.

The project in Toyota City, Aichi Prefecture, was promoted by Toyota, Denso, Sharp, Chubu Electric, Toho Gas Co., Fujitsu, Toshiba, KDDI, Mitsubishi Motors, Circle K, Lawson, Home Toyota, and Mitsubishi, and so on. The goals are to deploy 3100 next-generation vehicles, to improve energy efficiency by using a mix of different energy sources (electricity, heat, and unused energy), and to reduce CO_2 emissions by 20% in households and 40% in transport. The project in the Kansai Science City, Kyoto, is supported by Kyoto city, Keihanna Science City, Kyoto University, Kansai Electric Power Co, and Kyoto City Gas. The goals are to install PV systems in 1000 households, to build "nano-grids" in homes and buildings for the purpose of controlling power generation systems and electric storage systems, to propose an energy economy model based on "Kyoto eco-points," and to reduce CO_2 emissions by 20% in households and 40% in transport by 2030, compared to the level of 2005. The project in the Kitakyushu City, Fukuoka Prefecture, is supported by Kitakyushu City government, Nippon Steel, Fuji Electric Systems, IBM Japan, and so on. The goals are to create a city block where new energy (including wind power, solar power, waste heat, etc.) supports 10% of the total energy consumption, to deploy smart meters in 70 companies and 200 households, and to reduce CO_2 emissions by 50% in the residential/commercial and transport sectors by 2030, and 80% by 2050.

METI has issued 25 Smart Grid core research subjects: (i) monitoring and control of systems for wide area transmission system; (ii) optimal control of storage battery system; (iii) optimal control of the battery for power distribution; (iv) optimal control of cells in the region; (v) billing the contents of the theme standardization; (vi) power conditioners for a high efficiency battery and distribution automation system; (vii) power conditioners for distributed power; (viii) equipment power electronics for power distribution; (ix) demand response network; (x) HEMS (Home Energy Management System); (xi) BEMS (Building Energy Management System); (xii) FEMS (Factory Energy Management System); (xiii) CEMS (Cluster/Community Energy Management System); (xiv) stationary storage system; (xv) battery modules; (xvi) evaluation method of the residual value for automotive storage battery; (xvii) EV rapid charger for Electric Vehicles; and so on. In 2010, METI announced the Next Generation Vehicle Strategy 2010, which specifies the strategy for developing new-generation vehicles. It sets the diffusion target that the next generation vehicles will account for up to 50% in 2020. In order to attain this goal, each Japanese automobile company should develop 17 EVs and 38 hybrid-power vehicles at the latest by 2020.

In Japan, the electricity generation and distribution businesses are not separated. Following the Fukushima Daiichi accident in March 2011 caused by the earthquake in East Japan, development and research interests have shifted away from nuclear power to EVs, LED lamps, and smart home and building automation systems. NEDO under METI has launched several projects to promote the development of the Japanese Smart Grid. One example is the Japan-US collaborative Smart Grid project in Los Alamos, New Mexico, which was launched in 2009 with an investment of $10 billion. NEDO has established the JSCA to promote cooperation among various stakeholders to accelerate Smart Grid-related activities in Japan. In order for Japanese companies to participate in Smart Grid-related activities, Japan has become actively involved in expanding the Asian Smart Grid market. NEDO has conducted an investigation of the Smart Grid-related technology requirements in the industrial areas surrounding Jakarta, Indonesia in 2010. This is the first time that Japan launched investigations of the Smart Grid in Southeast Asia. NEDO has investigated the current situation of Java Island Power Company in Jakarta, electricity supply and demand, electricity quality, number of factories, conditions of power stations, and so on. On the basis of the investigation results, the Japanese government has cooperated with the Indonesian government for the development of a smart community in Indonesia. NEDO has commissioned the Sumitomo Corporation, jointly with Fuji Electric Co., Ltd., Mitsubishi Electric Corporation, and NTT Communications Corporation, to implement the Smart Industrial Park project in the Indonesian Island of Java from 2013 to 2016.

1.6   Status of South Korea

In 2008, South Korea enacted the Green Growth Basic Law, established the Green Growth Institute, and founded the Global Green Growth Institute (GGGI). The law explicitly states that the green growth development is the first priority among national issues, and lays the foundation for other relevant laws.

On 15 August 2008, the South Korea government has established the Presidential Committee on Green Growth, which is an organization directly responsible to the President for reducing CO_2 emissions and promote green growth. Its vision is to enable South Korea to become the seventh Green Power by 2020, and fifth Green Power by 2050 [54]. In 2009, the Presidential Committee on Green Growth issued the guideline, Building an Advanced Green Country, which specifies the contents of the South Korea Smart Grid. In August 2009, the Korea Smart Grid Institute (KSGI) was launched to promote the development of the Smart Grid in South Korea. KSGI issued Korea's Smart Grid Roadmap, which specifies five sectors for implementing the Smart Grid: smart power grid, smart consumer, smart transportation, smart renewable, and smart electricity service [55]. In the first stage (2010–2012), the implementation direction was to construct and operate the Smart Grid test bed in pilot projects. In the second stage (2012–2020), the implementation direction is to expand the Smart Grid into metropolitan areas. In the final stage (2021–2030), the implementation direction is to complete a nationwide intelligent power grid.

In order to ensure smooth implementation of smart grid, the South Korean Government developed several supporting policies. For instance, it supports research, development, and industrialization, propagates successful modes, builds infrastructures, and establishes related policies and regulations. More details are shown in Table 1.1.

In January 2010, the South Korea Ministry of Knowledge Economy (MKE) issued the Korea Smart Grid Development Directions 2030, which predicted a 27.5 trillion

Table 1.1 Policy directions and implementation plans

Policy direction	Implementation plans
Support research, development, and industrialization	Support activities in technology development, standardization, and commercialization, and reward companies and individuals for voluntary participation in the construction of the Smart Grid
Promote successful modes	Explore successful development modes and share the experience of the Jeju Smart Grid test bed
Build infrastructure	Make incentive plans for infrastructure constructions
Establish related policies and regulations	Refine and revise the Smart Grid-related laws and regulations

won investment in Smart Grid, from which 24.8 trillion won would be invested for private sectors [56]. In April 2011, the National Assembly passed the Smart Grid Stimulus Law, which specifies how to develop the Smart Grid and promote infrastructure construction, how to get returns on the investment and tax protection, and how to promote technology research and improve information supervision and security.

Among the pilot Smart Grid projects launched by the South Korea Government, the Jeju Smart Grid test bed is of the most successful. The Jeju Smart Grid test bed was launched in June 2009. The project started from 2009 to 2013, with a total investment of 2.4 trillion won. Its objective is to set up the world largest Smart Grid test bed at the initial stage. The whole project is divided into two phases. In the first phase (December 2009 to May 2011), the Smart Grid demonstration infrastructures were constructed, and the key areas were smart power grid, smart place, and intelligent transportation. In the second phase (June 2011 to May 2013), new Smart Grid services would be integrated with the existing infrastructures, and the key areas were renewable energies, and electricity services.

1.7 Status of China

China has achieved rapid economic growth, and at the same time, the electric demand has also increased dramatically. China has enough coal reserves to cover domestic needs. However, the self-sufficiency in crude oil is lower than 50%, and China is the second largest oil user after the United States. In addition, as the gap in economic development between eastern coastal areas and middle, western China expands, the Chinese Government has launched several projects to transmit the energy resources from the western region to the eastern region, due to the higher demand in the latter. Environmental pollution has already become a severe problem, that is, China is the world's second largest CO_2 emitting country after the United States.

China has separated the electricity generation and transmission businesses since 2002, in order to introduce competition in the electricity market. As a result, the assets of the State Electric Power Corporation were divided into 11 companies, including two transmission companies and five generation companies. Among them, the SGCC, the monopoly position in the electricity transmission and distribution market, and is the leading company to develop Smart Grid-related projects and standards.

On 7 June 2010, the Former President Hu Jintao gave a speech at the fifteenth Academician Conference of Chinese Academy of Sciences, in which he pointed out that China should build a smart, highly efficient, and reliable Smart Grid to cover both urban and rural areas. In July 2010, SGCC issued the SGCC Framework and Roadmap for Strong and Smart Grid Standards, which specifies that China would build a world-leading Strong and Smart Grid with Ultra High Voltage (UHV) grid as its backbone. Since 2009, SGCC has launched 228 projects of 21 categories in 26 provinces and cities [57]. The UHV transmission technology has been specified as one of the key technologies in the Outline of the National Long and Medium Term Program for Scientific and Technical Development (2006–2020), which was issued by the State Council. UHV demonstration projects were listed in the Key Projects List Of 2005–2006 National Energy Work Outline. SGCC specified three stages to realize the Strong and Smart Grid [59]:

- **2009–2010 Phase I Pilot Study**: issue technical and operational standards of the Smart Grid, develop technologies and equipment, and perform trial tests.
- **2011–2015 Phase II Construction and Development**: construct UHV urban/rural grids, establish the fundamental framework for Strong and Smart Grid operation control and interoperation, and achieve advancements in Smart Grid technologies and equipment productions.
- **2016–2020 year Phase III Upgrade**: enhance technologies and equipment development for the Strong and Smart Grids.

Furthermore, SGCC is actively involved in the standard development activities. China Electric Power Research Institute (CEPRI), which is a comprehensive research institute and subsidiary of SGCC, has been leading the development of specifications for IEC PC 118 Smart Grid User Interface. The goal of the IEC PC118 is to develop a standard and unified interface for information exchange between the demand-side equipment and/or systems and the Smart Grid. The simultaneous achievement of maintaining a stable power supply and reducing CO_2 emissions is an important issue in China's energy policy. The Chinese Government has launched numerous demonstration projects to develop the Smart Grid technologies. Among them, the Sino-Singapore Tianjin Eco-city is an international project launched by both China and Singapore Government in 2007. The Sino-Singapore Tianjin Eco-city has a total land area of $30\,km^2$, and is expected to be completed in around 2020. The science and technology project is promoted with focus on clean water, ecology, clean environment, green transport, clean energy, green building, and city management [58]. Experts from both countries gathered together and formulated the Key Performance Indicators (KPIs), which are used to guide the planning and development of the

Eco-city. Currently, there are 22 quantitative and four qualitative KPIs. It is expected that the Eco-city will have an estimated 350 000 residents in around 2020.

1.8 Conclusions

This chapter provided an introduction to various Smart Grid-related SDOs, Technical Consortia, Forums, and Panels, Political, Market, Trade, and Regulatory Organizations, Forums, and Alliances. It also provided an overview of the development of the Smart Grid in both developed and developing countries, such as the United States, the European Union, Japan, South Korea, and China. Through the NIST, the United States have been actively involved in promoting open and interoperable standards for the Smart Grid. Other SDOs such as IEEE, IEC, ISO, CEN, CENELEC, and IETF are cooperating closely with NIST to achieve interoperability among Smart Grid-related standards. Crude oil, natural gas, coal, and other natural resources, have a finite supply, and the world might run out of these natural resources in the future. Therefore, the use of renewable energy resources, including solar power, wind power, hydroelectric power, and geothermal power, is required for sustainable development. However, the renewable energy resources have some sort of seasonal and diurnal profile, which has been discussed in more detail in Chapter 9, and might cause the supply of energy to be unstable. Smart Grid technologies should be used to adapt energy production to energy consumption, to lower the total cost of energy production, and to increase the energy use efficiency. Bi-directional communication and information technologies should be integrated with the Smart Grid to enable customers to exchange information with electricity service providers or with each other. It is necessary to standardize all interfaces to ensure interoperability among different Smart Grid systems. It is also required for the Smart Grid to control a lot of energy information while ensuring the security and reliability at the same time. In conclusion, it is important to promote the development of international standards for the Smart Grid as early as possible.

References

[1] Green Living (2009) *15th Session of the Conference of the Parties to the UNFCCC (COP 15)*, The Copenhagen Climate Change Conference, www.cop15.dk/en (accessed 31 January 2013).
[2] Energy Community (2013) *Milestones*, www.etsi.org/about/our-role-in-europe/public-policy/ec-directives (accessed 11 March 2013).
[3] European Commission (2013) *Single Market for Gas and Electricity Internal Energy Market*, http://ec.europa.eu/energy/gas_electricity/legislation/legislation_en.htm (accessed 11 March 2013).
[4] European Commission (2013) *Single Market for Gas and Electricity Third Package*, http://ec.europa.eu/energy/gas_electricity/legislation/third_legislative_package_en.htm (accessed 11 March 2013).
[5] Agency for the Cooperation of Energy Regulators (2011) *The Agency*, www.acer.europa.eu/The_agency/Pages/default.aspx (accessed 11 March 2013).

[6] IEC SMB Smart Grid Strategic Group (2010) *IEC Smart Grid Standardization Roadmap Edition 1.0*, www.iec.ch/smartgrid/downloads/sg3_roadmap.pdf (accessed 27 December 2012).

[7] Institute of Electrical and Electronics Engineers (2010) *Smart Grid Information*, https://mentor.ieee.org/802-ec/dcn/10/ec-10-0013-00-00EC-smart-grid-information-update-july-2010.pdf (accessed 11 March 2013).

[8] International Organization for Standardization (2013) *About ISO*, www.iso.org/iso/about.htm. (accessed 18 February 2013).

[9] International Organization for Standardization (2013) *How to Use ISO Catalogue*, www.iso.org/iso/iso_catalogue/how_to_use_the_catalogue.htm. (accessed 18 February 2013).

[10] United Nations Development Group (2013) *UNDG Members*, www.undg.org/index.cfm?P=13 (accessed 18 February 2013).

[11] Internal Telecommunication Union (2011) *ITU Telecom World 2011*, http://world2011.itu.int/. (accessed 18 February 2013).

[12] Institute of Electrical and Electronics Engineers (2013) *IEEE Technical Activities Board Operations Manual*, www.ieee.org/about/volunteers/tab_operations_manual.pdf. (accessed 18 February 2013).

[13] Institute of Electrical and Electronics Engineers (2013) *IEEE at a Glance*, www.ieee.org/about/today/at_a_glance.html#sect1. (accessed 18 February 2013).

[14] Institute of Electrical and Electronics Engineers (2010) *IEEE Annual Report 2010*, www.ieee.org/documents/ieee_annual_report_10_1.pdf. (accessed 18 February 2013).

[15] European Committee for Electrotechnical Standardization (2013) *CENELEC Global Partners*, www.cenelec.eu/aboutcenelec/whoweare/globalpartners/nationalpartners.html. (accessed 19 February 2013).

[16] Telecommunication Industry Association (2013) *Technology and Standards*, www.tiaonline.org/standards/. (accessed 18 February 2013).

[17] Internet Engineering Task Force (2013) *About the IETF*, www.ietf.org/about/. (accessed 17 March 2013).

[18] SAE International (2013) *An Abridged History of SAE*, www.sae.org/about/general/history/. (accessed 17 March 2013).

[19] National Electrical Manufacturers Association *About the NEMA*, www.nema.org/About/Pages/default.aspx (accessed 17 March 2013).

[20] Organization for the Advancement of Structured Information Standards *About Us*, www.oasis-open.org/org. (accessed 17 March 2013).

[21] Hammerstrom, D.J., R Ambrosio, TA Carlon *et al.* (2007) *Pacific Northwest GridWise™ Testbed Demonstration Projects: Part I. Olympic Peninsula Project*, PNNL-17167, Pacific Northwest National Laboratory, Richland, WA.

[22] Cazalet, E.G. (2010) TeMIX: a foundation for transactive energy in a smart grid world. *Grid-Interop 2010, Chicago, IL*, www.pointview.com/data/files/2/1062/1878.pdf. (accessed 2 March 2013).

[23] Pacific Northwest Smart Grid (2013) *Pacific Northwest Smart Grid Demonstration Project*, www.pnwsmartgrid.org/. (accessed 2 March 2013).

[24] Smart Grid Interoperability Panel (2009) *About Us*, http://sgip.org/about_us/ .(accessed 2 March 2013).

[25] International Energy Agency (2013) *About Us*, www.iea.org/aboutus/. (accessed 11 March 2013).

[26] Clean Energy Ministerial (2009) *About the Clean Energy Ministerial*, www.cleanenergyministerial.org/About.aspx (accessed 11 March 2013).

[27] Clean Energy Ministerial (2013) *Initiatives*, www.cleanenergyministerial.org/OurWork/Initiatives.asp (accessed 11 March 2013).

[28] The Energy Library (2004) *Northeast Blackout of 2003*, www.theenergylibrary.com/node/13088 (accessed 11 March 2013).

[29] US Department of Energy (2002) *DOE National Transmission Grid Study*, www.ferc.gov/industries/electric/gen-info/transmission-grid.pdf. (accessed 12 March 2013).

[30] US Department of Energy (2003) *"Grid 2030" A National Vision for Electricity's Second 100 Years*, http://energy.gov/sites/prod/files/oeprod/DocumentsandMedia/Electric_Vision_Document.pdf (accessed 12 March 2013).

[31] Electric Power Research Institute (2003) *Electricity Technology Roadmap-Meeting the Critical Challenges of the 21st Century*, http://mydocs.epri.com/docs/CorporateDocuments/StrategicVision/Roadmap2003.pdf (accessed 12 March 2013).

[32] US Department of Energy (2006) *Roadmap to Secure Control Systems in the Energy Sector*, www.cyber.st.dhs.gov/docs/DOE%20Roadmap%202006.pdf. (accessed 12 March 2013).

[33] National Energy Technology Laboratory (2007) *A Vision for the Modern Grid V1.0*, www.bpa.gov/energy/n/smart_grid/docs/Vision_for_theModernGrid_Final.pdf (accessed 11 March 2013).

[34] National Energy Technology Laboratory (2009) *A Vision for the Smart Grid*, www.netl.doe.gov/smartgrid/referenceshelf/whitepapers/Whitepaper_The%20Modern%20Grid%20Vision_APPROVED_2009_06_18.pdf (accessed 11 March 2013).

[35] National Institute of Standards and Technology (2010) NIST Framework and Roadmap for Smart Grid Interoperability Standards Realease 2.0, http://www.nist.gov/smartgrid/framework-022812.cfm (accessed 26 November 2014).

[36] US Department of Energy (2005) *Energy Policy Act of 2005*, www1.eere.energy.gov/femp/regulations/epact2005.html. (accessed 11 March 2013).

[37] US Congress (2009) *The Recovery Act*, www.recovery.gov/About/Pages/The_Act.aspx. (accessed 11 March 2013).

[38] Austin Energy (2004) *Austin Energy Smart Grid Program*, www.austinenergy.com/about%20us/company%20profile/smartGrid/index.htm .(accessed 11 March 2013).

[39] Pacific Gas and Electric (2009) *The SmartMeter™ Deployment*, www.pge.com/myhome/customerservice/smartmeter/deployment/. (accessed 11 March 2013).

[40] Dominion Virginia Power (2009) *Dominion Virginia Power AMI Project*, www.sgiclearinghouse.org/ProjectList?q=node/1670&lb=1. (accessed 11 March 2013).

[41] Businesswire (2010) *Silver Spring and ClipperCreek Join PG&E and EPRI in Innovative Electric Vehicle Smart Charging Pilot*, www.businesswire.com/news/home/20100727005740/en/Silver-Spring-ClipperCreek-Join-PGE-EPRI-Innovative (accessed 11 March 2013).

[42] Stuart, B. (2012) *Japan-US Smart Grid Demonstration Project Announced.* PV-Magazine, www.pv-magazine.com/news/details/beitrag/japan-us-smart-grid-demonstration-project-announced_100006872/#axzz2MR2YKPus (accessed 11 March 2013).

[43] European Commission (2006) *European SmartGrids Technology Platform-Vision and Strategy for Europe's Electricity Networks of the Future*, ftp://ftp.cordis.europa.eu/pub/fp7/energy/docs/smartgrids_en.pdf. (accessed 11 March 2013).

[44] CEN/CENELEC/ETSI Joint Working Group (2011) *Final Report of the CEN/CENELEC/ETSI Joint Working Group on Standards for Smart Grids*, ftp://ftp.cen.eu/CEN/Sectors/List/Energy/SmartGrids/SmartGridFinalReport.pdf. (accessed 27 December 2012).

[45] European Commission (2012) *Smart Grids SRA (Strategic Research Agenda) 2035 Strategic Research Agenda Update of the Smart Grids SRA 2007 for the Needs by the Year 2035*, www.smartgrids.eu/documents/sra2035.pdf. (accessed 11 March 2013).

[46] European Network of Transmission System Operators for Electricity (2012) *Ten-Year Network Development Plan 2012*, www.entsoe.eu/fileadmin/user_upload/_library/SDC/TYNDP/2012/TYNDP_2012_report.pdf .(accessed 12 March 2013).

[47] Halper, M. (2011) *Construction of World's Biggest Solar Project Starts in 2012*, www.smartplanet.com/blog/intelligent-energy/construction-of-worlds-biggest-solar-project-starts-in-2012/10235 (accessed 12 March 2013).

[48] Federal Ministry of Economics and Technology (2008) *E-Energy-Smart Grids Made in Germany*, www.e-energy.de/en/. (accessed 12 March 2013).

[49] Creti, A., Pouyet, J., and Sanin, M.E. (2011) *The Law NOME: Some Implications for the French Electricity Markets*, CEPREMAP Working Papers (Docweb) 1102, CEPREMAP, www.cepremap .ens.fr/depot/docweb/docweb1102.pdf. (accessed 12 March 2013).

[50] Department of Energy and Climate Change (2005) *Electricity Networks Strategy Group ENSG*, www.decc.gov.uk/en/content/cms/meeting_energy/network/ensg/ensg.aspx. (accessed 12 March 2013).

[51] Richards, P. (2012) *Smart Meters*, www.parliament.uk/briefing-papers/sn06179.pdf. (accessed 12 March 2013).

[52] Japan National Press Club (2009) *Japan's Future Development Strategy and Growth Initiative towards Doubling the Size of Asia's Economy*, www.kantei.go.jp/foreign/asospeech/ 2009/04/09speech_e.html (accessed 14 March 2013).

[53] Ministry of Economy, Trade and Industry (2010) *Selection of Next-generation Energy and Social Systems Demonstration Areas*, www.meti.go.jp/english/press/data/pdf/N-G%20System.pdf. (accessed 14 March 2013).

[54] Ministry of Environment (2008) *The Presidential Committee on Green Growth*, http://eng .me.go.kr/content.do?method=moveContent&menuCode=pol_pol_edu_gov_growth (accessed 14 March 2013).

[55] Korea Smart Grid Institute (2010) *Korea's Smart Grid Roadmap*, www.smartgrid.or.kr/ 10eng4-1.php. (accessed 14 March 2013).

[56] Korea Smart Grid Institute (2010) *Korea, State of Illinois Conclude MOU on Smart Grid*, www.smartgrid.or.kr/view.php?id=10eng1&no=21. (accessed 14 March 2013).

[57] State Grid Corporation of China (2010) *SGCC Framework and Roadmap for Strong and Smart Grid Standards*, www.cspress.cn/u/cms/www/201208/16154808z5u9.pdf. (accessed 10 March 2013).

[58] Tianjin Eco-City *A Model for Sustainable Development*, www.tianjinecocity.gov.sg/ .(accessed 13 March 2013).

[59] Bojanczyk, K. (2012) *Reprint: China and the World's Greatest Smart Grid Opportunity*, www.greentechmedia.com/articles/read/enter-the-dragon-china-and-the-worlds-greatest-smart-grid-opportunity/. (accessed 13 March 2013).

2

Renewable Energy Generation

2.1 Introduction

The economic growth in the last few decades can be attributed to the improved ways of energy production, especially to the capability of electricity generation from various energy resources. The energy resources include fossil fuels, such as hydrocarbon and nuclear fuels, along with various renewable energy sources. Hydrocarbons are the major drivers of an increase in greenhouse gas emissions that contribute to climate change. On the contrary, nuclear energy, along with renewables, is one of the clean electricity sources. On the basis of the estimate of the United States Department of Energy, alternative energy sources, such as nuclear energy, have the potential to supply the energy demand for several decades [1]. The amount of energy contained in a mass of hydrocarbon fuel is considerably lower than the energy in a much smaller mass of nuclear fuel. However, there are various drawbacks associated with nuclear power. In addition to the long life span of the radioactive waste from nuclear fuel, the destructive potential of nuclear energy in a potentially unstable world will always remain a serious threat to peace and stability [2]. As a result, many countries and international bodies have shown overwhelming interest in renewable technologies for the production of the required clean energy.

Our planet has a large potential for diverse renewable energy resources. These resources include hydro, biomass, geothermal, ocean wave, solar, and windpower. These renewable resources are geographically constrained but widespread in various regions of the world. Commercially available renewable technologies among these broad categories of renewable sources have been deployed in various countries around the world. These technologies include stand-alone bio, geothermal, hydro, distributed photovoltaic (PV), utility-scale PV, Concentrated Solar Power (CSP), onshore wind, and fixed-bottom offshore windpower. As of now, 10% of the total US electricity supply is obtained from these resources. According to the report of the International Energy Agency's World Energy Outlook [3], the world electricity generation from

Smart Grid Standards: Specifications, Requirements, and Technologies, First Edition. Takuro Sato,
Daniel M. Kammen, Bin Duan, Martin Macuha, Zhenyu Zhou, Jun Wu, Muhammad Tariq and Solomon Abebe Asfaw.
© 2015 John Wiley & Sons, Ltd. Published 2015 by John Wiley & Sons, Ltd.

renewable resources is expected to be doubled over the next 25 years. Out of this generation, the share of the renewable energy is expected to be around 57% [3]. According to the Renewable Electricity Future Study (National Renewable Energy Lab, NREL) [4], which examines the implications and challenges of transitioning to a grid that relies on renewable electricity generation from now to the year 2050 by increasing renewable penetration from 30% to 90%, up to 80% of the 2050 US electricity demand could be supplied using renewable technologies. The study focuses on some key technical implications on environment, exploring whether the US power system can supply electricity to meet customer demand with high levels of renewable electricity, including the variable wind and solar energy-generating technologies.

Besides the desire to reduce pollution in the electricity industry, the increased interest in the renewable energy technologies has been motivated by the falling cost of renewable materials and an increase in the technical efficiency. However, the cost of energy generation per unit from renewable resources is still higher than that from conventional sources. The cost comparison of energy generated from different energy sources (according to the report of the United States Department of Energy Information Administration (EIA) [5]) is shown in Table 2.1.

In addition to cost, renewable energy sources have varying degrees of uncontrollability and uncertainty, and the output characteristics of the associated technologies vary considerably. Since renewable technologies provide greater levels of electricity to the grid, it is significant to consider variability, uncontrollability, uncertainty, and the output characteristics in grid planning and operations to ensure a real-time balance of electricity supply and demand over various timescales. For example, wind and solar PV are dependent on weather conditions, which are beyond the control of human beings. As a consequence, to reliably meet our electricity demand, power grids will have to be designed in a way that they will permit us to take advantage of the various advantages of these technologies [4, 6–8].

Table 2.1 Comparison of energy (generated from different sources) cost per unit (megajoule)

Sources	Cost $/megajoule (approximately)
Electricity	0.016
Natural gas	0.05
Barreled oil	0.013
Solar	0.10
Wind	0.10
Geothermal	0.03

2.2 Renewable Energy Systems and the Smart Grid

Distributed power generation is becoming an important part of the developing plans for most countries to meet the energy demand in the future [3]. The idea of distributed power generation is particularly attractive when different kinds of renewable energy sources are available. Distributed power generation of these renewable energy systems allows for the integration of renewable and nontraditional energy sources. With distributed energy generation and storage, the power and energy engineering will definitely face a new situation in which small-distributed power generators and scattered energy storage devices have to be integrated into a single grid commonly known as the Smart Grid. The Smart Grid will distribute electricity from suppliers to consumers using Information and Communication Technologies (ICT) to control appliances at consumers' homes and machines in the industries. In this way, it will reduce cost, save energy, and increase reliability with more transparency [9, 10].

With the growing electrification of various new applications, the demand for electricity is increasing every year. It is expected that, in the near future, more than 60% of all our energy need will be supplied by electricity [3]. Therefore, the efficient utilization of electricity, its production and distribution is important in order to optimize the overall energy consumption. In addition, recent challenges with nuclear power plants, such as the Fukushima Nuclear Power Plant disaster on 11th March 2011, are demanding more sustainable energy generation solutions. The Smart Grid is considered to be one of the key technologies to play an important role in solving parts of those current and future challenges. The Smart Grid will be helpful in changing the electrical power production from conventional fossil-based energy systems to the renewable energy systems. In addition, to make the system smarter, the new information technologies will be implemented in the existing power grid using highly efficient power electronics, communication technologies, and software-based systems in power generation, power transmission, distribution, and end-user applications.

In the following subsections, a detailed description of various renewable energy systems and technologies, their specification together with the relevant standards, which have been developed by different Standards Developing Organization (SDOs), is provided.

2.2.1 Hydroelectric Power

2.2.1.1 Historical Prospective of Hydroelectric Power

Hydroelectric power is the cheapest and widely available renewable energy source worldwide. It is traced centuries back when hydropower was used to turn wooden or stone waterwheels for grinding grain. Today, very big hydropower dams with compound equipment generate electricity to provide power for households, businesses, and industries. Commercially, hydroelectric power is still viable for a century or so [11]. This long lifetime of hydropower facilities reflects an important economic aspect. Hydropwer projects can recover capital investment much before than the end of their actual useful life. Once the debt services are paid, these projects have no fuel cost, the equipments are expected to work for a long period, while at the same time the operating costs are extremely low. A hydropower project developed privately will have a debt payment structure for 10–17 years, while a publicly funded project would have a bit longer term. Upon fulfilling the debt service, the only remaining costs are of Operation and Maintenance (O&M), and of the life extension of the equipment and structures. Once the debt is repaid the cost of power is reduced significantly. For example, the cost of power drops to less than $1/MW h for a small hydropower project, and for large-scale projects to less than $0.5/MW h. With the advancement in technology, upgradation of existing power facilities is important, which usually leads to an increased power output and energy production. Similarly, rehabilitation and upgrading of existing facilities can prove to be extremely cost effective, often ranging from approximately $200/kW to approximately $600/kW, which is a fraction of the cost of new facilities [4].

Hydropower is one of the oldest renewable energy sources used to generate electricity. For example, electricity has been generated from hydropower in the United States since 1880. However, since 1995, the induction of additional hydropower capacity in the United States has been small. In spite of this lower addition, hydropower is currently the largest source of renewable electricity generation in the United States, representing approximately 7% of total electricity generation. The historical growth in conventional hydropower capacity can be seen in Figure 2.1.

2.2.1.2   Introduction of Hydropower Technology

Water is the main resource behind hydroelectric power technology. Water is stored in reservoirs called dams. The stored water contains potential energy that can be converted to electricity in the hydropower plant. As the water passes from its source through a penstock, the potential energy is converted to kinetic energy. As the water spins a turbine, which may be a simple waterwheel, a reaction turbine, a propeller-like device, or a complex turbine with blades that can be adjusted during an operation, the kinetic energy of the water is converted to mechanical energy. The turbine is mechanically connected to a generator, which converts the mechanical energy into electrical energy as can be seen in Figure 2.2. In this way, the electricity is produced, commonly known as hydroelectricity. The hydroelectricity capacity is dependent on both the flow through the turbine and the hydraulic head. The hydraulic head is the height measured in feet or meters. The headwater surface behind the dam is above the tailwater surface directly downstream of the dam.

There are two main categories of conventional hydropower plants, namely, the run-of-river and the storage-based projects, commonly known as dams. The former

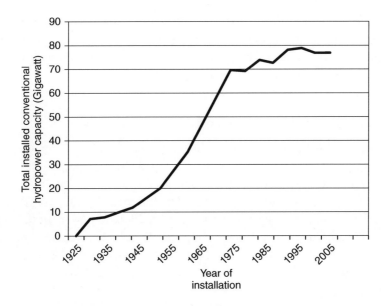

Figure 2.1   Historical development of hydroelectric power in the United States Source [4].

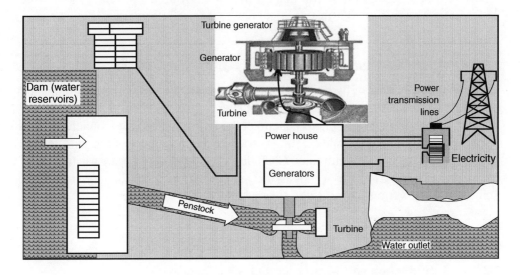

Figure 2.2   Sketch of the hydroelectric power plants

might or might not use a water reservoir to create hydraulic head for generating power. In such projects, the flow rate of water through the turbines is almost same as the rate at which water enters the reservoir from the river. The latter uses a water reservoir to store water to shift the generation of power to the times having the greatest need for electricity and also to increase the height of the water. The stored water in dams enables a project to vary electricity generation and send it to meet demand. In addition to the generation of electricity, the stored water commonly serves other useful functions such as flood protection, domestic water supply, irrigation, recreation, navigation, fishery, and protection of the environment.

2.2.1.3   Hydroelectric Power Standards

Despite its widespread use in various applications, about 73% of the world's hydroelectric energy potential still remains undeveloped. Much of this potential lies in Asia, Africa, and Latin America. Initially, various applications of hydroelectric power have benefited greatly from standardization. The International Electrotechnical Commission (IEC) began preparing standards for hydroelectric power. In the beginning, the IEC Technical Committee 4 was set up in 1911, which prepared standards and technical reports for designing, manufacturing, commissioning, testing, and operating hydraulic machines. Today, various SDOs have developed standards for turbines, storage pumps, and pump turbines for different types of wells, and for the related equipment such as speed governors and performance evaluation and testing [3, 11, 12].

Table 2.2 provides details about main hydropower standards. These standards are related to power generation, power plant installation, hydraulic turbines, monitoring and maintenance, power transmission, and power storage at power grid stations.

2.2.1.4   Advantages and Disadvantages of Hydroelectric Power

The production process of hydroelectric power is largely free from the environmental effects which are associated with traditional energy sources. However, the development of hydroelectric power plants can affect the environment by impounding water, flooding terrestrial habitats, and creating barriers to the movements of fish and aquatic organisms, sediments, and nutrients. Changing of water flows by constructing dams can affect aquatic and terrestrial habitats that are downstream of the dam reservoirs [4]. Table 2.3 shows various benefits and adverse effects of hydropower.

2.2.2   Solar Energy

2.2.2.1   Introduction and Background of Solar Energy

Humans have been using solar energy since ancient times using various forms of technologies over the years. These early technologies include architectural passive cooling and heating systems. In passive heating, the energy was stored and distributed without the need for complex controllers that are used today, thereby raising the indoor temperature level above the outdoor level. On the other hand, passive cooling systems transferred incident energy to heat sinks, such as the air, the upper atmosphere, water, and the earth. In this way, it decreased the temperature of the air inside the area [11].

Table 2.2 Standards of hydroelectric power systems developed by various SDOs

Standard	Main focus	Description
IEC 61850-7-410	Hydroelectric power plants – communication for monitoring and control	This standard is of high relevance to the Smart Grid. It specifies the additional common data class, the logical nodes, and data objects required for the use of the IEC 61850 standard in a hydropower plant
IEC-EN 61116	Electromechanical equipment guide for small hydroelectric installations	The standard applies to installations having outputs of less than 5 MW and turbines with diameters less than 3 m
IEC 60041	Field acceptance test	The standard is related to field acceptance tests in order to determine the hydraulic performance of hydraulic turbines, storage pumps, and turbines
IEC 60193	Hydraulic turbines, storage, and turbine pumps performance (model acceptance test)	The standard is related to hydraulic turbines, storage pumps, and pump turbines model acceptance tests
IEC 60308	Hydraulic turbine speed	The standard specifies the international code for testing of speed governing systems for hydraulic turbines
IEC 60994	Measurement of vibrations and pulsations in hydraulic machines	The standard is basically a guide for field measurement of vibrations and pulsations in hydraulic machines
IEC 60545	Testing of hydraulic turbine speed	The standard is a guide for commissioning, operation, and maintenance of hydraulic turbines
IEC 60609	Hydraulic machines' field measurement of vibrations and pulsations	The standard is related to cavitations pitting evaluation in hydraulic turbines, storage pumps, and pump turbines
IEC 60609-2	Hydraulic turbine operation, maintenance, and commissioning – related to cavitations pitting evaluation	The standard is a guide for commissioning, operation, and maintenance of storage pumps and of pump turbines operating as pumps (part 2 – evaluation in Pelton turbines)
IEC 60805	Equipment guide for small hydroelectric installations	It provides the electromechanical equipment guide for the small hydroelectric installations

(*continued overleaf*)

Table 2.2 (*continued*)

Standard	Main focus	Description
IEC 61116	Hydraulic turbine control systems	The standard is a guide to the specification of hydraulic turbine control systems
IEC 61362	Hydraulic turbine governing system	It provides guidelines for hydraulic turbine governing systems
IEC 61364	Nomenclature for power plant machinery	It provides specification for hydraulic turbine governing systems (the nomenclature for hydroelectric power plant machinery)
IEC 61366-1	General and annexes	The standard is related to hydraulic turbines, storage pumps, and pump turbines (part 1 – general and annexes)
IEC 61366-2	Guidelines for technical specifications for Francis turbines	Part 2 of IEC 61366 provides guidelines for technical specifications for Francis turbines
IEC 61366-3	Guidelines for technical specifications for Pelton and propeller turbines	Part 3 of IEC 61366 provides guidelines for technical specifications for Pelton turbines
IEC 61366-4	Guidelines for technical specifications for Kaplan and propeller turbines	Part 4 of IEC 61366 provides guidelines for technical specifications for Kaplan and propeller turbines
IEC 61366-5	Guidelines for technical specifications for tubular turbines	Part 5 of IEC 61366 standard provides guidelines for technical specifications for tubular turbines
IEC 61366-6	Guidelines for technical specifications for storage pumps	Part 6 of IEC 61366 standard provides guidelines for technical specifications for pump turbines
IEC 61366-7	Guidelines for technical specifications for pump turbines	Part 7 of IEC 61366 provides guidelines for technical specifications for storage pumps

Similarly, the sun emits radiant energy in the form of electromagnetic waves that heats the earth and gives green plants the energy they need for the process commonly known as photosynthesis.

The earth receives a huge amount of energy from the sun on its surface every day. For example, every year, the US population uses about 4000 TWh of electrical energy, which is equal to the amount of energy that the US land surface receives from the sun in a few hours of sunshine. Not only in the United States but in most of the regions of the world, solar energy technologies have access to an abundant energy resource, more than any other renewable energy technology. This is because, compared to other energy resources, solar energy is more equally distributed over the earth surface [4].

Table 2.3 Potential advantages and adverse effects of hydropower

Advantages	Adverse effects
Still the cheapest form of energy among the renewable energy systems	Building dams needs highest standards and quality construction materials
The dam can also be used for irrigation purposes	Production of electricity may not be continuous as drought or dry seasons would affect the production
Water dams are considered to last for generations	Building dams in high seismic areas are potentially very dangerous
Free from sulfur and nitrogen oxides emissions	Its impact on the nature has also been criticized by the ecologists and biologists, for example, in the cases of river flora and fauna
Very few solid wastes	Greenhouse gases emitted by the vegetables and plants that grow because of flooded water
Reservoir-based fisheries	Displacement of people and terrestrial wildlife
Negligible effects from the extraction of resources, its preparation, and navigation	Hydropower changes the quality and temperature of water considerably
Very effective in controlling floods	It changes the upstream and downstream passageway of aquatic organisms
Inland waterways, transportation, and navigation below the dam have been developed over the years	

2.2.2.2 Technology Overview

Solar Photovoltaic

Electricity production from a PV module is the result of charge carrier separation from a photon-absorbing medium by incoming photons from solar irradiation. The building blocks of PV module are solar cells, which are fabricated from wafers of crystalline silicon, or thin films created by depositing a photosensitive substance. These substances include amorphous silicon, copper indium gallium sulfide, and cadmium telluride. The basic components of a typical crystalline silicon cell are shown in Figure 2.3. Several PV cells are connected together to form PV modules. PV module project/plants typically include tens to thousands of PV modules connected electrically together in rows, known as a PV array. These arrays generate DC power which can be converted to AC power by using devices known as inverters.

A modern PV array consists of several PV modules that produce DC power, which is converted to AC using inverters in order to feed it into the AC power grid. It could also be used to charge a battery as given in Figure 2.4. PV arrays can be stationary, or mounted on sun-tracking stands. The tracking system can include optical concentration systems that will increase the amount of light under which the solar cell is being operated.

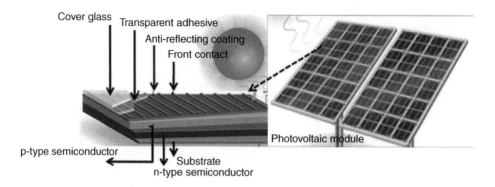

Figure 2.3   Components of typical crystalline silicon PV cell

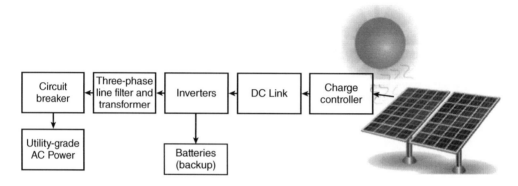

Figure 2.4   Block diagram of solar energy system

Concentrating Solar Power
CSP is another form of solar technologies to generate electricity. CSP technologies use lenses to focus sunlight onto a receiver. There is working fluid inside the receiver. The electrical generator is driven by the heat engine which is generated from thermal energy that is transferred from the receiver.

Other Solar Technologies
Other than Solar PV and CSP, there are several other solar technologies. These technologies include water heating, space heating, cooling, irrigation, transportation, and lighting. These technologies do not generate electricity but reduce the use of electricity and fossil fuel consumption. Although these technologies are not directly related to renewable energy systems, they are likely to be the important part to energy-efficiency investments for stabilizing or reducing end-use electricity demand in the future [4].

2.2.2.3 Solar Technologies – Growth, Cost, and Price Projection

Although both PV and CSP solar technologies were built on discoveries, which can be traced back to several centuries, the real development in electricity-generating technologies started in the 1970s and 1980s. The fraction of electricity generated by solar technologies in the United States at present is small, but it is growing rapidly. In 2011, the United States added just under 1500 MW of grid-tied AC-equivalent PV capacity, bringing the cumulative total to more than 3400 MW [13]. CSP capacity grew by about 100 MW from 2009 to 2011, bringing the cumulative total to approximately 520 MW [4, 13]. This reflects that about 0.2% of US electricity demand is being met by PV and 0.015% by CSP [14]. It should be kept in mind that the US PV market is responsible for only a small fraction of the total global PV market, which reached about 48 GW of grid-connected AC-equivalent capacity by the end of 2011 [15, 16, 29]. In comparison with cumulative installed global CSP capacity, the US CSP market made up approximately one-third of the cumulative installed global CSP capacity by 2011. Outside the United States, most of the CSP technology is installed in Spain [4, 17, 30–32].

The history of the operating mechanism that enables PV cells to generate electricity has been traced back to the mid-1800s. However, the first cell based on silicon using this mechanism was developed in the mid-1900s, while manufacturing techniques for bulk electricity-generating PV modules were developed in the late 1970s and 1980s. Thin-film PV technologies first came to the market in the 1970s, many of them being non-silicon based. The production on commercial scale of bulk electricity-generating PV modules began over the last quarter century. Similarly, CSP technology was used in the late 1800s in different agricultural applications; however, it was utilized for electricity generation in the 1980s. The trend in the growth of solar PV and CSP markets in the United States is illustrated in Figure 2.5.

The cost of a solar energy project is not only based on the cost of PV/CSP modules but also on other costs such as that of inverters, batteries, mounting or tracking structures, wiring, site-specific installation, and the indirect costs. The indirect cost includes engineering, procurement and construction, land, and project management. Direct and indirect price projection is shown in Figure 2.6.

2.2.2.4 Solar Technologies Standards

Various SDOs have developed standards for solar energy systems. For example, the IEC Technical Committee 82 (Solar photovoltaic energy systems, PVESs) has developed standards for both crystalline silicon and thin-film PV and CSP modules. In addition, it develops standards for certification of PV products and accreditation of PV test laboratories and balance-of-system components such as charge controllers, inverters, and power conditioners.

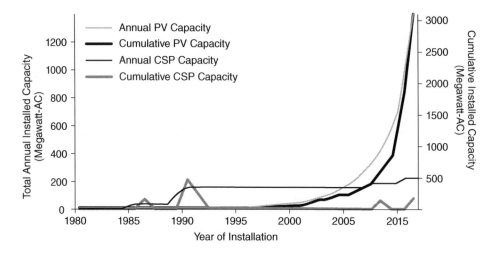

Figure 2.5 Growth of Solar PV and CSP in the United States. Source [4].

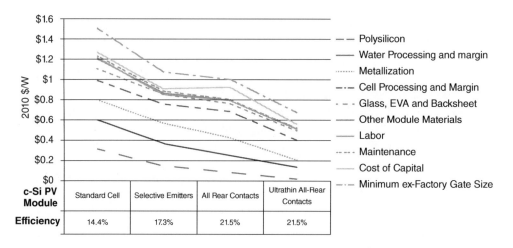

Figure 2.6 Price project of module by component for monocrystalline silicon PV. Source: [16].

Table 2.4 provides details about various standards developed for solar and PVESs by different organizations. These standards include the information related to materials of solar panels, batteries, inverters, and other objects used in solar energy systems. These standards also include specifications about installation, safety requirements, maintenance, and storage of solar energy.

2.2.2.5 Advantages and Disadvantages of Solar Energy

Table 2.5 lists the advantages and disadvantages of solar energy systems.

Table 2.4 Standards of solar PV and CSP systems developed by various SDOs

Standard name	Main focus	Descriptions
IEC-EN 61427	the Secondary cells and batteries for solar PV energy systems (general requirements and method of test)	The standard basically provides information about secondary cells and batteries for solar PV energy systems. If storage is used then the secondary cells are the rechargeable batteries
IEC-EN 61724	PV system performance monitoring (guidelines for measurement, data exchange, and analysis)	The standard recommends procedures for the monitoring of energy-related PV system characteristics, and for the exchange and analysis of monitored data. The main objective of designing this standard is to assess the overall performance of PV systems
IEC-EN 61727	Characteristics of the utility interface of a PV system	The specifications in this standard apply to utility-interconnected PV power systems operating in parallel with the utility and utilizing static non-islanding inverters for the conversion of DC to AC. It lays down requirements for interconnection of PV systems to the utility distribution system
IEC/EN 61215	Crystalline silicon terrestrial PV modules (design qualification and type approval)	The standard lays down requirements for the design qualification and type approval of terrestrial PV modules suitable for long-term operation, as defined in IEC 60721-2-1. It determines the electrical and thermal characteristics of the module, such as its capability to withstand prolonged exposure in certain climates
IEC 61646	Thin-film terrestrial PV modules (design qualification and type approval)	The IEC 61646 standard lays down requirements for the design qualification and type approval of terrestrial, thin-film PV modules suitable for long-term operation as defined in IEC 60721-2-1. This standard applies to all terrestrial, flat-plate module materials not covered by IEC 61215. The significant technical change with respect to the previous edition deals with the pass/fail criteria

(continued overleaf)

Table 2.4 (*continued*)

Standard name	Main focus	Descriptions
IEC/EN 61730	PV module safety qualification (part 1 – requirements for construction and part 2 – requirements for testing)	The standard describes the fundamental construction requirements for PV modules in order to provide safe electrical and mechanical operation during their expected lifetime. It addresses the prevention of electrical shock, fire hazards, and personal injury due to mechanical and environmental stresses. It pertains to the particular requirements of construction and is to be used in conjunction with IEC 61215 or IEC 61646
IEC 60891	PV devices – procedures for temperature and irradiance corrections to measure I–V (current–voltage) characteristics	IEC 60891 defines procedures to be followed for temperature and irradiance corrections to the measured I–V (current–voltage) characteristics of PV devices. It also defines the procedures used to determine factors relevant for these corrections. Requirements for I–V measurement of PV devices are laid down in IEC 60904-1
IEC 60904-1	Measurement of PV current–voltage characteristics	The standard is related to measurement of PV current–voltage characteristics
IEC 61194	Characteristic parameters of stand-alone PV systems	The standard defines major electrical, mechanical, and environmental parameters for the description and performance analysis of stand-alone PV systems
IEC 61215	Crystalline silicon terrestrial PV modules	It is related to crystalline silicon terrestrial PV modules (mainly deals with the design qualification and type approval)
IEC 61345	Ultraviolet (UV) test for PV modules	The standard determines the ability of a PV module to withstand exposure to UV radiation from 280 to 400 nm
IEC 61427	Secondary cells and batteries for PV energy systems	It provides recommendations for secondary cells and batteries for PV energy systems (PVES). It deals with general requirements and methods of test revision to include latest battery technology
IEC 61646	Thin-film terrestrial PV	The standard is related to thin-film terrestrial PV modules (mainly design qualification and type approval)

Table 2.4   *(continued)*

Standard name	Main focus	Descriptions
IEC 61701	Corrosion testing of PV	The standard is related to salt mist corrosion testing of PV modules
IEC 61730-1	PV module safety requirement	The standard defines PV module safety qualification (part 1 – requirements for construction). In this standard, some amendments have been made and published in 2010
IEC 61702	Rating of direct coupled PV pumping systems	The standard defines predicted short-term characteristics of direct coupled PV water-pumping systems
IEC 61829	Crystalline silicon PV array – on-site measurement of I–V characteristics	It describes procedures for on-site measurement of crystalline silicon PV array characteristics and for extrapolating these data to standard test conditions (STCs) or other selected temperatures and irradiance values
IEC 61853-1	Module performance testing and energy rating	It defines PV module performance testing and energy rating (part 1 is related to irradiance and temperature performance measurements and power rating)
IEC 61853-2	Module performance testing and energy rating	The standard is related to PV module performance testing and energy rating. Part 2 – spectral response, incidence angle, and module operating temperature measurements
IEC/TS 62257-7-1	Generators-PV arrays	The standard provides recommendations for small renewable energy and hybrid systems for rural electrification (part 7-1 is related to generators-PV arrays)
IEC/TS 62257-7-3	Generator set – selection of generator sets for rural electrification systems	Part 7-3 is related to generator sets, that is, selection of generator sets for rural electrification systems)
IEC/TS 62257-8-1	Selection of batteries and battery management systems for stand-alone electrification systems	Part 8-1 mainly deals with the selection of batteries and battery management systems for stand-alone electrification systems. In addition, there are specific cases of automotive flooded lead-acid batteries, which are available in developing countries
IEC/TS 62257-9-1	Micropower systems	Part 9-1 of IEC/TS 62257 is related to PV micropower systems
IEC/TS 62257-9-2	Microgrids	Part 9-2 of IEC/TS 62257 is also related to PV micropower systems (microgrids)

(continued overleaf)

Table 2.4 (*continued*)

Standard name	Main focus	Descriptions
IEC/TS 62257-9-3	Integrated system user interface	Part 9-3 of IEC/TS 62257 is related to integrated system (user interface)
IEC/TS 62257-9-4	Integrated system user installation	Part 9-4 of IEC/TS 62257 is related to integrated system (user installation)
IEC/TS 62257-9-5	Integrated system – selection of portable PV lanterns for rural electrification projects	Part 9-5 mainly deals with integrated system specifically in selection of portable PV lanterns for rural electrification projects
IEC/TS 62257-9-6	Integrated systems – selection of photovoltaic individual electrification systems (PV-IES)	Part 9-6 is related to integrated systems; specifically it deals with the selection of PV individual electrification systems (PV-IES)
IEC/TS 62257-12-1	Selection of self-ballasted lamps (CFL) for rural electrification systems	Part 12-1 deals with the selection of self-ballasted lamps (CFL) for rural electrification systems and recommendations for household lighting equipment
IEC 62108	CSP module	The standard is related to CSP modules and assemblies, that is, its design qualification and type approval
IEC/TS 62257-1	General introduction to rural electrification	The standard provides recommendations for small renewable energy and hybrid systems for rural electrification (part 1 – general introduction to rural electrification)
IEC/TS 62257-2	Requirements to a range of electrification systems	Part 2 of this standard is about requirements to a range of electrification systems
IEC/TS 62257-3	Project development and management	Part 3 is related to project development and management
IEC/TS 62257-4	System selection and design	Part 4 is related to the system selection and design
IEC/TS 62257-5	Protection against electrical hazards	Part 5 is related to protection against electrical hazards
IEC/TS 62257-6	Acceptance, operation, maintenance, and replacement	Part 6 is related to acceptance, operation, maintenance, and replacement
IEC/TS 62257-7	Generators	Part 7 of this standard is related to generators
IEC62108	CSP module	It specifies characteristics, installation, and relevant information for CSP solar cells

Table 2.5 Advantages and disadvantages of solar energy systems

Advantages	Disadvantages
With no combustion and greenhouse gas emission, solar energy is the cleanest form of the renewable energy	It is still an expensive renewable energy source, until and unless the price of solar panels reduces considerably
There is abundant fuel supply in the form of sunlight	It is not a continuous source of energy as sunlight is not always available
It is available everywhere, that is, wherever the sun shines	Needs expensive inverters to convert DC current to AC current
It is suitable for distributed power generation	It requires storage batteries or grid connection for continuous supply of electricity
In solar energy, no moving parts are required	Expensive material is required in many thin-film systems
Power generation is silent, that is, no noise or pollution	Compared to other energy sources, it relatively requires a large amount of open space
It requires very negligible or minimal maintenance	Efficiency of solar panels is still low (17–40%)
Land requirement for centralized generations	Fragile materials are used in solar energy, which require intensive care

2.2.3   Wind Energy

2.2.3.1   Introduction of Wind Energy

Historians trace utilization of wind as a source of energy as early as 3200 BC, when sailors were using it for transporting goods. Today, wind is considered as one of the most promising renewable energy sources and it is expected to be an important source of electricity in the near future. The incentives of various national programs worldwide for advanced technology research and markets is helping considerably in promoting the widespread deployment of wind energy [11, 18–21]. For example, over 20% of Denmark's energy comes from wind energy. Denmark's wind energy share is about 27% of the global market today [4].

Electricity generated in bulk from the wind generators received attention in the 1970s when the energy crisis started because of oil embargoes. Out of these early installations, some of them were under-designed and did not sustain the starting years of operations. However, some of the best technology continues to operate some 30 years later, generating electricity as part of profitable commercial businesses. For example, after gaining a momentum during the early to mid-1980s, the US wind industry slipped as favorable policy and tax provisions incentives were gradually taken back in the late 1980s and 1990s. As a result, very little investment was made in the wind energy industry. However, with the reduction in cost and the announcement of a more favorable policy regime at both the state and federal levels, the US wind industry regained momentum back in the late 1990s. Today, wind energy constituted 35% (approximately) of annual electric power capacity installations [22]. This recent growth can be attributed to the abundant resource, continuous reducing cost of turbines, and favorable government policies.

The total installed wind capacity in the United States exceeded 40 000 MW by the end of year 2010 [22], generating 2.4% of the US electricity production [23]. At the end of 2010, the global wind capacity added during 2010 was more than 38 000 MW, representing an estimated $71.8 billion in asset investments.

2.2.3.2 Technology Overview

Wind turbine is the main component of a wind machine. The rotation of wind turbines converts the kinetic energy of wind into mechanical energy. The mechanical energy is subsequently converted into electrical energy. The available wind power increases as the cube of the wind speed. The Lanchester–Betz limit [24, 25, 27] defines the ability of a wind turbine to capture and convert wind power. The power has an upper bound, which is based on a simple theoretical model of energy extraction from an uninterrupted flow.

Utility-scale wind turbines are shown in Figure 2.7. The three-bladed pitch-controlled wind turbines are upstream rotor machines which are horizontally configured. Current power ratings extend above 3 MW for models under development having two to three times higher current ratings. The diameter of the upstream rotor ranges from 80 to 100 m with the supporting towers having a comparable height. The largest wind machines currently under design having a capacity of 5–10 MW are primarily intended for offshore installations. The extensive growth of land-based turbines greater than 4 MW is constrained by the logistic challenges associated with overland transport. However, latest technologies continue to overcome these challenges, allowing for the installation of more land-based wind turbines [4].

The block diagram of a wind energy conversion system (WECS) is shown in Figure 2.8.

2.2.3.3 Wind Power Generation Performance

Over the past 20 years, the wind energy generation plant's performance and efficiency has been improved significantly. The feet-averaged capacity factors increased to nearly 35% in 2008 for wind plants from about 25% in 1999 [23, 28]. This improvement in efficiency has been observed during these years despite an array of factors of variability, such as wind resource variability, congestion in transmission, and variability in the quality of the wind resource at sites where new projects are installed. Figure 2.9 shows that a relatively continual incremental improvement in fleetwide capacity factor has been observed in the last decade. For projects built since 2004, the average capacity factors have been above 30%. Various factors contribute to the capacity factors. For example, the increase in turbine hub heights that allow access

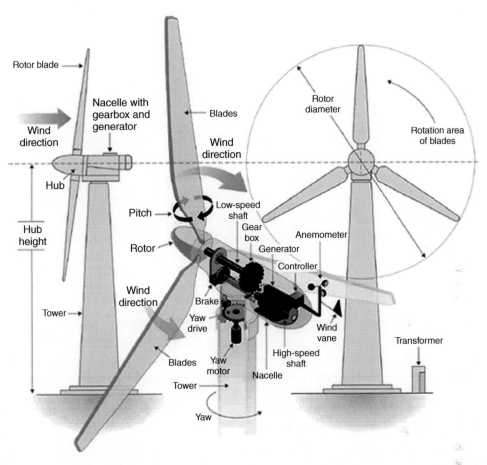

Figure 2.7 Modern horizontal axis wind turbine with gearboxes. Reproduced from R. Thresher, *et al.*, "Renewable electricity generation and storage technologies", Copyright © 2012 National Renewable Energy Laboratory, with permission from National Renewable Energy Laboratory [4]

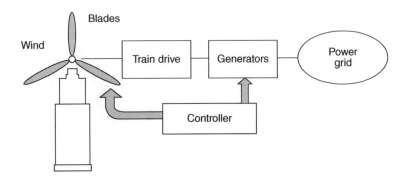

Figure 2.8 Block diagram of energy generation through a wind turbine

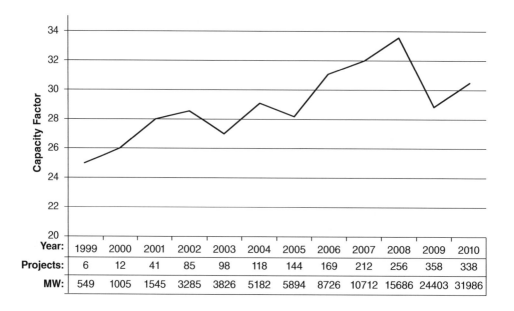

Year:	1999	2000	2001	2002	2003	2004	2005	2006	2007	2008	2009	2010
Projects:	6	12	41	85	98	118	144	169	212	256	358	338
MW:	549	1005	1545	3285	3826	5182	5894	8726	10712	15686	24403	31986

Figure 2.9   Total average sample-wide capacity factor per year. Source [4].

to better wind resources, and larger rotor diameters relative to generator capacity are among the reasons behind the increase in the capacity factor.

2.2.3.4   Wind Energy Standards

Wind energy standards are mainly related to the maintenance, certification, specification, turbines, gearboxes, storage, power station equipment, offshore structures, and the interconnection of equipments. Throughout the world, whether it is a national or local government, test houses, or industry in general, all of them use these standards. With the maturity of wind turbine industry, it is expected that the user groups and retract committees will be compelled to work in coordination in preparing these standards [9]. Table 2.6 provides details about various standards developed for wind energy technologies by different organizations. The organizations include American Gear Manufacturers Association (AGMA), British Standards Institution (BSI), Canadian Standards Association (CSA), Deutsches Institut für Normung (DIN), Danish Standards Association (DS), IEC, and Institute of Electrical and Electronics Engineers (IEEE).

2.2.3.5   Advantages, Deployment Opportunities, and Potential Barriers of Wind Energy

The diverse resource and relatively low cost of wind are key factors that have resulted in the popularity of wind as a favorable form of renewable energy generation today.

Table 2.6 Standards of wind energy systems

Standard	Main focus	Description
AGMA 6006-A03	Design and specification of gearboxes for wind turbines	The standard supersedes AGMA 921-A97. It is developed for design and specification of gearboxes of wind turbines
BSI BS EN 45510-5-3	Procurement of power station equipment	Part 5-3 of this standard is a guide for procurement of power station equipment
BSI BS EN 50308	Protective measures requirements for design, operation, and maintenance	The standard is a guide of protective measures requirements for design, operation, and maintenance of wind turbines
BSI PD CLC/TR 50373	Wind turbines' electromagnetic compatibility	The standard is related to wind turbines' electromagnetic compatibility
BSI BS EN 61400-12	Wind turbine generator systems	Part 12 of this standard is related to generator systems of wind turbines' power performance testing
BSI PD IEC WT 01	Conformity testing and certification	The standard defines rules and procedures for conformity testing and certification of wind turbines
CSA F417-M91-CAN/CSA	Wind energy conversion systems (WECS)	The standard defines the performance and general instructions of WECS
DIN EN 61400-25-4	Communications for monitoring and control of wind power plants	Part 25-4 of 61400 is related to communications for monitoring and control of wind power plants
DS DS/EN 61400-12-1	Power performance measurements of electricity-producing wind turbines	Part 12-1 of 61400 is related to wind turbines' performance measurements of electricity production
DNV DNV-OS-J101	Design of offshore wind turbine structures	This is a design of offshore wind turbine structures
GOST R 51237	Terms and definitions	The standard defines terms and conditions of nontraditional power engineering
IEC 60050-415	Wind turbine generator systems	Part 415 of IEC 60050 is related to wind turbine generator systems' international electrotechnical vocabulary
IEC 61400-1	Wind turbine design requirements	The standard is related to design requirements of wind turbines

(*continued overleaf*)

Table 2.6 (*continued*)

Standard	Main focus	Description
IEC 61400-2	Wind turbine generator systems	The standard is related to design requirements for small-scale wind turbine systems
IEC 61400-11	Wind turbine generator system (acoustic noise measurement)	Part 11 of this standard deals with the acoustic noise measurement techniques
IEC 61400-13 TS	Wind turbine generator systems (measurement and assessment of power quality)	It deals with measurement and assessment of power quality characteristics of grid-connected wind turbines
IEC 61400-21	Measurement and assessment of power quality characteristics	It deals with the full-scale structural testing of rotor blades
IEC 61400-22	Wind turbine certification	The standard is related to wind turbine's conformity testing and certifications
IEC 61400-23 TS	Wind turbine generator systems	The standard is related to a wind turbine generator systems. Specifically, it deals with full-scale structural testing of rotor blades
IEEE 1547	Interconnecting distributed resources with electric power systems	This IEEE standard is related to interconnecting distributed resources with electric power systems including wind power systems [36]
IEEE P2032.2	Guide for the interoperability of energy storage systems	This IEEE standard is a guide for the interperability of energy storage systems integrated with the electric power infrastructure including wind power systems [35]

It is expected that wind will continue to play a leading role in the supply of renewable power for many decades to come. Wind technologies have various benefits; however, there are various disadvantages as well, which are mentioned in Table 2.7. Table 2.8 lists out wind energy system deployment opportunities to enable high penetration of wind energy technologies, and the potential barriers in the way of wind technology to becoming a favored form of renewable energy.

2.2.4   Fuel Cell

2.2.4.1   Introduction of Fuel Cell

Besides other renewable energy sources such as solar and wind energy, hydrogen is an important source of clean and renewable energy. Hydrogen is abundantly available in the universe. It is also found in small quantities in the air. It is basically a colorless

Table 2.7 Advantages and disadvantages of wind energy

Advantages	Disadvantages
Just like solar energy, wind energy is clean and safe from the greenhouse gas emission	Wind turbines can only be installed in areas where wind often blows as all geographical locations are not appropriate for wind turbines
Investment on wind turbines can be returned in a few years	For home users, the wind turbine may not be suitable as it makes a noise and disturbing people in the surroundings
It is a completely renewable source of energy	Wind turbines can be damaged in thunderstorms
Compared to other renewable energy systems, wind energy is a cheaper option	

Table 2.8 Wind deployment opportunities, barriers, and responses

Research and development area	Barrier	Responses
Onshore turbines	Marginal competitiveness exists with conventional resources	To overcome the barrier, advanced technology solutions need to be applied to increase reliability and reduce the cost of technology including turbines, logistics, and installation
Offshore turbines	Offshore-specific design needs and challenges	To address this barrier, dedicated offshore equipment to maximize work at sea is required. In addition, easy maintenance and accessibility from offshore vessels and appropriate sea transport are also required
Offshore foundations with support structures	Current foundation structures add to costs and limit the depth of water for offshore installations	To minimize the foundation costs via standardization and design refinement
Wind resource assessment	A sophisticated understanding of both onshore and offshore wind resources and flow through plants is lacking. Limited wind-forecasting capabilities inhibit grid operations and dispatch planning	To develop a network of resource assessment facilities to better characterize the wind resource. In addition, it requires concurrent efforts to develop and implement improved wind plant-modeling and forecasting capabilities

(*continued overleaf*)

Table 2.8 (*continued*)

Research and development area	Barrier	Responses
Market and regulatory	**Barrier**	**Responses**
Market design and structure	The small operations areas are the barrier, which increase the cost of integrating wind energy into the grid	To rectify this issue, it requires a policy and market design that allow smaller operating areas to function in a consolidated manner
	Transmitting wind energy to population centers requires simple transfer of power over long distances	To resolve the limit on long-distance power transfer, including cost allocation for new transmission projects
	Restriction with low marginal costs might become an issue with high renewable electricity penetration	To ensure grid market access to plants with operational characteristics of renewable energy
Operational value	Instead of being recognized for grid service capabilities of wind energy, it is criticized for its variable output due to the nature of the wind	To develop techniques and methods for accurately valuing the additional costs as well as the value of grid services, which can determine issues of wind energy evaluation
Workforce development	Skilled manpower is required to support a rapid expansion of the wind industry	To facilitate worker training and encourage committed interest in the industry, including the establishment of a strong university/industry linkage
Environmental and siting	**Barrier**	**Responses**
Wildlife impacts	Rare birds can restrain the deployment of wind energy	It needs regular monitoring of wildlife impacts, development of impact mitigation strategies, and standardization
	Regular permit requirements will increase deployment costs	Intensive study is required to identify the impacts to habitat and resulting wildlife displacement, which can result in policy solutions to constant industry challenges
Siting policy	Host communities could oppose new installation	In both cases a proper awareness program about wind energy is required, which can help policymakers in weighing the trade-offs of wind energy. It helps them in making such policy which protects the interests of locals while facilitating deployment
	Poor land sectoring power-using policy increases developer risk	

Table 2.8   (*continued*)

Research and development area	Barrier	Responses
Radar and communications	Wind turbines can also affect the aviation, military radar systems, and communications	Technological advancement in wireless systems is required, which can address radar challenges. In addition, software solutions and system upgrades can alleviate the radar and communications interference

Source: Reproduced from R. Thresher, *et al.*, "Renewable electricity generation and storage technologies", Copyright © 2012 National Renewable Energy Laboratory, with permission from National Renewable Energy Laboratory [4]

and odorless gas. When hydrogen gas is used as a source of energy (fuel cell), it gives away only water and heat with no carbon emissions [33].

2.2.4.2   Technology Overview

In a hydrogen fuel cell, hydrogen atoms are divided into protons (ions+) and electrons (ions−). The negatively charged electrons from the hydrogen atom generate an electric current with water (H_2O) as a by-product. In other words, fuel cells are electrochemical devices that convert hydrogen gases or hydrocarbons and oxygen (O_2) from air into electrical power and heat. Fuel cells need either hydrogen gas or hydrocarbons to operate. This process is shown in Figure 2.10.

Basically, fuel cells are not truly renewable energy producers by themselves as they require a fuel supply. However, hydrogen gas is easily available on electrolyzing

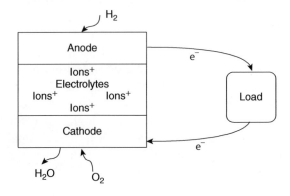

Figure 2.10   Energy generation process through a fuel cell

water into hydrogen and oxygen. The gas can then be fed into a fuel cell to generate electricity whenever there is a shortage. Therefore, a fuel cell can be used to balance the supply of electricity from intermittent renewable energy sources [2, 3, 12]. The advantage of fuel cells over solar and wind energy is that they can generate electricity continuously unlike the latter whose electricity generation depends on availability of sunshine or wind. When extra electricity is produced, it can be used to generate hydrogen gas.

2.2.4.3   Fuel Cell Standards

It should be noted that there still exist some basic challenges in introducing fuel cell technology into the consumer market, such as ensuring overall safety, inter compatibility with existing systems, and the establishing of standards. Standards are indeed one of the most important steps toward the commercialization of fuel cells. IEC TC 105, which was formed for fuel cell technologies, was established in 1998. The task of TC 105 is to prepare the necessary safety and interface standards. It is expected to design a standard in such a way that leaves enough space for further development in this area in the future. Table 2.9 provides details about various standards developed for fuel cell systems by different SDOs.

2.2.4.4   Advantages and Disadvantages of Fuel Cells

Table 2.10 provides details about advantages and disadvantages of the fuel cell.

2.2.5   Geothermal Energy

2.2.5.1   Introduction of Geothermal Energy

The heat energy which is stored in rock and fluids in the Earth's crust, is the main source of geothermal energy. This form of heat is found in hot springs and geysers that come to the surface of the earth or in the reservoirs found deep under the ground. The core of the earth is made of iron, which is surrounded by a layer or molten rocks. To generate energy, geothermal power plants are built on these heat energy reservoirs. This kind of renewable energy is primarily used in heating homes and commercial industries. The temperature of the thermal energy determines the amount of electricity generated. The higher-temperature resources will have higher potential of generating electricity and vice versa. With the best resources (high temperatures, large amounts of in-site fluids, and high reservoir permeability), the geothermal resource forms a continuum. Depending on the nature of the resource, the technology used to recover the subsurface thermal energy varies in geothermal energy systems [2–4, 11].

Table 2.9 Standards of fuel cell systems, developed by various SDOs

Standard	Main focus	Description
IEC/PAS 63547	System aspects for electric energy supply	It is an IEEE standard for interconnecting distributed resources with electric power systems
IEC 60079-29-1	Equipment for explosive atmospheres	It is related to explosive atmospheres, gas detectors, that is, performance requirements of detectors for flammable gases
IEC 60079-29-2	Explosive atmosphere	Part 29-2 deals with detectors selection, installation, use, and maintenance of detectors for flammable gases and oxygen
IEC/TS 62282-1	Terminology	It specifies fuel cell terminologies
IEC/TS 62282-2	Modules	It is related to different fuel cell modules
IEC 62282-3-100	Safety	It is related to safety of stationary fuel cell power systems
IEC 62282-3-200	Performance test methods	It mainly deals with how to measure the performance of stationary fuel cell power systems designed for residential, commercial, and agriculture systems
IEC 62282-3-3	Installation	It is related to installation of stationary fuel cell power systems
IEC 62282-5-1	Safety – fuel cell appliances	It is related to safety of portable fuel cell appliances
IEC 62282-6-100	Safety – micro fuel cells	It is related to safety of micro fuel cell power systems
IEC/PAS 62282-6-150	Safety – water-reactive compounds	It is related to the safety of water-reactive (UN Division 4.3) compounds in indirect PEM fuel cells
IEC 62282-6-200	Performance – micro fuel cells	It is related to the performance of micro fuel cell power systems
IEC 62282-6-300	Interchangeability	It is related to micro fuel cell power system's fuel cartridge interchangeability
IEC 62282-7-1	Cell testing method	It is related to the single cell test method for polymer electrolyte fuel cells
OIML R 81	Dynamic measuring devices and systems	It is related to dynamic measuring devices and systems for cryogenic liquids
OIML R 139	Metrological and technical requirement	It is related to metrological and technical requirements of compressed gaseous fuel measuring systems for vehicles
ISO 23273-1	Safety specifications	It is related to fuel cell road vehicles safety specifications (part 1 – vehicle functional safety)
ISO 23273-2	Safety specification	Part 2 of ISO 23273 is related to the protection against hydrogen hazards for vehicles fueled with compressed hydrogen gas

(continued overleaf)

Table 2.9 (*continued*)

Standard	Main focus	Description
ISO 23273-3	Safety specifications	Part 3 of ISO 23273 is related to protection of persons against electric shock
ISO 23828:2008	Energy consumption measurement	It is related to fuel cell road vehicle's energy consumption measurement. Part 1 deals with vehicles fueled with compressed hydrogen gas
ISO/TR 11954	Road speed measurement	It deals with fuel cell-based vehicle's maximum speed measurement
ISO 6469-1	Safety specifications – RESS	It deals with safety specifications of electrically propelled road vehicles (part 1 – on-board rechargeable energy storage systems (RESSs))
ISO 6469-2	Safety specifications – vehicle operational safety	Part 2 of this standard is related to vehicle operational safety means and protection against failures
ISO 6469-3:2011	Safety specifications – electric shocks	Part 3 specifies methods for protection of persons against electric shock
ISO/TR 8713:2012	Vocabulary	It deals with vocabulary of electrically propelled road vehicles
ISO 13985	Liquid hydrogen-fuel tanks	It deals with liquid hydrogen in land vehicle fuel tanks
ISO/PAS 15594	Liquid hydrogen-fueling facility	It deals with the airport hydrogen fueling facility operation
ISO 17268:2006	Liquid hydrogen-refueling connecting devices	It deals with the compressed hydrogen surface vehicle (refueling connection devices)
ISO/TS 15869	Liquid hydrogen land-vehicles fuel tanks	It deals with the gaseous hydrogen blends and hydrogen fuels (land vehicles fuel tanks)
ISO TR 15916:2004	Safety – hydrogen systems	It deals with the basic considerations for the safety of hydrogen systems
ISO 22734-1 2008	Hydrogen generators	It deals with hydrogen generators using water electrolysis process (part 1 – industrial and commercial applications)
ISO 22734-2	Hydrogen generators	Part 2 of this standard is related the residential applications
ISO 16110-1	Safety – generators	It is related to hydrogen generators using fuel processing technologies (part 1 – safety)
ISO 16110-2	Test method for performance	It is related to hydrogen generators using fuel processing technologies (part 2 – test method for performance)
ISO 16111	Storage devices	It is related to transportable gas storage devices (hydrogen absorbed in reversible metal hydrides)

Table 2.9 (*continued*)

Standard	Main focus	Description
ISO TS 20100	Gaseous hydrogen	It deals with the service stations of gaseous hydrogen
ISO/TR 14687-2	Hydrogen fuel	It deals with the product specification (part 2 – PEM fuel cell applications for road vehicles)
ISO 26142	Hydrogen fuel	It deals with the hydrogen detector apparatus (stationary applications)

PEM, proton exchange membrane.

Table 2.10 Advantages and disadvantages of fuel cells

Advantages	Disadvantages
Fuel cells have high efficiency (maximum theoretical efficiency is 83% at 250 °C [26])	The cost of fuel cells very high as they use expensive materials such as platinum
Fuel cells are a clean form of energy (using only H2 and O2)	Fuel cell reliability is still an issue
Fuel cells can run continuously as long as fuel is available	Fuel cells are still not durable and robust, particularly high temperatures
Their fuel can be produced from water (H2O), which is abundant in the universe	For fuel cells, hydrogen fuel is not readily available
Fuel cells are well suited for distributed generation	Fuel cells have low density of fuel, compared to gasoline
Fuel cells can be run in reverse for energy storage, that is, hydrogen gas can be produced from electricity and water	

2.2.5.2 Technology Overview

Table 2.11 categorizes geothermal resources based on the technology and methods used to develop the resource.

2.2.5.3 Geothermal Energy Standards

In order to extract energy from the heat generated beneath the earth's surface, heat pumps are used. Therefore, heat pumps are the important components of geothermal energy systems. Most of the geothermal standards are related to heat pumps, such as EN 378, EN 12263, and EN 14511. For example, a common standard, that is, EN 15450 for the design of the complete heat pump system was published toward the end of 2007. On the ground side in geothermal energy systems, the standards are mostly related to maintenance and safety of drilling rigs. Table 2.12 provides complete details about major geothermal standards developed by various SDOs.

Table 2.11 Geothermal resources, technologies, and methods

Geothermal resources	Description
Hydrothermal	Hydrothermal is a traditional and commercially available geothermal technology. Its reservoirs are enough for development of geothermal electricity at economically feasible rates
Enhanced geothermal system (EGS)	EGS has enough resources with a large amount of thermal energy but lacking reasonable in-site water and permeability
Co-production from oil and gas wells	In this case, the electricity is generated from the energy which resides in fluids produced along with oil and gas using organic Rankine-cycle power plants
Geo-pressured	Geo-pressured means a very highly pressured shale and sandstone formation that have high temperature with dissolved methane brine
Direct use	It contains those applications that use thermal energy from hydrothermal reservoirs directly instead of using it to generate electricity. Examples are space heating and cooling as well other heating and cooling applications such as greenhouse operations, irrigation, lighting, and recreation
Geothermal ground source heat pumps	Ground source heat pumps use the relatively constant temperature of the earth near the surface as a heat source for heating or cooling industrial and residential buildings. It is a widely available resource which can be used everywhere

Source: Reproduced from R. Thresher, *et al.*, "Renewable electricity generation and storage technologies", Copyright © 2012 National Renewable Energy Laboratory, with permission from National Renewable Energy Laboratory [4]

2.2.5.4 Advantages and Disadvantages of Geothermal Energy

Geothermal power is an emerging form of renewable technology, which has a number of advantages over existing technologies. However, despite its advantages, it faces a variety of market barriers such as risk and long development timelines in the early project stages of leasing, permitting, and exploration; insufficient understanding of the resource; few demonstration projects to confirm their technical feasibility; and incomplete basic research and development. Table 2.13 lists out advantages and disadvantages of geothermal energy.

2.2.6 Biomass

2.2.6.1 Introduction of Biomass

Biomass is another type of renewable energy source, in which fuel is produced from organic matter. The organic matter which is used in biomass is usually vegetable oil, rotting waste, animal fat, or industrial waste. Currently, biofuels such as biodiesel mixed with gasoline are used as fuel to run automobiles or to produce heat as fuel (wood and straw) in power stations to produce electric power, known as biopower. Methane gas generated from sewage and industrial waste is also a source of biomass energy [2, 4].

Table 2.12 Standards developed by various SDOs for geothermal systems

Standard name	Main focus	Description
EN 255-3	Heat pumps – testing for hot water units	The standard mainly deals with testing of hot water units such as air conditioners, liquid chilling packages, and heat pumps with electrically driven compressors (heating mode)
EN 378	Heat pumps – safety and environmental protection	The standard is related to refrigerating systems and heat pumps. Part 1-4 deals with safety and environmental requirements
EN 14511	Heat pumps in general, requirements and testing	The standard is mainly related to air conditioners, liquid chilling packages, and heat pumps with electrically driven compressors for space heating and cooling
EN 15450	Heat pumps in general (specifically system design)	The standard deals with heating systems in buildings (design of heat pump heating systems)
ISO 5149	Heat pumps in general (specifically safety)	The standard is related to mechanical refrigerating systems used for cooling and heating (safety requirements)
ISO 5151	Heat pumps in general, (specifically testing and rating)	The standard is related to the non-ducted air conditioners and heat pumps (testing and rating for performance nationalized) (e.g., in Great Britain)
ISO 13256	Heat pumps in general (specifically, testing and rating)	The standard deals with water-source heat pumps (testing and rating for performance), particularly the rating and testing of the heat pumps that are used in Denmark and Netherlands
Vereinung Deutche Ingenerung (VDI) 2067 Blatt6	Heat pumps (economic calculations)	The standard is related to the budget calculation of heat-consuming installations of heat pumps
VDI 4650 Blatt 1	Heat pumps (efficiency calculations)	The standard deals with calculation of heat pumps (short-cut method for the calculation of the annual effort figure of heat pumps)
ONORM M 7755-1	Design and installation of heat pump systems	Part 1 of ONORMM7755 standard deals with general requirements for design and construction of heat pump heating systems

(*continued overleaf*)

Table 2.12   (*continued*)

Standard name	Main focus	Description
DIN 8901	Protection of groundwater and soil against pollution (heat pump)	The standard is related to refrigerating systems and heat pumps (protection of soil, ground, and surface water)
VDI 4640 Blatt 1-4	Geothermal heat pump systems (design and installation)	The standard is related to design and installation of heat pump systems (thermal use of the underground heat system)
ONORM M 7753	Geothermal testing, and rating	It is related to the heat pumps with electrically driven compressors for direct expansion, ground coupled (testing and indication of the producer)
ONORM M 7755-2+3	Electrically driven heat pumps	It is related to the design and installation of ground source heat pump systems
OWAV RB 207	Avoiding risks to underground and groundwater	It is related to the systems for the exploitation of geothermal heat
Swedish Heat Pump Association (SVEP) standard	Installation standard for ground heat collectors	It is related to the correct installation of geothermal systems (ground heat collectors)
Deutsche Vereinigung des Gas- und Wasser- faches (DVGW) DVGW W 110	Investigations for geological reconnaissance inside a borehole	This standard deals with the investigations in boreholes and wells sunk to tap groundwater, compilation of methods
DVGW W 115	Well drilling	The standard is related to well drilling, that is, boreholes for exploration, capture, and observation of groundwater
DVGW W 116	Selection of drilling fluids in order to protect the groundwater	The standard is related to the use of mud additives in drilling fluids for drilling in groundwater
PN-G-08611	Work safety	The standard is related to drilling machinery (requirements of safety)

DVGW, Deutsche Vereinigung des Gas- und Wasserfaches.

Biopower is currently estimated to be the third largest form of renewable electricity generation after hydropower and wind energy in the United States [14]. In 2010, 56.2 TWh of biopower generation came from 10.7 GW of capacity [14]. The breakdown of this capacity was such that 7.0 GW was based on forest product industry and agricultural industry residues, and 3.7 GW was based on municipal solid, including landfill gas. The 5.8 GW of biopower capacity in the electric power sector represents

Table 2.13   Advantages and disadvantages of geothermal energy

Advantages	Disadvantages
Geothermal energy systems are entirely emission free	In geothermal energy, major sites are very location specific
For geothermal energy, no fuel is required for mining or transportation	Geothermal sites are often in remote locations far away from population centers
It is a very simple and reliable form of renewable energy	Owing to long distance transmission of electricity, it faces power losses
It is already a cost, competitive source of renewable energy in some areas	Sulfur dioxide and silica emissions are observed in geothermal energy systems
New technologies in geothermal energy systems assure to utilize lower temperatures	Construction cost is high in geothermal energy systems
	Very high temperature is required

about 0.56% of the total electric sector generating capacity in 2010 while the 5.1 GW of end-use generation capacity exhibits about 17% of total end-use capacity.

In the United States major growth of the biopower industry started in the 1980s when the Public Utilities Regulatory Policies Act (PURPA) of 1978 was passed. The PURPA guaranteed small generators, that is, generators with less than 80 MW capacity that the regulated utilities would buy electricity at a price equal to the cost of electricity avoided by the utilities. Many utilities offered PURPA contracts in anticipation of increasing fuel prices and for avoiding high costs, for example, the Four Standard Offer contracts in California, which made biopower projects economically attractive [4]. However, the biopower projects became less attractive in the early 1990s because of deregulation of the electric industry in combination with increased natural gas supplies and reduced fuel costs. Some variation in capacity and generation has occurred as older PURPA contracts expired, during the past 15 years [4].

Comparing the size of the United States biopower industry to that of the European Union, the biopower generation in the European Union was approximately 62.2 TWh in 2009. Out of this, the 23.3 TWh was generated from electricity-only plants and 38.9 TWh from combined heat and power plants. The top four European Union countries contributing biopower were Germany (11.4 TWh), Sweden (10.1 TWh), Finland (8.4 TWh), and Poland (4.9 TWh) [4].

2.2.6.2   Technology Overview

Co-Firing with Coal

In coal boilers, co-firing is the method of introducing biomass as a supplementary energy source. In the existing boiler's co-firing is the lowest-cost biopower option. In co-firing, only the existing boilers and generating equipment are used, while the major investment is in feed systems. The typical co-firing systems include the feed handling and preparation important for separate addition of biomass into a coal boiler [4].

Direct Combustion

Most biopower plants use solid biomass residues by employing the direct-fired systems. In this method, combustion involves the oxidation of biomass with excess air to give hot flue gas, which produces steam in the heat exchange sections of boilers. In order to produce electricity in a Rankine cycle, the steam contained in the heat exchange section of boilers is used. All of the steam is condensed in the turbine cycle in the electricity-only processes, while in the co-firing heat operation, only a portion of the steam is extracted to provide process heat. The direct firing process is shown in Figure 2.11.

Gasification

In the process of gasification, the biomass in an atmosphere of steam or sub-stoichiometric air/oxygen is converted to a medium- or low-calorific gas to produce a gas rich in carbon monoxide (CO) and hydrogen (H_2) plus other gases such as methane (CH_4) and carbon dioxide (CO_2). The heating value of a medium-calorific-value gas is 10–20 MJ/m^3 (270–540 (British Thermal Unit (BTU)/ft^3), and a low-calorific gas has a heating value of 3.5–10 MJ/m^3 (100–270 BTU/ft^3) [4].

2.2.6.3 Biomass Standards

In April 2007, the standard roadmap for biofuels was developed, which outlined the necessary steps that needed to be undertaken by the European Union, the

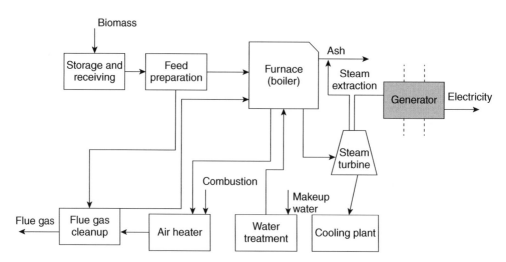

Figure 2.11 Schematic of direct fire system. Reproduced from R. Thresher, *et al.*, "Renewable electricity generation and storage technologies", Copyright © 2012 National Renewable Energy Laboratory, with permission from National Renewable Energy Laboratory [4].

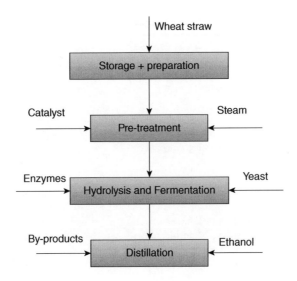

Figure 2.12 Block diagram of the ecological concept symbolizing bio energy (ethanol)

United States, and Brazil to achieve greater compatibility among existing biofuel standards [26]. A list of the major biofuel standards is provided in Table 2.14.

2.2.6.4 Ethanol

Ethanol is an important biofuel, which can be produced from sugarcane, wheat straws, or corn as well as by other similar other means. The process of producing ethanol from wheat straws can be explained with the block diagram shown in Figure 2.12. Firstly, dry wheat straws are stored for some pretreatment. During pretreatment, steam, and catalysts are provided. Then hydrolysis and fermentation are performed. After the distillation process, ethanol is produced along with some by-products.

Ethanol is an important form of energy production in emerging economies such as Brazil. However, the analysis of using fossil fuels in the production of agriculture and the carbon cycle leaves many open questions, which need to be answered. Energy generated per unit area from solar panels and wind form is 100 times more than corn ethanol, which leaves enough doubts on the viability of ethanol as an economically feasible form of renewable energy [4, 33].

2.2.6.5 Capital and Operating Costs of Biopower Systems

Table 2.15 provides some insights into the initial investment and operating costs of various representative biopower systems.

Table 2.14 Standards for biomass systems, developed by various SDOs

Standard name	Main focus	Description
British Standards (BS) European Norms (EN) 14774-1	Determination of moisture content – total moisture reference method	The standard deals with the determination of moisture content (oven dry method). It explains the total moisture reference method
BS EN 14774-2	Determination of moisture content – moisture simplified method	This standard also deals with the determination of moisture content (oven dry method). It explains the moisture-simplified method
BS EN 14774-3	Determination of moisture content	The standard explains the moisture in the general analysis sample
BS EN 14775	Determination of ash content	It is related to extracting of ash content
BS EN 14918	Determination of calorific value	The standard is related to extracting of calorific value
BS EN 14961-1	Fuel specifications and classes	The standard is related to specification and classes of biofuel (general specifications)
BS EN 15103	Determination of bulk density	The standard is related to biofuel bulk density
BS EN 15148	Determination of the content of volatile matter	The standard deals with extraction of the content of volatile matter
BS EN 15210	Determination of mechanical durability of pellets and briquettes	The standard deals with extracting of mechanical durability of pellets and briquettes
CEN/TS 14588	Terminology, definitions, and descriptions	The standard is related to various terminology, definitions, and descriptions of biofuels
CEN/TS 14778-1	Biofuel sampling (methods)	The standard is related to biofuel sampling (part 1 – methods for sampling)
CEN/TS 14778-2	Biofuel sampling (method for sampling materials)	The standard is related to biofuel sampling (part 2 – methods for sampling particulate material transported in lorries)
CEN/TS 14779	Instrumental methods	The standard is related to biofuel sampling (methods for preparing sampling plans and sampling certificates)
CEN/TS 14780	Sampling preparation of biofuel	The standard is related to the methods for sample preparation
CEN/TS 15104	Sampling (determination of total content of biofuel)	The standard deals with the determination of total content of carbon, hydrogen, and nitrogen (instrumental methods)
CEN/TS 15105	Determination of particle size distribution	The standard deals with the determination of the water-soluble content of chloride, sodium, and potassium

(continued overleaf)

Table 2.14 (*continued*)

Standard name	Main focus	Description
CEN/TS 15149-1	Determination of particle size distribution	The standard deals with the methods for the determination of particle size distribution – (part 1 – oscillating screen method using sieve apertures of 3.15 mm and above)
CEN/TS 15149-2	Determination of particle size distribution	Part 2 is basically related to the vibrating screen method using sieve apertures of 3.15 mm and below
CEN/TS 15149-3	Rotary screen method	Part 3 is related to the rotary screen method
CEN/TS 15150	Determination of particle density	The standard is related to the particle density of biofuels
CEN/TS 15210-2	Determination of mechanical durability	The standard is related to the determination of mechanical durability
CEN/TS 15234	Fuel quality	The standard is related to quality of biofuels
CEN/TS 15289	Determination of total content of sulfur and chlorine	The standard is related to the total content of sulfur and chlorine that are used in biofuels
CEN/TS 15290	Determination of major elements	The standard deals with the determination of major elements of biofuels
CEN/TS 15296	Analysis to different bases	The standard is related to the analysis of different bases in biofuels
CEN/TS 15297	Determination of minor elements	The standard is related to the determination of minor elements in biofuels
CEN/TS 15370-1	Determination of ash melting behavior	The standard is related to the determination of ash melting behavior of biofuels

2.2.6.6 Advantages and Disadvantages of Biomass

Biomass is considered to be a useful form of renewable energy. However, there are various controversial issues associated with the use of biomass. It is a reality that production of biofuels may involve cutting trees, which is against the concept of green energy. In addition, transforming the organic matter into energy may be expensive with higher agriculture products and higher carbon footprints, which may be redirected instead of being used for food purposes. Table 2.16 provides details of various advantages and disadvantages associated with biomass.

Table 2.15 Capital and operating costs of representative biopower systems

Representative technology	Year	Plant size (MW)	Capital cost Overnight w/AFUDC (1000 $/MW)	Capital cost w/AFUDC (Allowance for Funds used during construction) (1000 $/MW)	Operating cost Fixed ($/kW-year)	Operating cost Variable ($/MWh)	Heat rate Feed ($/tonne)	($/MWh)	$\frac{\mathrm{MMBtu}}{\mathrm{MWh}}$
Combustion, stoker	2010	50	3657	3794	99	4	82.60	59	12.50
Combustion, stoker	2010	50	3742	4092	99	5	82.60	68	14.48
Combustion, circulating fluidized bed	2010	50	3771	3911	102	6	82.60	59	12.50
Combustion bubbling fluidized bed	2010	50	3638	–	94	5	82.60	63	13.50
CHP	2010	50	3859	4002	101	4	82.60	67	14.25
Gasification, base	2010	75	4149	4417	94	7	82.60	44	9.49
Gasification, advance	2010	75	3607	3795	60	7	82.60	38	8.00
Gasification, (=Integrated Gasification Combined Cycle (IGCC))	2010	20	7498	–	332	16	82.60	58	12.35
Co-firing, pulverized coal, co-feed	2010	20	449	555	13	2	82.60	47	Coal heat rate +1.5%
Co-firing, cyclone co-feed	2010	20	353	353	13	1	82.60	47	Coal heat rate
Composite	2010	50	3872	–	95	15	82.60	68	14.50
Composite	2030	50	3872	–	95	15	82.60	63	13.50
Composite	2050	50	3872	–	95	15	82.60	59	12.50
Composite	2010	50	3865	–	103	5	82.60	59	12.50
Composite	2020	50	3864	–	102	5	82.60	59	12.40
Composite	2030	50	3843	–	89	5	82.60	52	11.10
Composite	2040	50	3822	–	76	6	82.60	46	9.70
Composite	2050	50	3811	–	63	7	82.60	39	8.40

Source [23].

Table 2.16 Advantages and disadvantages of biomass

Advantages	Disadvantages
Biofuels are carbon neutral	Gases such as CO_2, which are emitted into the atmosphere during biomass processing, may damage the ozone layer
Biofuel is cost effective	Building a biomass power plant takes considerable space and the processing also requires a large amount of water
Unlike wind and solar energy, biomass can be stored for future usage	Usage of corn and soybeans, and so on, has resulted in hiking food prices
Biomass is available everywhere and can be easily collected	Raw materials of biomass, which are agricultural wastes of certain crops, are not available all year around
Compared to fossil fuels, which takes million of years to stock up, biomass takes only as much time as it is required to grow or gather the source which can takes months up to maximum of a year	The characteristics of biomass are geographic dependent
Its processing is identified as an excellent energy-efficient waste disposal by converting biological waste to energy, which can be used for diverse applications	Excessive use of biomass may lead to environmental issues and deforestation
It can be used as manure in agriculture	

2.3 Challenges of Renewable Energy Systems

In last few decades, renewable energy systems have shown rapid growth. However, there are many challenges which could affect large-scale deployment of these technologies. In addition, there are various technical challenges associated with renewable technologies which must be addressed to make these technologies more reliable to deliver uninterrupted supply of electricity. Currently, the main challenges associated with renewable technologies are high capital cost, integrating renewable technologies to existing grid (on-grid), transmission, distribution, storage, and reliability [34, 37–40].

2.3.1 High Capital Cost

The cost of building and maintaining renewable energy technologies has decreased significantly in recent years, however, compared to per unit cost of the traditional energy technologies, it is still higher. Most of the materials used in renewable technologies, their construction, and maintaining costs are still high. In order to bring renewable material, construction, and maintain costs to a reasonable level, further improvement is required in renewable research and development and the commercialization of these technologies. Owing to the increasing interest of various countries in renewable energy technologies, it is expected that the overall cost will be very competitive in the coming years.

2.3.2   Integrating Renewable to the On-Grid

The off-grid system is simple as it does not require integration with the existing power grid. However, most of the renewable technologies, such as small-scale wind and solar PV systems are connected with the existing grid (on-grid system), which has the ability to add thousands of power units to the nation's power system. In order to maintain the balance between demand and response, smart metering is required, which will keep proper record of both demand and response and thus ensure the reliability and integrity of the power system[34]. For this purpose, the deployment of Advanced Metering Infrastructure (AMI) on wider scale is required.

2.3.3   Reliable Supply of Power

The main issue with renewable energy systems is to provide uninterrupted power supply according to the demand. For the said purpose, a reliable supply of electricity must be available around the clock in all weather and seasons. However, the most popular renewable resources, such as wind and solar energy depend on the wind power and irradiation from the sun. It means that both solar and wind energy systems are variable and are not always available. Both of these systems always require a backup power supply.

2.3.4   Power Transmission

Renewable energy resources are also location dependent. The availability of renewable resources varies from region to region. For example, coastal areas are more suitable for wind energy; however, it may be far away from the area where power supply is required. In addition, significant transmission systems will be needed to supply power to the areas where it is needed. A bulk investment is also required in building a transmission system to provide power supply from remote areas to the populated areas where power is required.

2.3.5   Power Distribution

Just like power transmission, power distribution is also a challenging issue. In peak hours, industrial areas require more power supply then residential areas. Smart power distribution and effective load management is required to supply power according to the demand of specific area and applications.

2.4 Conclusion

Global warming and environmental changes have forced the human beings to reexamine the excessive use of hydrocarbon and look for alternative renewable sources of energy. As a result, various efforts have been made to generate energy from unconventional sources in the last few decades, which are environmental friendly, long lasting, economically feasible, and easily accessible. With the increasing demand for renewable energy generation, the power and energy engineering face the problem of integration and interoperability so that the small distributed power generators and scattered energy storage devices are integrated together into a single grid, called the Smart Grid. The Smart Grid will distribute electricity from suppliers to consumers using a variety of ICTs to control appliances at the consumer's home and industries to reduce cost, increase reliability with more transparency, reduce power loss, avoid power outage, avoid complete blackout and thus save overall energy.

In this chapter, we have introduced various forms of renewable energy sources and their utilization. We have discussed the main standards which have been developed by various SDOs for various renewable systems and technologies. These standards have been designed for energy generation, storage, commissioning, operation, and the overall maintenance of the systems. We have also provided a technology overview, discussed different types, benefits, potential barriers, and drawbacks of each technology in detail.

References

[1] US Department of Energy http://energy.gov/data/open-energy-data/ (accessed November 17, 2014).
[2] Keyhani, A. (2011) *Design of Smart Power Grid Renewable Energy Systems*, John Wiley & Sons, Inc., Hoboken, NJ.
[3] International Energy's Agency World Energy Outlook (2006) *World Energy Outlook 2013*, www.worldenergyoutlook.org/ (accessed November 17, 2014).
[4] Thresher, R. *et al.* (2012) *Renewable Electricity Generation and Storage Technologies*, National Renewable Energy Laboratory, Golden, CO.
[5] Tonn, B. and Peretz, J.H. (2007) State-level benefits of energy efficiency. *Energy Policy*, **35** (7), 3665–3674.
[6] Blaabjerg, F. and Guerrero, J.M. (2011) Smart grid and renewable energy systems. *Proceeding of the International Conference on Electrical Machines and Systems (ICEMS)*, pp. 1–10.
[7] Guerrero, J.M., Blaabjerg, F., Zhelev, T. *et al.* (2010) Distributed generation: toward a new energy paradigm. *IEEE Industrial Electronics Magazine*, **04** (01), 52–64.
[8] Blaabjerg, F., Teodorescu, R., Liserre, M. and Timbus, A.V. (2006) Overview of control and grid synchronization for distributed power generation systems. *IEEE Transactions on Industrial Electronics*, **53** (05), 1398–1409.
[9] IEC Renewable Energies www.iec.ch/about/brochures/pdf/technology/renewable_energies_2.pdf (accessed November 17, 2014).
[10] Guerrero, J.M., Vasquez, J.C., Matas, J. *et al.* (2011) Hierarchical control of droop-controlled DC and AC microgrids – a general approach towards standardization. *IEEE Transactions on Industrial Electronics*, **58** (01), 158–172.

[11] ESkam IDM Report (2012) *Small-Scale Renewable Energy Standards and Specifications*, www.eskomidm.co.za/docs/renewable_energy/20120531_Standards_and_Specs.pdf (accessed November 17, 2014).

[12] Timbus, A., Liserre, M., Teodorescu, R. *et al.* (2009) Evaluation of current controllers for distributed power generation systems. *IEEE Transactions on Power Electronics*, **24** (03), 654–664.

[13] Solar Energy Industries Association/GTM Research (2012) *U.S. Solar Market Insight*, www.seia.org/cs/research/SolarInsight (accessed November 17, 2014).

[14] Energy Information Administration (EIA) (2012) *Annual Energy Outlook 2012: With Projections to 2035*, Washington, DC.

[15] Photon Consulting (2012) *Solar Annual 2012: The Next Wave.*

[16] Goodrich, A., James, T. and Woodhouse, M. (2012) *Residential, Commercial, and Utility-Scale Photovoltaic (PV) System Prices in the United States: Current Drivers and Cost-Reduction Opportunities*, National Renewable Energy Laboratory.

[17] EPIA (2009), *Global Market Outlook for Photovoltaics Until 2013*, European Photovoltaic Industry Association.

[18] Tenca, P., Rockhill, A.A., Lipo, T.A. and Tricoli, P. (2008) Current source topology for wind turbines with decreased mains current harmonics, further reducible via functional minimization. *IEEE Transactions on Power Electronics*, **23** (03), 1143–1155.

[19] El-Moursi, M.S., Bak-Jensen, B. and Abdel-Rahman, M.H. (2010) Novel STATCOM controller for mitigating SSR and damping power system oscillations in a series compensated wind park. *IEEE Transactions on Power Electronics*, **25** (02), 429–441.

[20] Li, R., Bozhko, S. and Asher, G. (2008) Frequency control design for offshore wind farm grid with LCC-HVDC link connection. *IEEE Transactions on Power Electronics*, **23** (03), 1085–1092.

[21] Grabic, S., Celanovic, N. and Katic, V.A. (2008) Permanent magnet synchronous generator cascade for wind turbine application. *IEEE Transactions on Power Electronics*, **23** (03), 1136–1142.

[22] Bolinger, M. and Wiser, R. (2011) *Understanding Trends in Wind Turbine Prices Over the Past Decade. LBNL-5119E*, Lawrence Berkeley National Laboratory, Berkeley, CA.

[23] REN21 (Renewable Energy Policy Network for the 21st Century) (2011) *Renewables 2011 Global Status Report*, Paris: REN21 Secretariat, www.ren21.net/Portals/97/documents/GSR/REN21_GSR2011.pdf (accessed November 17, 2014).

[24] Cuerva, A. and Sanz-Andres, A. (2005) *Renewable Energy*, Elsevier, San Francisco, CA.

[25] Iov, P. Soerensen, A. Hansen, F. Blaabjerg, (2006) Modelling, analysis and control of DC-connected wind farms to grid, *International Review of Electrical Engineering*, Praise Worthy Prize, Vol. 0. n. 0, pp. 14–22 .

[26] Bright Hub www.brighthub.com/environment/renewable-energy/articles/7730.aspx (accessed November 17, 2014).

[27] Twidell, J. and Weir, T. (2006) *Renewable Energy Sources*, Taylor & Francis, London and New York.

[28] Hansen, A.D., Iov, F., Blaabjerg, F. and Hansen, L.H. (2004) Review of contemporary wind turbine concepts and their market penetration. *Journal of Wind Engineering*, **28** (3), 247–263.

[29] Asiminoaei, L., Teodorescu, R., Blaabjerg, F. and Borup, U. (2005) A digital controlled PV-inverter with grid impedance estimation for ENS detection. *IEEE Transactions on Power Electronics*, **20** (06), 1480–1490.

[30] Svrzek, M. and Sterzinger, G. (2005) *Solar PV Development: Location of Economic Activity*, Renewable Energy Policy Report.

[31] IEA-International Energy Agency (2007) *Trends in Photovoltaic Applications: Survey Report of Selected IEA Countries between 1992 and 2006*. Report IEA-PVPS T1-16:2007.

[32] IEA-International Energy Agency (2008) *Trends in Photovoltaic Applications: Survey Report of Selected IEA Countries between 1992 and 2007*, Report IEA-PVPS T1-17:2008.

[33] Tripartite Task Force (2007) *A White Paper on Internationally Compatible Biofuel Standards*, Biofuel Roadmap.

[34] Mohd, A., Ortjohann, E., Schmelter, A. *et al.* (2008) Challenges in integrating distributed Energy storage systems into future smart grid. *Proceeding of the IEEE International Symposium on Industrial Electronics*, pp. 1627–1632.

[35] IEEE IEEE P2032.2. *Draft Guide for the Interoperability of Energy Storage Systems Integrated with the Electric Power Infrastructure*, IEEE Standards Association, http://grouper.ieee.org/groups/scc21/2030.2/2030.2_index.html (accessed November 17, 2014).

[36] IEEE Standards Association *1547 Series of Interconnection Standards*, http://grouper.ieee.org/groups/scc21/dr_shared/ (accessed November 17, 2014).

[37] Layton, B.E. (2008) A comparison of energy densities of prevalent energy sources in units of joules per cubic meter. *International Journal of Green Energy*, **5**, 438–455.

[38] World Energy Council www.worldenergy.org/focus/fuel_cells/377.asp (accessed 21 February 2013).

[39] AWEA (American Wind Energy Association) (2008) *Wind Energy Siting Handbook*, www.awea.org/sitinghandbook/ (accessed November 17, 2014).

[40] EIA (2010) *Annual Energy Outlook 2010: With Projections to 2035*, U.S. Energy Information Administration, Washington, DC.

3

Power Grid

One of the major differences between the Smart Grid and conventional power grid systems is the integration of modern Information and Communication Technologies (ITC) with the existing electric systems. In today's utility enterprise where information exchanges among the various generation, transmission and Distribution Management Systems (DMSs) and other Information Technology (IT) systems is not only desirable but necessary in most cases, each system plays the role of either supplier or consumer of information, or more typically both. This means that both data semantics and syntax need to be preserved across system boundaries. The system boundaries mentioned in this chapter are interfaces where data are made publicly accessible to other systems or where requests for data residing in other systems are initiated. In other words, Unified and consistent specifications and standards are required to enable seamless integration of various subsystems.

Most previous efforts to define system architectures have dealt primarily with the definition of protocols and profiles. The increasing use of object modeling schemes make the standard development process much easier than ever before. However, the side effect is that each organization has development its own specific modeling language and definitions, given the limited scope of applications and users. The consequence is that instead of one object model for each physical entity in the generation, transmission, and distribution operations domain being standardized, at least two or more object models exist in most cases with different definitions for classes, attributes, data types, and relationships between classes. What is worse, even different modeling languages have been used to define the same object. Therefore, it is important to have an overview of current existing power grid standards and specifications before addressing the interoperability issues.

Table 3.1 contains all these standards, which are categorized according to their function fields. Each standard is then explained one by one in detail.

Smart Grid Standards: Specifications, Requirements, and Technologies, First Edition. Takuro Sato,
Daniel M. Kammen, Bin Duan, Martin Macuha, Zhenyu Zhou, Jun Wu, Muhammad Tariq and Solomon Abebe Asfaw.
© 2015 John Wiley & Sons, Ltd. Published 2015 by John Wiley & Sons, Ltd.

Table 3.1 Standard list of advanced metering infrastructure

Function field	Standard name	Short introduction
Product	IEC/TR 62357	Service-oriented architecture (SOA)
Data exchange	IEC 61850	Substation automation
Service	IEC 61968	Common information model (CIM)/distribution management
Product	IEC 60870-5	Telecontrol equipment and systems – Part 5 – transmission protocols
Data exchange	IEC 61334	Distribution automation using distribution line carrier systems
Product	IEC 60834	Teleprotection equipment of power systems – performance and testing
Service	IEC 61970	Common information model (CIM)/energy management
Data exchange	IEC 61400-25	Wind turbines – Part 25: communications for monitoring and control of wind power plants
Data exchange	IEC 60255-24	Electrical relays – Part 24: common format for transient data exchange (COMTRADE) for power systems
Electrical transmission	IEC 61954	Power electronics for electrical transmission and distribution systems

3.1 Power Grid Systems

Figure 3.1 shows a basic model of smart power grid architecture. Staff in the power control center should analyze the data and make corresponding changes in operation to ensure the implementation of Supervisory Control and Data Acquisition (SCADA), which provides efficiency, reliability, and safety control and monitor the Smart Grid. There are mainly four types of staff in a control center: dispatcher, check engineer, planner, and relay protection engineer. These staff have different responsibilities, and are interested in different aspects of the Smart Grid data. The dispatcher is the commander of the power grid operation and fault handling. He is responsible for real-time monitoring and control of the network, keeping real-time balance of power production, and regulating the power flow on lines and voltages on buses to prevent limit violation. The security check engineer prepares evaluates and checks the yearly, monthly, weekly, and daily operation modes and other special operation modes required. He is responsible for verifying the security, presenting stability measures, and providing stability analysis or fault analysis. The planner issues the schedules of generation and power supplies, including the power exchange of the tie-line in an interconnected grid. The relay protection engineer is responsible for the setting of relay protection and verifying the protection settings when the operation mode changes.

Market services staff can perform the suitable and dynamic services for the power uses according to the situation of the Smart Grid. Power user can also access some of the data in Smart Grid, such as smart meter data, power price, and so on. The two

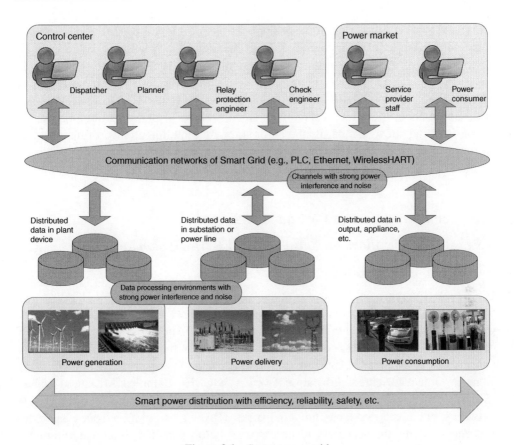

Figure 3.1 Smart power grid

kinds of data users can also affect the running of the Smart Grid. For example, the power price can be decided by the services market for power, which have effect with power consumption of the power users each other.

3.2   An Overview of the Important Key Standards for the Power Grid

In the substation, IEC 61850 provides the communication standards among different devices, such as the transformer and the switch gear. In addition, this standard series provides communication interface between the substation and the control center. This standard series is proposed to design electrical substation automation. The IEC 60870 standards is a set of standards used for telecontrol. On the basis of the protocol of these standards, equipment from many different suppliers can interoperate. The fifth part of IEC 60870 defines the systems used for telecontrol (i.e., SCADA) in electrical engineering and power system automation applications. Moreover, the IEC 60870–6 standards define the communication interface between different control centers. The IEC 61970 standards can provide Application Program Interface, which can perform the integration of applications developed by different suppliers in the control center environment and the exchange of information to systems external to the control center environment. Meanwhile, IEC 61968 standards can perform the software inter-application integration. In addition, IEC 60834 standards can be applied to narrowband and wideband teleprotection systems used to convey analog information about the primary quantities such as phase or phase and amplitude. On the basis of the above standards, the automation and intelligentialized functions can be performed for the power grid. The organization and relations among the above standards in a power grid is shown in Figure 3.2.

3.3   Communications in the Smart Grid

3.3.1   Communications for Substations: IEC 61850 Standards

3.3.1.1   Overview of IEC 61850

IEC 61850 (International Electrotechnical Commission) standards [1–13] is proposed to design electrical substation automation. IEC 61850 is a part of the IECs Technical Committee 57 (TC57) reference architecture for electric power systems. The abstract data models which correspond to IEC 61850 can be mapped to many protocols. Current mappings in IEC 61850 are to Manufacturing Message Specification (MMS), Generic Object Oriented Substation Event (GOOSE), Sampled Measured Values (SMVs), and soon to Web Services (WSs). These protocols can run over TCP/IP networks or substation local area networks (LANs) using high speed switched Ethernet to obtain the necessary response times below 4 ms for protective relaying.

IEC 61850 consists of the following parts detailed in separate IEC 61850 standard documents.

- **IEC 61850-1**: Introduction and overview
- **IEC 61850-2**: Glossary
- **IEC 61850-3**: General requirements
- **IEC 61850-4**: System and project management – Ed.2
- **IEC 61850-5**: Communication requirements for functions and device models
- **IEC 61850-6**: Configuration language for communication in electrical substations related to intelligent electronic devices (IEDs) – Ed.2
- **IEC 61850-7**: Basic communication structure for substation and feeder equipment
  - **IEC 61850-7-1**: Principles and models – Ed.2
  - **IEC 61850-7-2**: Abstract communication service interface (ACSI) – Ed.2
  - **IEC 61850-7-3**: Common data classes – Ed.2
  - **IEC 61850-7-4**: Compatible logical node classes and data classes – Ed.2
  - **IEC 61850-7-10**: Communication networks and systems in power utility automation – Requirements for web-based and structured access to the IEC 61850 information models (Approved new work)

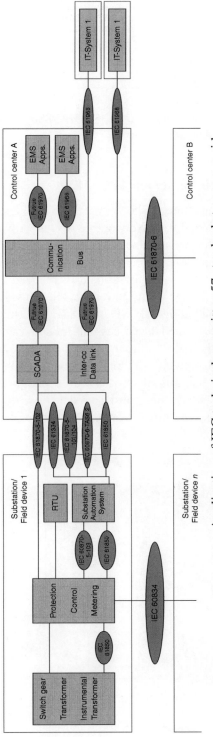

Figure 3.2 Application of IEC technical committee 57 standards to power grid

- **IEC 61850-8**: Specific communication service mapping (SCSM)
  - **IEC 61850-8-1**: Mappings to MMS (ISO/IEC9506-1 and ISO/IEC 9506-2) – Ed.2
- **IEC 61850-9**: SCSM
  - **IEC 61850-9-1**: Sampled values over serial unidirectional multidrop point-to-point links
  - **IEC 61850-9-2**: Sampled values over ISO/IEC 8802-3 – Ed.2
  - **IEC 61850-10-**: Conformance testing.

3.3.1.2   Substation Architecture and IEC 61850

A typical substation architecture is shown in Figure 3.3. The substation network is connected to the outside Wide Area Network (WAN) via a gateway. Outside, remote operators and control centers can use the ACSI defined in IEC 61850-7-2 to query and control devices in the substation. There are one or more substation buses connecting all the IEDs inside a substation. A substation bus is realized as a medium bandwidth Ethernet network, which carries all ACSI requests/responses and generic substation events messages (GSEs, including GOOSE and GSSE generic substation status event). There is another kind of bus called the process bus for communication inside each bay. A process bus connects the IEDs to the traditional dumb devices (merge units, etc.) and is realized as a high bandwidth Ethernet network. A substation usually has only one global substation bus but multiple process buses, one for each bay.

The IEC 61850 based substations can be modeled as a three-layer architecture: substation layer, bay layer, and process layer. The substation layer usually includes

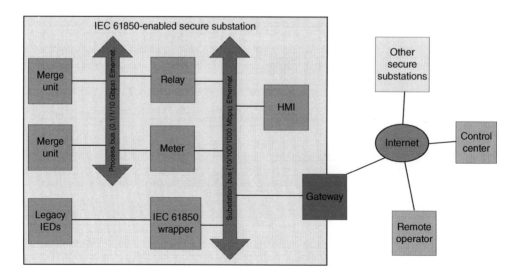

Figure 3.3   Substation structure

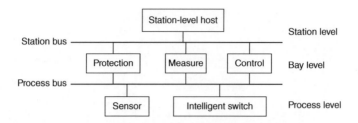

Figure 3.4   Three-layer architecture of IEC 6185 substation

Human Machine Interface (HMI), monitoring host, and so on. The bay layer usually includes the protection, control, and meter equipment of the primary equipment. The process layer usually includes the smart primary equipment, such as smart switch, voltage/current transformer, and so on. The three-lay architecture of IEC 61850 is shown in Figure 3.4. The equipment in the three layers are connected by the substation bus and process bus. The communications among smart sensors, meters, and protection and control equipment in the process layer are based on Sampling Value (SV) messages through the process bus. Smart switches in the process layer communicate with the protection equipment in the bay layer by using the GOOSE message through the process bus. The communications among various equipment in the process layer are based on MMS, GOOSE message, and so on, through either the process bus or the substation bus. The records, reports, logs, and so on, between bay-layer equipment and substation-layer equipment are transmitted in the substation bus based on MMS.

3.3.1.3   Data Model of Substation

On the basis of IEC 61850, when data modeling needs to be performed for the equipment in the substation, each equipment should be layered. The top layer of an equipment data model is the Logic Device (LD), which includes a number of Logic Nodes (LNs). And an LN includes a number of Data Objects (DOs), which covers many Data Attributes (DAs). Figure 3.5 shows the data modeling of a substation [14]. The names of the LD, LN, DO, and so on, can be obtained from the LD Directory, LN Directory, DO Directory, Get DO Definition services, and so on. For example, when the data model of a breaker needs to be established, the name of the LD equipment is IED1. LD includes the LNs of XCBR (Logical Node Name of Circuit Breaker), LLN0 (Logical Node Zero), LPHD (Logical Node Physical Device), and so on. Here, LLN0 usually is the description of the LN of each equipment, and LPHD is the LN of the physical state of the equipments. The important data of the equipment are stored in the LN of the XCBR. The XCBR LN includes the DO (Data Object) of Pos, mod, and so on. Here Pos is the DO of state of the switch. Pos DO includes DA of q, t, ctlnum, stal, and so on. The state information of the switch is stored in stal DA. Therefore, if the state information of the equipment needs to be read, the value of IEDq.XCBR.Pos.stVal must be read on the basis of the data modeling method.

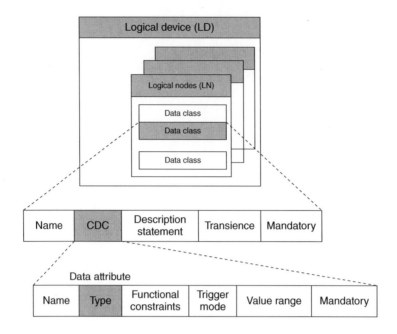

Figure 3.5   Multilayer data model

3.3.1.4   Communication Model of Substation

ACSI requests/responses, GSE messages, and sampled analog values are the major active data in the substation network. Since we are less interested in communication on the process buses (such as sampled value multicasting), we focus on the activities on the substation bus in this chapter, especially the ACSI activities. Interactions inside a substation automation system mainly fall into three categories: data gathering/setting, data monitoring/reporting, and event logging. The former two kinds of interactions are the most important – in the IEC 61850 standard, all inquiries and control activities toward physical devices are modeled as getting or setting the values of the corresponding data attributes, while data monitoring/reporting provides an efficient way to track the system status, so that control commands can be issued in a timely manner.

To realize the above kinds of interaction, the IEC 61850 standard defines a relatively complicated communication structure, which is shown in Figure 3.6. Five kinds of communication profiles are defined in the standard: the ACSI profile, the GOOSE profile, the GSSE profile, the SMV multicast profile, and the time synchronization profile. ACSI services enable client–server style interaction between applications and servers. GOOSE provides a fast way of data exchange on the substation bus and GSSE provides an expressway for substation status exchange. SMV multicast provides an effective way to exchange data on the process bus.

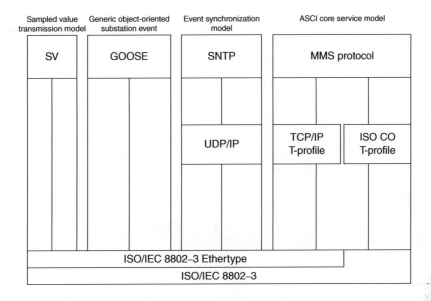

Figure 3.6   Communication stacks of IEC 61850

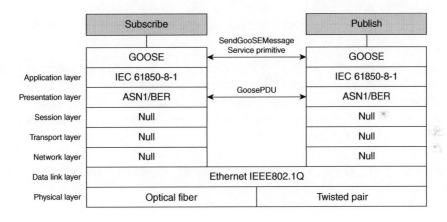

Figure 3.7   GOOSE message communication stack

IEC 61850 mentions that some types of messages (e.g., bay blocking, real-time trip) have very high real-time requirement. In order to satisfy the real-time requirement, GOOSE is used to transfer-related information. The communication stack of a GOOSE message is shown in Figure 3.7.

The GOOSE message communication stack consist of four layers, excluding the session layer, transport layer, and network layer in the ISO/OSI (Open System Interconnection) seven-layer model. GOOSE messages are mapped directly from the application layer to the data link layer. The most important feature of this structure is the

```
IEC 61850   DEFINITIONS : : = BEGIN
IMPORTS   Data  FROM IOS—IEC—9506—2
IEC 61850—8—1 Specific  Protocol : : =  CHO ICE{
gseMngtPdu [ APPLICATION 0 ]   IMPLICIT  GSEMngt Pdu ,
  goosePdu [ APPLICATION 1 ] IMPLICIT IECGoosePdu
  ...
}
IECGoosePdu : : = SEQUENCE{
gocbRef [ 0 ]   IMPLICIT  VISIBLE-STRING ,
timeAllowedtoLive [ 1 ]  IMPLICIT  VISIBLE-STRING OPTIONAL ,
datSet [ 2 ]   IMPLICIT  VISIBLE-STRING ,
goID [ 3 ]   IMPLICIT  VISIBLE-STRING  OPTIONAL
t   [ 4 ]   IMPLICIT   UtcTime ,
stNum[ 5 ]   IMPLICIT   INTEGER ,
sqNum[ 6 ]   IMPLICIT   INTEGER ,
test [ 7 ]   IMPLICIT   BOOLEAN  DEFAULT  FALSE ,
confRev  [ 8 ]   IMPLICIT   INTEGER ,
ndsCom   [ 9 ]   IMPLICIT BOOLEAN DEFAUL T FALSE ,
numDatSetEntries [ 10 ]   IMPLICIT INTEGER ,
allData [ 11 ]   IMPLICIT  SEQUENCE OF Data ,
security[ 12 ]   ANY  OPTIONAL ,
}
END
```

Figure 3.8   ASN.1 code of a GOOSE message

low time delay in the networks. For example, the delay caused by unpacking packets can be reduced in order to improve the data transfer speed and decrease the congestion possibility. In addition, the router cannot be performed because there is no network layer, so the GOOSE message cannot be used in the Internet. It is usually just used in the LAN which has high real-time requirement.

Sending GOOSE Message services at the presentation layer use abstract syntax notation 1 (ASN.1)/Basic Encoding Rules (BER). The ASN.1 encode of a GOOSE message is shown in Figure 3.8.

The GOOSE message' PDU (Protocol Data Unit) includes 13 parts: *gocbRef, timeAllowedtoLive, datSet, goID, t, stNum, sqNum, test, confRev, ndsCom, numDat SetEn-tries, allData, and security*. Here *gocbRef* is the control reference; *timeAl-lowedtoLive* is the max period of a message; *sqNum* is the counter of the GOOSE message sequence; *StNum* is the counter of the stage change of a message; *t* is the time of the state change; *datSet* is the name of the data set; *allData* is information of the switch state, which is the core data; *security* is the security extension.

3.3.2   Communications for Telecontrol: IEC 60870-5 Standards

3.3.2.1   Overview of IEC 60870 Standards

The IEC 60870 standards is a set of standards used for telecontrol (i.e., SCADA). Such SCADA systems are applied to control electric power transmission grids and other geographically distributed control systems. On the basis of the protocol of these standards, equipment from many different suppliers can perform interoperations.

IEC 60870 has six parts, which define general contents related to the standard, operating conditions, electrical interfaces, performance requirements, and data transmission protocols. Of these, IEC 60870-5 [15–25] is usually used in the Smart Grid.

The IEC TC57 (Working Group 03) has developed a protocol standard for telecontrol, teleprotection, and associated telecommunications for electric power systems. The result of this work is IEC 60870-5, which is one of the IEC 60870 set of standards that define systems used for telecontrol (i.e., SCADA) in electrical engineering and power system automation applications. This part of the standard provides a communication profile for sending basic telecontrol messages between two systems. In IEC 61850-5, the following documents are included: IEC 60870-5-1 Transmission Frame Formats, IEC 60870-5-2 Data Link Transmission Services, IEC 60870-5-3 General Structure of Application Data, IEC 60870-5-4 Definition and Coding of Information Elements, IEC 60870-5-5 Basic Application Functions, and IEC 60870-5-6 Guidelines for conformance testing for the IEC 60870-5 companion standards.

The IEC TC57 has also generated companion standards, which are as follows:

- IEC 60870-5-101 Transmission Protocols, companion standards especially for basic telecontrol tasks
- IEC 60870-5-102 Companion standard for the transmission of integrated totals in electric power systems
- IEC 60870-5-103 Transmission Protocols, Companion standard for the informative interface of protection equipment
- IEC 60870-5-104 Transmission Protocols, Network access for IEC 60870-5-101 using standard transport profiles.

IEC 60870-5-101/102/103/104 are companion standards generated for basic telecontrol tasks, transmission of integrated totals, data exchange from protection equipment, and network access of IEC 101, respectively.

3.3.2.2 Architecture of the Protocol

The architecture of the protocol of the terminal nodes of the communication systems is shown in Figure 3.9. The transport interface, which means the interface between the user and TCP, is a stream-oriented interface. This interface does not define any start or stop mechanism for the Application Service Data Unit (ASDU) of IEC 60870-5-101. The Application Protocol Control Information (APCI) includes three delimiting elements: (i) a start character, (ii) a specification of the length of the ASDU, and (iii) a control field. The start and end of the ASDU can be detected by checking these delimiting elements.

User process	Initialization	Selection of application function of IEC 60870-5-5 according to IEC60870-5-101
Application layer	Selection of ASDUs from IEC60870-5-101 and IEC 60870-5-104	
	APCI (Application Protocol Control Information) Transport Interface (user to TCP interface)	
1-4 Layers of OSI model	Selection of TCP/IP protocol suite (RFC 2200)	

Figure 3.9   Selected standard provisions of the defined telecontrol companion standard

3.3.2.3   Mapping to TCP Services

The station which performs the connection establishment is referred to as the controlling station (station A) in the subsequent description, while the partner station is referred to as the controlled station (station B). The release of connections may be initiated by either a controlling or a controlled station. Connection is performed based on (i) the controlling station if a controlled station exists as a partner and (ii) a fixed selection in the case of two equivalent controlling stations or partners.

Figure 3.10 shows the controlling station which establishes a connection by giving an active open call to its TCP. Then the controlling station sends Reset_Process to the connected controlled station, which confirms back the Reset_Process and gives an active close call to its TCP. The connection will be closed after the controlling station has given a passive close call to its TCP. Next, the controlling station tries to connect the controlled station by giving cyclic active opens to its TCP.

3.3.2.4   Redundant Connections

Through providing the possibility to establish more than one connection between two stations, redundant communication in a system using IEC 60870-5-104 can be performed. A logical connection is defined by a unique combination of two IP addresses and two port numbers.

The following general rules apply to the clause concerning redundant connections: Firstly, the controlling and controlled station shall be able to handle multiple

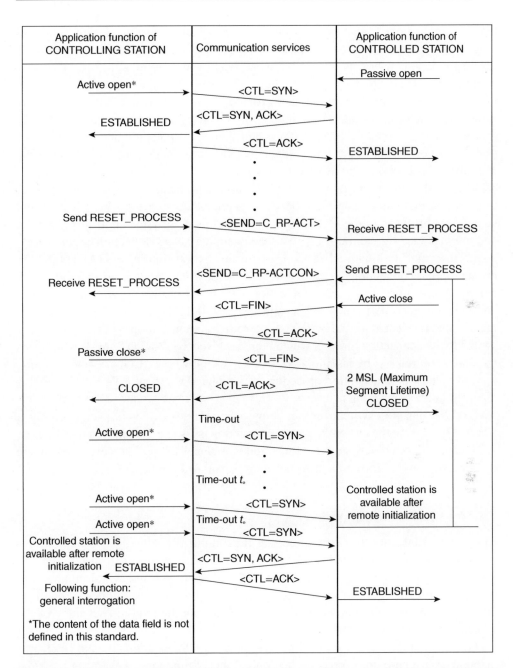

Figure 3.10 Remote initialization of the controlled station. Reproduced from IEC 60870-5-104 ed. 2.0 copyright © 2006 IEC Geneva, Switzerland. www.iec.ch [25] with permissions from International Electrotechnical Commission (IEC)

logical connections. Secondly, a total of N logical connections represent one redundancy group. Thirdly, only one logical connection is in the start state and is sending/receiving user data at a time for one redundancy group. Fourthly, all logical connections of a redundancy group shall be supervised by test frames. Fifthly, if more than one controlling station needs to access the same controlled station at the same time, each controlling station must be assigned to a different redundancy group (process image). Sixthly, a redundancy group shall rely upon only one process image (database/event buffer). And finally, the controlling station decides which one of the N connections is in the started state.

The logical connection which is enabled for user data exchange at any time is defined to be the started connection the others' connections are stopped. The selection and switchover of the started connection are always initiated by the controlling station, and are managed by the transport interface or higher layers. Selection of the started connection after station initialization is performed by transmitting a STARTDT_ACT on the desired connection, where STARTDT denotes "start data transfer" and _ACT denotes "activation." Similarly, connection switchover in the case of a failure (connection failover) is performed by transmitting a STARTDT_ACT on the stopped connection that is selected to take over. The controlled station (station B) should be able to switch to the connection through which it has received a STARTDT_ACT. It confirms the activation request by issuing a STARTDT_CON, where _CON denotes "confirmation." The whole activation procedure is completed when the STARTDT_CON is received in the controlling station.

Manual connection switchover can be performed by the following steps. Firstly, a STOPDT_ACT on the currently started connection is issued. Secondly, a STARTDT_ACT on the selected new started connection is issued. This will gracefully terminate data transfer on the first connection before it is resumed on the new connection.

3.3.2.5 Clock Synchronization

Clock synchronization is a very important function of IEC 60870-5. The clock synchronization function enabled by IEC 60870-5 avoids the necessity of installing additional clock synchronization receivers or similar equipment in several hundreds or thousands of controlled stations. Clock synchronization may be applied in configurations where the maximum network delay is less than the required accuracy of the clock in the receiving station.

The clocks of the controlled stations and the controlling stations must be synchronized in order to provide correct chronological sets of time-tagged events or information objects. After system initialization, the clocks are initially synchronized by the controlling station and then resynchronized periodically. Controlled stations expect the reception of clock synchronization messages within a legal time interval. In case the synchronization command does not arrive within this time interval, the controlled station sets all time-tagged information objects with a mark to indicate that the time tag may be inaccurate (invalid).

3.3.3 Inter-Control Center Communications: IEC 60870-6 Standards

3.3.3.1 Overview of IEC 60870-6

IEC 60870 part 6 [26–35] is one of the IEC 60870 set of standards which define systems used for telecontrol (SCADA) in electrical engineering and power system automation applications. The IEC TC57 (Working Group 03) have developed part 6 to provide a communication profile for sending basic telecontrol messages between two systems, which is compatible with ISO standards and ITU-T (ITU Telecommunication Standardization Sector) recommendations.

The Inter-Control Center Communications Protocol (ICCP or IEC 60870-6/TASE.2 (Telecontrol Application Service Element 2)) is being specified by utility organizations throughout the world to provide data exchange over WANs among utility control centers, utilities, power pools, regional control centers, and Non-Utility Generators.

Inter-utility real-time data exchange has become critical to the operation of interconnected systems in most parts of the world. For example, the development of electricity markets has seen the management of electricity networks by a functional hierarchy that is split across boundaries of commercial entities. At the top level, there is typically a system operator with coordination responsibilities for dispatch and overall system security. The lower layer is composed of regional transmission companies, which are linked with distribution companies and generating companies. In continental power systems, there is now considerable interconnection across national borders. ICCP allows the exchange of real-time and historical power system information including status and control data, measured values, scheduling data, energy accounting data, and operator messages.

The following documents are included in IEC 61850-6:

- IEC 60870-6-1 Application context and organization of standards
- IEC 60870-6-2 Use of basic standards (OSI layers 1–4)
- IEC 60870-6-501 TASE.1 Service definitions
- IEC 60870-6-502 TASE.1 Protocol definitions
- IEC 60870-6-503 TASE.2 Services and protocol
- IEC 60870-6-504 TASE.1 User conventions
- IEC 60870-6-601 Functional profile for providing the connection-oriented transport service in an end system connected via permanent access to a packet switched data network
- IEC 60870-6-602 TASE transport profiles
- IEC 60870-6-701 Functional profile for providing the TASE.1 application service in end systems
- IEC 60870-6-702 Functional profile for providing the TASE.2 application service in end systems.

3.3.3.2 Architecture and Network Model

The TASE.2 protocol relies on the use of MMS services (and hence the underlying MMS protocol) to implement data exchange among control center. Figure 3.11 shows

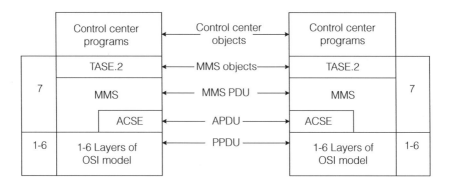

Figure 3.11 Protocol relationships

the relationship of TASE.2, the MMS provider, and other components of the protocol stack. In most cases, the values of objects being transferred are translated from/to the local machine representation automatically by the local MMS provider. Some TASE.2 objects require a common syntax (representation) and a meaning (interpretation) to constitute a form of protocol. The control center applications are not part of this standard. It is assumed that these applications should be able to communicate with TASE.2 through a specific interface. The specific interface between TASE.2 and the control center applications is a local issue and not a part of this standard.

As shown in Figure 3.11, the protocol architecture for TASE.2 requires the use of ISO protocols in layers 5–7 of the OSI reference model. The Transport Profiles (layers 1–4) may use virtually any standard or de facto standard (including TCP/IP) connection-mode transport layer and connectionless-mode network layer services over any type of transmission media.

The TASE.2 Data Exchange network may be either a private or public packet-switched or mesh network connecting communications processors which provide adequate routing functionality to allow for redundant paths and reliable service. Figure 3.12 shows a typical network topology using a router-based WAN.

The WAN provides routing and reliable services between control centers (which may include internal networking and routing capabilities).

The mesh network consists of redundant paths (Figure 3.13). Each control center maintains its own series of direct circuits, and also provides a mechanism for routing between those direct circuits. Control center C provides an alternate routing path for network traffic going from Control center A to B.

3.3.3.3   Informal TASE.2 Model Description

The model of a control center includes several different classes of applications, such as SCADA/EMS (Energy Management System), DSM (Demand Side Management)/ Load Management, Distributed Applications, and Man/Machine Interface. In inter-actions with other computing elements, a control center may act as a client, server, or both. As a server, a control center appears as a singular entity to interact with a

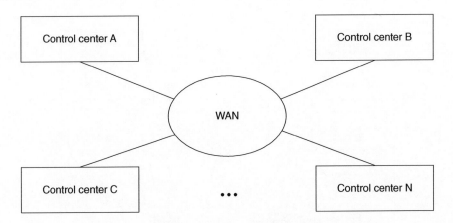

Figure 3.12 Router-based WAN

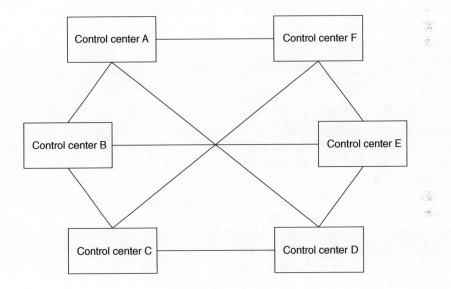

Figure 3.13 Mesh network

number of clients. Controls stations follow the Bilateral Agreement, which lists the data required to make and validate a connection between two control centers and the data to be exchanged. Control stations compare their respective interpretations of the Bilateral Agreement before a connection can be established and data exchanged.

TASE.2 is modeled as one or more processes operating as a logical entity which performs certain communications that allow the control centers to acquire or change data, and control devices. This specification defines the services and protocols for performing the communications between these processes. It also uses object models to define the data types and devices on which the TASE.2 services perform. TASE.2 is defined in terms of the client/server model of ISO 9506 (MMS).

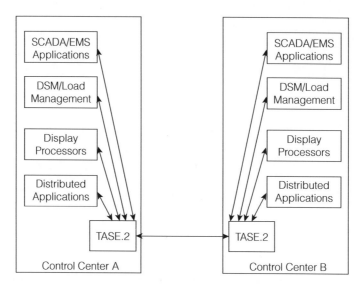

Figure 3.14 Informal TASE.2 model. Reproduced from IEC 60870-6-503 ed.2.0 Copyright © 2002 IEC Geneva, Switzerland. www.iec.ch [30] with permissions from International Electrotechnical Commission (IEC) [30]

The TASE.2 specification defines a number of operations and actions. TASE.2 operations are associated with a TASE.2 client. TASE.2 actions are associated with a TASE.2 server. There are two TASE.2 services that are considered to be both an operation and an action because either a TASE.2 client or a TASE.2 server may invoke them. These services are Conclude and Abort. Each TASE.2 operation begins with a local TASE.2 instance, acting as a TASE.2 client, invoking an MMS service. This invocation causes the local MMS provider to make use of the MMS protocol to communicate with the remote MMS server associated with the TASE.2 server. The remote MMS server may deliver indications to the remote TASE.2 server, which in turn responds appropriately, invoking one or more MMS responses and/or services as defined in this standard.

Figure 3.14 shows the logical relationships of TASE.2 to the control center applications. The local TASE.2 instance uses the services of the local MMS provider to communicate with the remote TASE.2 instance. It should be noted that the actual relationships, structure, location, connectivity, and interfaces between TASE.2 and the rest of the control center are local to the control center and outside the scope of this standard.

3.3.3.4 General Access Control Requirements

A Bilateral Table is required if a control center is going to serve any data values to any other control center. A control center shall have a Bilateral Table for each remote control center it serves. The Bilateral Table has a conceptual entry for each data object and data set included in the Bilateral Agreement. Each attribute in the object model is required to exist in some locally defined representation. Each data object specified in the Bilateral Table shall have an identifier which uniquely identifies the object.

The Client Control Center Designation identifies the TASE.2 client control center which is requesting an association with the TASE.2 server control center. Upon the initiation of the association, the TASE.2 server shall check the Client Control Center Designation that is provided in the association request to ensure that a Bilateral Agreement exists with this particular control center. If one exists, the TASE.2 server shall continue with further checks in the association establishment process. If it does not exist, then the TASE.2 server shall refuse the association request.

The List of Application References in the Bilateral Table are the Application References that are identified with the control center designated in the Client Control Center Designation. For TASE.2, each Application Reference maps to a unique AE-title (Application Entity Title). Upon the initiation of the association, after the TASE.2 server has checked the validity of the Client Control Center Designation, it shall check the List of Application References in that particular client control center Bilateral Table to ensure that the Application Reference provided in the association request is on the list. If it is on the list, the TASE.2 server shall continue with further checks in the association establishment process. If it is not on the list, then the TASE.2 server shall refuse the association request.

3.4   Energy Management Systems

3.4.1   Application Program Interface: the IEC 61970 Standards

3.4.1.1   Overview of IEC 61970

IEC 61970 standards [36–46] are based to a large extent upon the work of the Electric Power Research Institute (EPRI) Control Center Application Program Interface (CCAPI) research project (RP-3654-1). The principal objectives of the EPRI CCAPI project are as follows:

- to reduce the cost and time overheads to add new applications to an EMS or other system;
- to protect the investment in existing applications that are working effectively;
- to improve the capability to exchange information between disparate systems both within and external to the control center system;
- to provide an integration framework for interconnecting existing applications/systems that are based on a common architecture and information model;
- provide information model that is independent of the underlying technology.

The principal task of the IEC 61970 standards is to develop a set of guidelines and standards to facilitate, the following, among others:

- the integration of applications developed by different suppliers in the control center environment;
- the exchange of information to systems external to the control center environment.

The scope of these specifications includes other transmission systems as well as distribution and generation systems external to the control center that need to exchange real-time operational data with the control center. Therefore, another related goal of these standards is to enable the integration of existing legacy systems as well as new systems to conform to these standards in these application domains.

IEC 61970 consists of the following parts, under the general title Energy Management System Application Program Interface (EMS-API):

- **IEC 61970-1**: Guidelines and general requirements
- **IEC 61970-2**: Glossary
- **IEC 61970-3**: Common Information Model (CIM)
  - **IEC 61970-301**: CIM base
  - **IEC 61970-302**: CIM financial, energy scheduling, and reservations
- **IEC 61970-4**: SCSM
  - **IEC 61970-401**: Component Interface Specification (CIS) framework
  - **IEC 61970-402**: CIS – Common services
  - **IEC 61970-403**: CIS – Generic data access
  - **IEC 61970-404**: CIS – High speed data access
  - **IEC 61970-405**: CIS – Generic eventing and subscription
  - **IEC 61970-407**: CIS – Time series data access
  - **IEC 61970-453**: Exchange of Graphics Schematics Definitions (Common Graphics Exchange)
- **IEC 61970-501**: SCSM CIM eXtensible Markup Language (XML) Codification for Programmable Reference and Model Data Exchange.

3.4.1.2   EMS–API Reference Model

The EMS–API reference model is an abstract architecture which can provide following services. Firstly, it provides a visualization of the problem space being addressed. Secondly, it provides a language for describing and discussing solutions. Thirdly, it defines terminology. And finally, it provides other similar aids toward achieving a mutual understanding of the problem being solved with the EMS-API standards.

Figure 3.15 is a diagram of the reference model, with the shaded areas representing those portions of the reference model that are the subject of the IEC 61970 standards. The primary goal is to show clearly which parts of the problem space are the subject of the EMS-API standards, and by contrast, which are outside the domain of

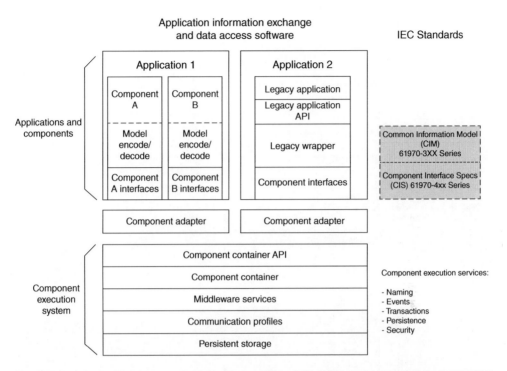

Figure 3.15 EMS-API reference model. Reproduced from IEC 61970-1 ed.1.0 Copyright © 2005 IEC Geneva, Switzerland. www.iec.ch [36] with permissions from International Electrotechnical Commission (IEC) [36]

the EMS-API project and the reasons. It is also intended to show how different parts of the standard relate to each other. The reference model is not a design, nor is it intended to describe software layers, although a layering approach is hard to avoid and is implied.

The reference model is specifically intended to apply to control center environments, which typically comprise networks of computers connected by LANs and sometimes WANs. A control center supports many different user groups and organizational functions, including operators, supervisors, operator training, operations planning, database maintenance, and software development. In an EMS, many applications are applied in a number of these contexts. Also, it is important that an application can be easily configured (preferably automatically) for use in multiple contexts.

In Figure 3.16, an EMS-API reference model based EMS system is shown. On the basis of a component execution system and component adapters that provide the infrastructure services, individual application components are interconnected. Real-time operational SCADA data are obtained from two components which are (i) a legacy SCADA system that are wrapped with a legacy wrapper and (ii) an ICCP data server. A DMS system external to the control center is also shown which could receive the same SCADA data updates.

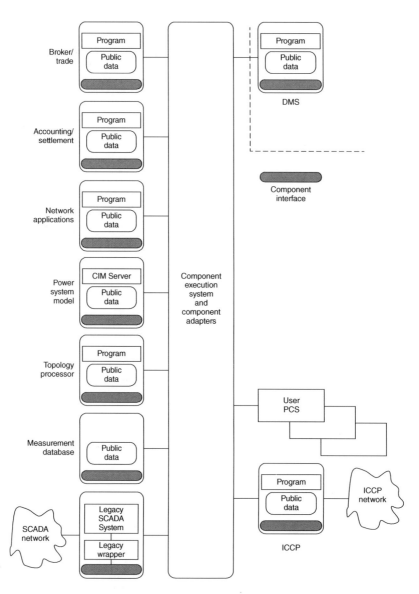

Figure 3.16 EMS using EMS-API component standard interfaces. Reproduced from IEC 60870-5-104 ed.2.0 Copyright © 2006 IEC Geneva, Switzerland. www.iec.ch [25] with permissions from International Electrotechnical Commission (IEC) [36]

3.4.1.3 Common Information Model (CIM)

The CIM provides an elaborate model for SCADA/EMS/DMS application components including measurements, network connectivity, and device characteristics. The CIM is part of the overall EMS-API framework. The CIM supplies these application components with a comprehensive logical view of a power system. This model is described in the Unified Modeling Language (UML) and maintained as single unified Rational ROSE model file. The CIM is an abstract model that represents all the major objects that are needed by multiple applications in an electric utility enterprise typically contained in an EMS information model. This model includes public classes and attributes for these objects, as well as the relationships between them. The objects represented in the CIM are abstract in nature and may be used in a wide variety of applications. The use of the CIM goes far beyond its application in an SCADA/EMS/DMS. This model can be regarded as a tool to perform integration in any domain where a common power system model is needed to facilitate interoperability and plug compatibility between applications and systems independent of any particular implementation.

CIM provides a standardized method to represent Smart Grid resources as object classes and attributes, along with their relationships. Also, the CIM performs the integration of EMS applications proposed independently by different vendors. This is accomplished by defining a common language (i.e., semantics and syntax) based on the CIM to enable these applications or systems to access public data and exchange information independent of how such information is represented internally. Specific anticipated uses of the CIM include the following:

- Initialize application components with configuration data.
  - Typically, before an application component can be made operational, it shall be initialized with current state and event information in a configuration phase (sometimes called data engineering) based on this model.
  - During the lifetime of an application component, extensions are made to the power system, requiring changes to the model and thus to the configuration data.
- Reuse existing configuration data from legacy systems.
- Incorporate existing configuration data from foreign systems.
- Provide base data to support on-line exchange of data between application components.
  - Data produced by application components during online operation is presented to operators and made available as input to other applications.
  - Application components are also receivers of data from other application components.

3.4.1.4 Component Interface Specification (CIS) and Technology Mappings

On the basis of CIS, the component vendors can transfer different collections of component interfaces into component packages while still remaining compliant with the EMS-API standards. The purpose of CIS in the IEC 61970-4XX is to specify the interfaces that a component shall use to facilitate integration with other independently developed components.

The CIS specifies the two major parts of a component interface: Firstly, a component (or application) must be able to access publicly available data through a standard interface. Also, these components (or applications) must be able to exchange information with other components (or applications). Secondly, the information contents or messages are defined to enable a component exchange information with other components.

CIS shall be mapped to specific technologies for implementation purposes because the CIS documents are independent of the underlying technology. To ensure interoperability, there shall be a standard mapping for each interface to each technology. Similarly, the event definitions compiled in the Information Exchange Model (IEM) from the CIS documents for each application category need to be mapped to the specific language used for transmission of the information. For instance, in the web-service communications, related CIS events need to be mapped to XML if a message broker is to be used to deliver XML messages.

Next, the general mappings will be discussed. It is anticipated that the following mappings or specializations will become companion standards in this series as these standards are deployed. Some are language-specific and some are middleware-specific, such as C++, C, CORBA, DCOM, Java, and XML.

3.4.1.5   Communication and Utility-Specific Services

IEC 61970 requires that the Communication Profile services perform the following functions. Firstly, they must guarantee delivery of network messages to their network destination (if active). Secondly, they must provide guaranteed delivery, ensuring that network messages are delivered exactly once, regardless of network failures or changes. Thirdly, they must provide guaranteed ordering, preserving the sending sequence of the source when delivering messages, regardless of network failures or changes. Fourthly, they must guarantee that if a network message cannot be delivered to a network destination, the network source will receive a message indicating the nondelivery. Fifthly, they must provide a selectable quality of service for prioritization of network messages or delivery via specific network paths. And finally, they must provide dynamic adaptation to the processing speed of network messages to ensure fairness.

On the basis of the above functions, many utility-specific services may be needed to support components in an EMS that are not available from commercially available component execution systems.

3.4.2   Software Inter-Application Integration: the IEC 61968 Standards

3.4.2.1   Overview of IEC 61968

Intra-application integration is aimed at programs in the same application system, usually communicating with each other using middleware that is embedded in their underlying runtime environment. Different from intra-application integration, the IEC 61968 series [47–53] has proposed to realize inter-application integration of the various distributed software application systems supporting the management of utility electrical distribution networks. As used in IEC 61968, a DMS consists of various distributed application components for the utility to manage electrical distribution networks. These capabilities include monitoring and control of equipment for power delivery, management processes to ensure system reliability, demand-side management, voltage management, outage management, automated mapping, and facilities management. IEC 61968 recommends that system interfaces of a compliant utility inter-application infrastructure be defined using UML. Standards interfaces are to be defined for each class of applications identified in the Interface Reference Model (IRM).

Communication between application components of the IRM requires compatibility on two levels. The first level is message formats and protocols. And the second level is that message contents including application-level issues of message layout and semantics that must be mutually understood.

Published IEC 61968 document consist of the following parts detailed in separate IEC 61968 standard documents.

- **IEC 61968-1** – Interface architecture and general requirements
- **IEC 61968-2** – Glossary
- **IEC 61968-3** – Interface for Network Operations
- **IEC 61968-4** – Interface for Records and Asset Management
- **IEC 61968-9** – Interface Standard for Meter Reading and Control
- **IEC 61968-11** – CIM Extensions for Distribution
- **IEC 61968-13** – CIM RDF (Resource Description Framework) Model exchange format for distribution.

3.4.2.2　Interface Reference Model

In IEC 61968, the distribution management domain covers all aspects of management of utility electrical distribution networks. A distribution utility will have some or all of the responsibility for monitoring and control of equipment for power delivery, management processes to ensure system reliability, demand-side management, voltage management, work management, outage management, automated mapping, facilities management, and so on.

In the Smart Grid, the distribution management domain may be organized as two interrelated types of business, namely, (i) electricity distribution and (ii) electricity supply. Electricity distribution covers the management of the physical distribution network that connects the producers and consumers. Electricity supply is concerned with the purchase of electrical energy from bulk producers for sale to individual consumers.

Various departments within a utility cooperate to perform the operation and management of a power distribution network; this activity is termed as distribution

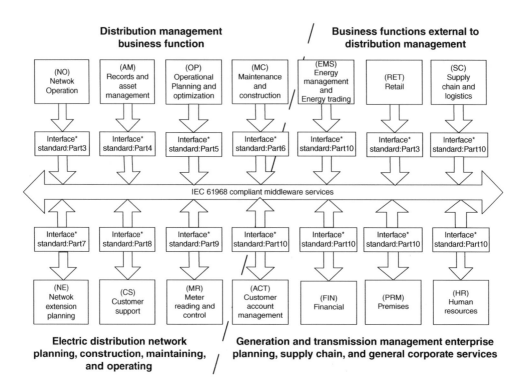

Figure 3.17   Typical applications mapped to interface reference model. Reproduced from IEC 61968-1 ed.1.0 Copyright © 2003 IEC Geneva, Switzerland.www.iec.ch [47] with permissions from International Electrotechnical Commission (IEC) [47]

management. Other departments within the organization may support the distribution management function without having direct responsibility for the distribution network. Here, a utility domain includes the software systems, equipment, staff, and consumers of a single utility organization which can be uniquely identified.

The use of a business-related model should ensure independence from vendor-produced system solutions. It is an important test of the viability of this standard that the IRM be recognizable to utility staff as a description of their own distribution network operation and management. Major utility business functions, which provide the top-level categories of the IRM, are shown in Figure 3.17.

3.4.2.3   Interface Profile

The communications of either a piece of data or the result of an execution of functionality cause information exchange among components. If the result of an execution of functionality causes information exchange among components, this case can be called a services exchange. Components can be distributed across the communication networks of the Smart Grid.

There is a middleware adapter in IEC 61968 which is profile-compliant software that increases existing middleware services so that the inter-application infrastructure supports required services. Information exchanged among components can be performed within the same process, across processes on the same machine (local), and across machines (remote). Usually, an object request brokers support of different communication patterns, such as asynchronous and synchronous interaction. Subscription refers to the ability to read or modify objects at cyclic or event-driven times. Messaging covers more of the features of current messaging middleware, such as store-and-forward, persistence of messages, and guaranteed delivery.

The middleware services shall provide a set of Application Program Interfaces (APIs) so that the previous layers in the interface profile among others can realize the following functions. Firstly, they locate transparently across the network, and interact with other applications or services. Secondly, they scale up in capacity without losing functionality. Thirdly, they are reliable and available. Fourthly, they are independent of communication profile services. Fifthly, they provide the ability to support Business-to-Business (B2B) transactions where needed.

A connection is required when integration is performed between two components. As there is more than one kind of network, different resources use different protocols, such as IIOP (Internet Inter-ORB Protocol) and HTTP (Hypertext Transfer Protocol). In order to connect multiple components, an integration system must reconcile the network and protocol differences transparently to the components.

Services are based on hardware and software standard platforms. As a matter of fact, the interoperability issues of different platforms of the hardware and operating systems from different vendors needs to be addressed. In other words, it cannot be expected that a component running in a dedicated hardware environment will be able to also run on another hardware environment without adjustment. The hardware environment of the IEC 61968 series requires the following items. Firstly, it shall support inter-process communication between concurrent processes. Secondly, it shall support multiple local processes running concurrently and it does not matter if this is achieved on a single processor or multiprocessor hardware. Thirdly, all other specifics of the hardware environment shall be shielded by the other layers in the interface profile.

3.4.2.4　Information Exchange Model

In IEC 61968 series, the following items from a compliant utility inter-application infrastructure are needed. Firstly, it shall have one logical IEC 61968 IEM and its implementation may be physically distributed. This facility allows information exchanged among components to be declared in a publicly accessible manner. Secondly, the IEM shall be accessible in machine-readable and platform-independent form. Thirdly, information is exchanged between components via one or more events whose types are defined in the IEM. Fourthly, the IEM shall maintain descriptions of the contents, syntax, and semantics (i.e., meaning) of the information exchanged between components. Such descriptions are commonly referred to as metadata (or a data dictionary). Fifthly, the IEM shall be capable of containing the following: named business object types such as breaker, outage schedule, and network diagram; name and data type of the attributes of the business objects such as "inService" and "voltage," names of primitive data types and their mapping to standard data types such as float, integer; named event types which act on objects, for example, object attribute update, object creation, object deletion; names of relationships between business objects such as "owns" and "connectedTo." The IEM may be capable of containing named datasets (i.e., sets of business object types, object attributes, event types, or object instances), and the IEM shall support services.

3.4.2.5 Maintenance Aspects

The IEC 61968 specification of component interfaces does not place requirements on how each component should be designed internally. However, the design is encouraged to be modular and decoupled from other component designs. In other words, components should be largely self-contained and have little interdependence.

Maintenance is an important part of the life cycle, which will reflect the quality of the design and implementation of the integrated components, each of which is produced by different sources. Reduced reliability, increased executable size, and reduced performance are among the likely consequences of a poor implementation. Reduced testability, reduced usability, and reduced modifiability are important primary causes. Secondary causes include increased link time, reduced comprehension, and increased compile time.

3.5   Teleprotection Equipment

3.5.1   An Overview of the IEC 60834

International Standard IEC 60834 standards [54, 55] has been developed by IEC TC57: power system control and associated communications. As a matter of fact, digital teleprotection systems such as microprocessor-based Phase Change Material (PCM), current differential relaying systems employing microwave, or fiber optic links such as telecommunication media have been widely used. IEC 60834 standards can be applied to narrowband and wideband teleprotection systems used to convey analog information about primary quantities such as phase or phase and amplitude. The teleprotection equipment can either be separate or integrated in one unit with the protection equipment or the telecommunication equipment. Narrowband systems include systems operating within a 4 kHz band (for each direction of transmission). Wideband systems include systems occupying more than 4 kHz bandwidth (for each direction of transmission). Broadband communication systems are not dealt with in this standard.

Published IEC 60834 documents consists of the following parts detailed in separate IEC 60834 standard documents.

- **IEC 61968-1**: Teleprotection equipment of power systems – Performance and testing – Part 1: Command systems
- **IEC 61968-2**: Teleprotection equipment of power systems Performance and testing of – Part 2: Analog comparison system.

IEC 60834-1 applies to teleprotection command systems used to convey command information, generally in conjunction with protection equipment. It aims at establishing performance requirements and recommended testing methods for command-type teleprotection equipment. The information conveyed by the teleprotection equipment can be in analog or digital form. The command-type teleprotection equipment referred to in this standard can be Power Line Carrier (PLC) equipment or voice frequency equipment which is used in connection with various telecommunication systems, such as PLC, radio links, optical fiber, rented circuits, leased or privately owned cables. In addition, the command-type teleprotection can be digital equipment which are used with a digital telecommunication system or media such as optical fibers, radio links, leased, or privately owned digital links.

The object of IEC 60834-2 is to establish performance requirements and recommended testing methods for analog comparison teleprotection equipments used in connection with power network protection systems and to define the associated terminology. The information conveyed and compared such as phase and amplitude quantities can be in either analog or digital form. In addition to the power supply and the interfaces belonging to the teleprotection equipment, the performance of the teleprotection equipment in conjunction with the protection equipment shall be tested.

3.5.2 Types of Teleprotection Command Schemes

1. **Permissive tripping schemes**
   This term refers to schemes where the received command initiates tripping in conjunction with a local protection equipment. Command channels of this type can operate in an audiofrequency band, a PLC frequency band, or at a digital bit rate. The channel is often designed with the premise that dependability of operation should be high even under conditions when, because of a power system disturbance, the telecommunication medium may be adversely affected.

2. **Intertripping schemes (direct or transfer tripping)**
   This term refers to schemes where the received command initiates tripping without qualification by local protection. Intertrip channels utilize principles similar to those of permissive trip channels. However, security against unwanted operation and dependability of correct operation are prime requirements. Speed of operation is usually sacrificed to meet security and dependability requirements, particularly in analog systems.

3. **Blocking protection schemes**

This term refers to schemes where the received command blocks the operation of local protection. These channels utilize principles similar to tosse of permissive trip channels; however, dependability of operation and speed are prime requirements.

3.5.3 Requirements for Command-Type Teleprotection Systems

The following requirements apply to the interface between protection equipment and teleprotection equipment as well as to the interface between teleprotection equipment and the telecommunication system. These protection architectures are defined in Figure 3.18. The requirements apply equally when the various types of equipment are integrated as well as separated from each other.

If the protection equipment and the teleprotection equipment form a combined system installed in the same enclosure in the same location, the requirements for interface (a) may not be applicable. If the teleprotection equipment and the telecommunication equipment are part of a common apparatus and are installed in the same bay in the same location, the requirements for interface (b) may not be applicable.

3.5.4 Teleprotection System Performance Requirements

Dependability, security, and transmission time of a given teleprotection command system are interdependent parameters; for a constant bandwidth, for instance, security can only be improved at the expense of dependability or transmission time.

The requirements are met by teleprotection command systems, and therefore the optimum compromise of the above parameters depends on the particular application (permissive tripping, intertripping, or blocking) and on the type of transmission path used.

The design of teleprotection systems and the way in which information links are used need to take account of the practical limitations arising from the fact that the influence of interference, noise, and communication failures cannot be completely avoided.

While transmitting and receiving either guard or command signals, the teleprotection equipment shall monitor the transmission path and as much of the terminal equipment as possible. Failure to receive the transmitted signal, whether by failure of the transmission path or the terminal equipment, shall be detected by monitoring circuits associated with the receiver and transmitter. In addition, an alarm shall be given, provided that the period of failure exceeds a specified time (usually settable within a defined range). The monitoring circuits may further respond to excessive interference and noise at a level that could impair the correct operation.

Again, an alarm shall then be given, should the period of interference exceed a specified time. Furthermore, for digital teleprotection systems, the command or guard messages transmitted by a particular equipment shall not be received as valid guard or command messages by the originating equipment or by the wrong equipment. If the

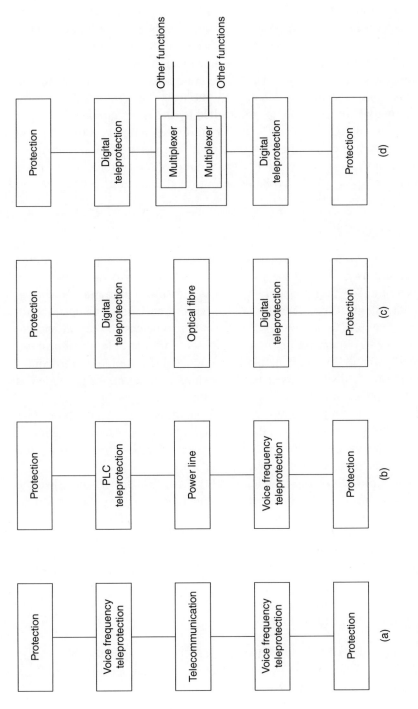

Figure 3.18 Four kinds of configuration

(a) Voice frequency configuration
(b) Power line carrier frequence configuration
(c) Directly connected digital teleprotection
(d) Directly teleprotection connected via a multiplexed communication system

communication system mistakenly directs messages back to the originating equipment or to the wrong equipment, then an alarm shall be given. It should be possible to disable such a mechanism for test purposes. Facilities shall be provided for clamping or inhibiting the receiver output once the monitoring circuits have responded to abnormal conditions. The receiver output shall be clamped or inhibited either in the state that existed before the signal failure, or in a steady "command-off" or "command-on" state. The various options shall be selectable by the user. The clamping or inhibit action may be immediate or delayed (e.g., controlled by the alarm circuit).

When a digital teleprotection system is used in a digital network, measurements shall be carried out to ensure that jitter at the output of the teleprotection transmitter does not affect the network and that jitter at the input of the teleprotection receiver does not result in any type of malfunctioning or unwanted operation.

3.5.5   Teleprotection System Performance Tests

A teleprotection channel may be exposed to various kinds of noise depending on the transmission medium used. Testing the security and to some degree also testing the dependability is a time-consuming operation. It is therefore very important to choose a procedure which will give a good measure of the performance of the channel within a reasonable time frame. The chosen procedure shall also be easy to repeat and the test instruments easy to obtain. Furthermore, the tests represent only a compromise on actual operating conditions and it should be recognized that more onerous conditions will occur in practical situations. The results should, therefore, be presented as indicative of comparative, and not absolute, performance.

One way to fulfill all the requirements for analog systems is to adopt a test procedure using white noise. Both the security and the dependability can be tested in relation to the S/N ratio when the noise is white noise. In some special cases, the results are not comparable with measurements made with impulse noise.

On the other hand, nowadays, digital teleprotection is a popular way to realize the teleprotection. Noise on a digital teleprotection channel tends to corrupt the information being transferred by causing bit errors. The bit errors may delay reception of a command, or may cause an unwanted command. Generally, therefore, the BER will alter the levels of security and dependability of the system. The main aim of the performance tests, therefore, is to examine the security and dependability characteristics of the system in relation to the BER.

For security and dependability tests, it is preferable to use random bit errors. This approach is chosen for the following reasons:

1. The use of a random variable for the BER aligns with the main methodology used for mathematical analysis of the protocol, which generally relies on probabilistic techniques utilizing random error variables.

2. It is often difficult to introduce nonrandom bit errors into the digital transmission path.

  The statistical properties of the disturbed digital communication channel shall correspond to those of the Binary Symmetric Channel model (BSC). Bit errors with these properties may be introduced by injecting direct bit errors, provided the bit errors can be introduced randomly and the BSC properties apply. Bit errors may also be applied by injecting white noise into the input of the line terminal of the telecommunication system used.

3.6   Application Cases of Related Standards in the Power Grid

3.6.1   Case 1: Engineering Process in Smart Substation Automation

3.6.1.1   Configuration in Electrical Substations Related to IEDs

1. **Use of Substation Configuration Language (SCL) in the Engineering Process**
   Figure 3.19 explains the usage of SCL data exchange in an engineering process. To make the engineering tasks and responsibilities clear, tool roles are introduced for an IED configureurator and a system configureurator. The tool provides IED-specific settings and generates IED-specific configuration files, or it loads the IED configuration into the IED.
   The IED configureurator is a manufacturer-specific, may be even IED-specific, tool that shall be able to import or export the files introduced in the next section. The tool then provides IED-specific settings and generates IED-specific configuration files, or it loads the IED configuration into the IED.
   The system configurator is an IED-independent system-level tool that shall be able to import or export configuration files. It shall be able to import configuration files from several IEDs, as needed for system-level engineering, and used by the configuration engineer to add system information shared by different IEDs. Then the system configureurator shall generate a substation-related configuration file, which is fed back to the IED configurator for system-related IED configuration. The system configuration should also be able to read a system specification file, for example, as a base for starting system engineering, or to compare it with an engineered system for the same substation.

2. **SCL description file types**
   SCL files are used to exchange the configuration data between different tools, possibly from different manufacturers. As already shown in Figure 3.23, there are at least five different purposes of SCL data exchange, and therefore five kinds of SCL files to be distinguished for the data exchange between tools. Nevertheless, the contents of each file shall obey the rules of the ConfigurationSCL defined in the next section.

The following types of SCL files are distinguished:
a. Firstly, this file describes the functional and engineering capabilities of an IED-type. It shall contain exactly one IED section for the IED-type whose capabilities are described. Furthermore, the file shall contain the needed data type templates inclusive of logical node-type definitions, and may contain an optional substation section, where the substation name shall be TEMPLATE.
b. Secondly, the file extension shall be .ICD for IED Capability Description.
c. Thirdly, this file describes the single line diagram and functions of the substation and the required logical nodes. It shall contain a substation description section and may contain the needed data-type templates and logical node-type definitions. This file extension shall be .SSD for System Specification Description.
d. Fourthly, this file extension shall be .SSD.
e. Fifthly, this file contains all IEDs inclusive of the needed DataTypeTemplates, a communication configuration section, and a substation description section.
f. Sixthly, this file extension shall be .SCD for Substation Configuration Description.
g. Seventhly, it describes the communication-related part of an instantiated IED within a project. The communication section contains the address of the IED. The substation section related to this IED may be present and then shall have name values assigned according to the project-specific names. It is an SCD file, possibly stripped down to what the concerned IED shall know (restricted view of source IEDs).
h. Eighthly, this file extension for the SCL part shall be .CID for Configured IED Description.
i. Ninthly, this file describes the interfaces of one project to be used by the other project, and at reimport the additionally engineered interface connections between the projects. It is a subset of a SCD file, containing the interfacing parts of the IEDs to which connections between the projects shall be engineered, and fix IEDs referenced by them to not loose, already defined references. Therefore additionally to an SCD file, it states at each IED the engineering rights and the owning project from the view of the using (importing) project at each IED.
j. Finally, this file extension shall be SED for System Exchange Description.

3. **The SCL language**

The SCL language is based on XML.

The SCL element shall contain a header section, and at least one of the following sections: Substation, Communication, IED, DataTypeTemplates. The UML diagram given in Figure 3.20 gives an overview of how the SCL schema is structured.
a. *Header:* The header serves to identify an SCL configuration file and its version, and to specify options for the mapping of names to signals.
b. *Substation section:* It serves to describe the functional structure of a substation, and to identify the primary devices and their electrical connections.

c. *IED section:* It describes the (pre-) configuration of an IED: its access points, the logical devices, and the logical nodes instantiated on it. Furthermore, it defines the capabilities of an IED in terms of communication services offered and, together with its LNType, instantiated data (DO) and its default or configuration values.

d. *Communication section:* It describes the direct communication connection possibilities between logical nodes by means of logical buses (SubNetworks) and IED access points. The communication section now describes which IED access points are connected to a common subnetwork. This is done in a way that reflects the hierarchical name structure within the IED, which is based on IED relative names for access points, LDs, and LNs.

e. *Data type templates:* This section defines instantiable logical node types. A logical node type is an instantiable template of a logical node. An LNodeType (elsewhere also called LN type) is referenced each time that this instantiable type is needed within an IED. A logical node type template is built from DATA (DO) elements, which is derived from the DATA classes (Common Data Class (CDC)) defined in IEC 61820-7-3.

4. **Configuration tools**

   a. *IED Configurator*

      The task of the IED configureurator is to create the ICD file, and to modify the data model, parameter, and configuration values for a new ICD file. Further, the IED configureurator is responsible to bind incoming data from other IEDs as defined within an imported SCD file to internal signals, for example, by means of the SCL input section, and to generate and load the IED instance specific configuration data, where a CID file could be a part of.

   b. *System Configurator*

      The tasks of the system configuration tool are to create IED instances from IED templates, engineer the data flow between the IEDs, give addresses to them, and bind the logical nodes to the primary system. Therefore beneath instantiating IED templates, the system tool handles the following SCL sections:

      i. substation section, including references to logical nodes on IEDs;

      ii. communication section inclusive of project-specific instance addresses;

      iii. data sets and control blocks, as allowed by the IED capabilities;

      iv. allocation of data flow and report control block instances to clients, as allowed by the IED capabilities;

      v. creating IED input sections as seen from system engineering point of views, however without binding to IED internal signals;

      vi. reorganizing the DateTypeTemplate section to keep the type identifiers unique and the template section short.

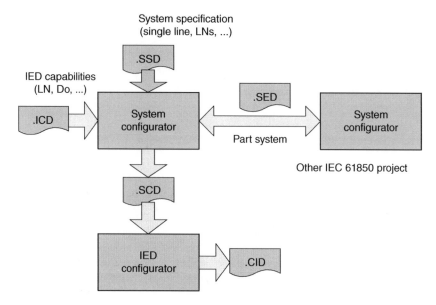

Figure 3.19 Information flow in the configuration process with information exchanged between projects

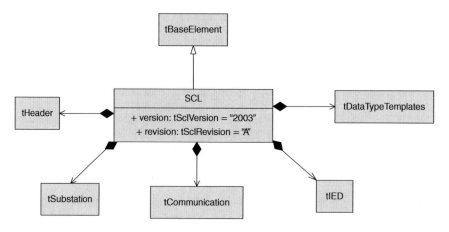

Figure 3.20 UML diagram overview of SCL schema

3.6.1.2 Application in Configuration of Interlocking between Substations

1. **Interlocking between Substations**

   The interlocking of the line earth switch depends on whether there is voltage on the line or not. To be able to detect this, the states of the earthing switch and the line disconnector switch of the other line side should be transferred and used. Interlocking is shown in Figure 3.21.

2. **Engineering process between substations**

   As normally for a certain engineering purpose, the concerned IEDs only have to know a part of the other system, it seems to be appropriate to use system exchange descriptions (SED) which contain only a part of the overall system to be engineered for a certain purpose. This SED file is then transferred from one substation project to the other project to include the needed DATA into the data flow. The engineered data flow is then transferred back to the originating system/project by means of an enhanced SED file, so that both systems then have the same consistent state regarding the data to be exchanged between them across the connecting Ethernet link (see Figure 3.20).

In Figure 3.22, the IEDs IED1_1, IED1_2, and IED1_3 are owned by the system tool for handling the A1 project (substation), while IED2_1, IED2_2, IED2_3 are owned by the system tool for handling the A2 project. IED1_3 is a "boundary" IED for A1 project, while IED2_1 is a boundary IED for the A2 project.

At export, a system tool can decide which kind of engineering rights another system tool might have. The engineering rights machine state is shown in Figure 3.27.

3. **State machine in data exchange**

   The right of data flow engineering can be formally transferred from one project to another project by means of an SED file. The concerned IEDs are marked with the *engRight* attribute value *dataflow*. To not lose already predefined references on these IEDs, all referenced IEDs have also to be exported at least partly, that is, just the LDs and data set references, with *engRight=fix*.

   The transfer and handling of the *dataflow* and *fix* rights is defined in the state diagrams in Figure 3.23 for the project owning the IEDs as well as for the receiving project. It may be observed that the *full* engineering right always stays in the owner project.

   The state diagram (a) shows the internal states in the IED owner project, when exporting and importing owned IEDs via SED file. The diagram (b) shows the IED states of imported/exported IEDs at the receiving project, which is not owner of the IEDs.

   When an IED is exported to an SED file, then the IED ownership is marked by means of the (new) owner attribute of the IED, which is set to the SCD header identification respective to the project identification, and an IED exported as data flow is blocked for changes within the owning system tool, for example, by setting it to fix.

   When the data flow IED with same owner ID as the project ID is re-imported, then the change block is taken away again. The owner attribute at IEDs with fix and data flow engineering right marks to which project they belong, to block changes at the owner after data flow export.

4. **Configuration example**

   This example is based on the system configuration of Figure 3.24. The engineering process of interlocking works as follows.

   a. Project A1 exports an SED file for IED IED1_3, with IED1_3 as data flow IED, and all needed source IEDs as fix IEDs.

   b. Project A2 imports the SED file, and engineers the data flow between IED1_3 and IED2_1.

c. Project A2 now can configure IED2_1 with its IED tool; then it exports and SED file with same identification as the imported SED, containing IED2_1 as fixed IED and IED1_3 as original data flow IED (plus all needed source IEDs as fix) with the added data flow definitions.

d. Project A1 imports the SED file, and finalizes any engineering on IED1_3 based on its now complete system description; now IED1_3 can be configured by its IED tool.

In fact, the SED file example can be imported in above engineering step (d). The SED file example is shown in Figure 3.22. It also contains the modeling of the line connection between the two substations, to make the primary system modeling complete, and shows that the two bays in the substations are connected by a common line. Further, the protection IED IED2_1, which, in principle, is the same IED type as IED 1_3, is exported as fix, and therefore only the relevant (interface) parts for IED2_1 are contained in it.

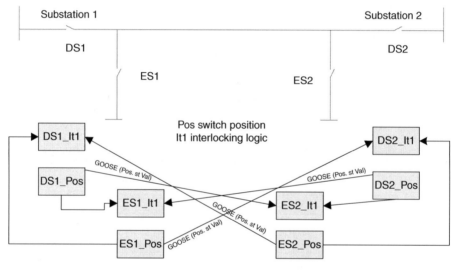

Figure 3.21   Interlocking

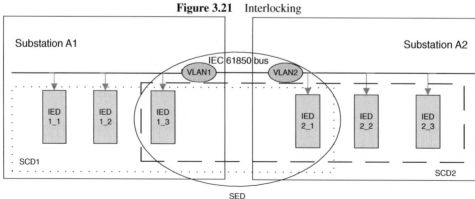

Figure 3.22   SCD files and SED region for SS-to-SS communication

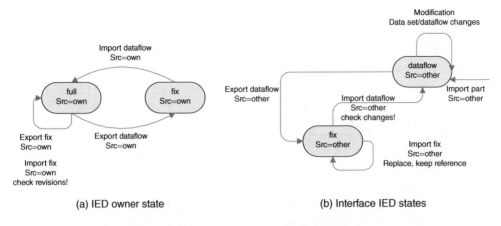

(a) IED owner state　　　　　　　　(b) Interface IED states

Figure 3.23　IED states when exchanging SED files

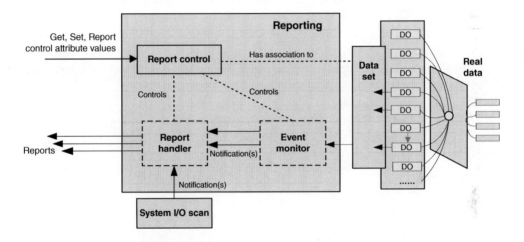

Figure 3.24　Basic building blocks for reporting

3.6.2　Case 2: Information Exchange Services and Service Tracking

3.6.2.1　The Introduction of Reporting

Reporting meets a number of crucial requirements for event-driven information exchange. Reporting provides mechanisms to report packed values of instances of a data object immediately or after some buffer time. The format of reporting is shown in Table 3.2.

1. **The model of Reporting**
   The reporting model is composed of three building blocks, as shown in Figure 3.24. There are Report control, Report handler, and Event monitor.

The data set (referencing data object) represents the values of data objects. The values are conceptually monitored by the event monitors. An event monitor determines, on the basis of the state of the real data and the attributes of the control class, when to generate a notification to the appropriate handlers (for example, report handler). The notification includes the data object values and reasons for data inclusion.

2. **The procedures for report generation**

   The procedures for report generation are shown in Figure 3.25. The parameter Opt-Flds shall specify the information to be included in the report. The corresponding relevance shall be as specified in the following.

3. **The composition of Entry**

   The report handler assigns Entry(s) and TimeOfEntry(s) to the values contained within a set of notifications. The number of notifications combined into a single EntryID is determined by the RCB (Report Control Block) control parameters (for example, BufTm). The value of the EntryID is local issue but it shall be a unique arbitrary OCTET STRING whose value is unique within the scope of entries for a specific RCB. The value of the TimeOfEntry shall be the timestamp representing the time at which the report handler received the first notification that is used to form an EntryID. The report handler decides when and how to send a report to the subscribed client.

4. **The storage of reporting**

   The general queue of entries for a report handler are shown in Figure 3.26.

Table 3.2 The format of reporting

Attribute name			BRCB Opt Flds
Rpt ID			—
Opt Flds			—
Sq Num			sequence-number (BOOLEAN) = TRUE
Sub Sq Num			
More Segments Follow			
Dat Set			data-set-name = TRUE
Buf Ovfl			buffer-overflow = TRUE
Conf Rev			conf-revision = TRUE
Entry	Time Of Entry		report-time-stamp = TRUE
	Entry ID		entry ID = TRUE
	Entry Data [1..n]	Data Ref	data-reference = TRUE
		Value	—
		Reason Code	reason-for-inclusion

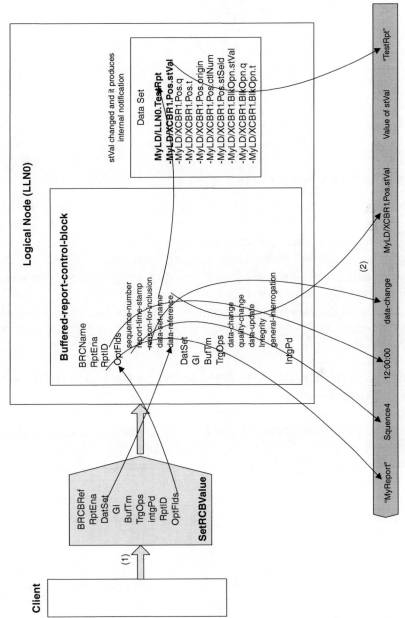

Figure 3.25 Procedures for report generation

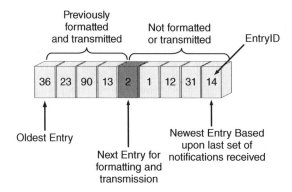

Figure 3.26 General queue of entries for report handler

3.6.2.2 Service Tracking for Control-Block-Related Services

This clause introduces a general model for control block, as used later on for service definitions of control-block-based services. Further it models control block attributes and services-related attributes into common data classes, which allow using a data model for service tracking by providing service-related events. The events can be monitored by using the already defined reporting and logging services. This reporting and logging of service-related events are called service tracking.

The service-tracking function for a CB (control block) class can use a single instance of such a data object to drive the reporting or logging process for all control block objects of the same class in the order of their events.

1. **Tracking of service for BRCB (buffered report control block) (BTS [buffered report tracking service])**

   The tracking service for a BR (Buffered Report) control block is shown in Figure 3.27. All these tracking-related CDCs inherit the attributes from the base-tracking CDC CST (Common Service Tracking) and add the relevant attributes from the control blocks. To better distinguish these from the control block attributes, the attributes within the tracking CDCs start with a lower case letter, while those used in the control block start with an upper case letter.

   When the BRCB is notified by an internal notification of a data-change, quality-change, or data-update event of a member(s) of the referenced data set whose values are to be reported, the BRCB shall include the values of the member(s) of the references data set that produced the internal notification in the report. The value to be reported shall be the value that produces the internal notification.

2. **Tracking of service for log control block (LTS) (Figure 3.28)**

   ServiceType=SetLCBValues

   The attributes LogEna, datSet, optFlds, bufTm, trgOps, intPd, logRef mirror the service parameters of the service SetLCBValues of the LCB (Log Control Block) referenced by the attribute objRef.

In case one of the attributes is not written while using the SetLCBValues service (LogEnable[0,1]), the value that is returned in the LTS reflects the value that is set in the LCB referenced by objRef.

3. **Tracking of service for GOOSE control block (GTS) (Figure 3.29)**

   The attributes goEna, goID, and datSet mirror the service parameters of the service SetGoCBValues of the GoCB referenced by the attributes objRef.

   The attribute originatorID contains the ID of the client. In case one of the attributes is not written while using the SetGoCBValues service (GoEnable [0,1], …), the value that is returned in the GTS is the value that is set in the GoCB referenced by objRef.

   The mechanism of GOOSE and Reporting can take the same data set. The GOOSE service and its protocol are summarized in Table 3.3.

4. **Tracking of service for MSVCB (Multicast Sampled Value Control Block) control block – MSVCB tracking service (MTS) (Figure 3.30)**

   ServiceType=SetMSVCBValues

   The attributes svEna, datSet, msvID, smpMod, smpRata, and optFlds mirror the service parameters of the service SetMSVCBValues of the MSVCB referenced by the attribute objRef. The attribute errorCode returns the error that occurred during the service or the value no error if it was successful. The attribute originatorID contains the ID of the client. In case one of the attributes is not written while using the SetMSVCBValues service (SvEnable[0,1], …), the value that is returned in the MTS reflects the value that is set in the MSVCB referenced by objRef.

5. **Tracking of service for SGCB (Setting Group Control Block) (STS) (Figure 3.31)**

   ServiceType=SelectEditSG,ServiceType=SelectActiveSG

   The service SelectActiveSG determines which values shall be copied to the "active buffer."

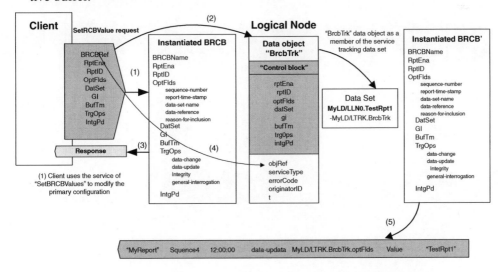

Figure 3.27   Tracking service for BR (buffered report) control block

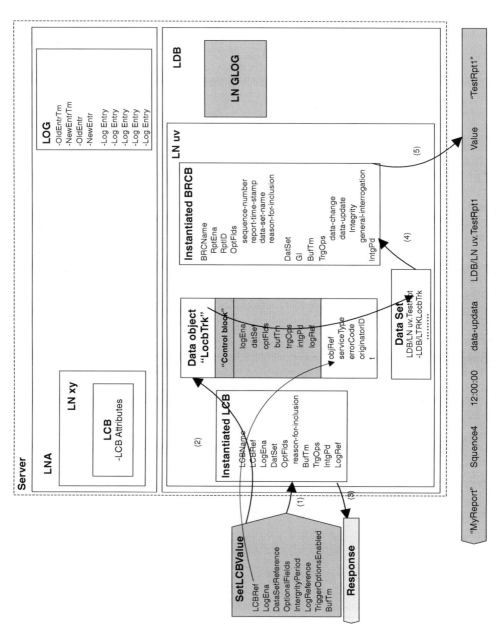

Figure 3.28 Tracking service for log control block

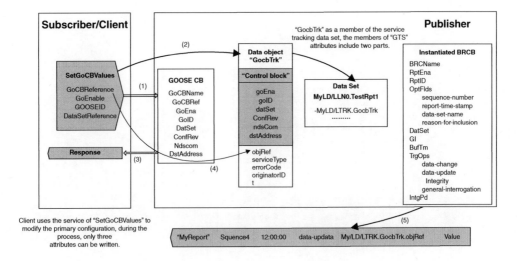

Figure 3.29   Tracking service for GOOSE control block

Table 3.3   GOOSE service and its protocol

GOOSE service and its protocol	
SendGooseMessage	MultiCast
GetGoReference	UniCast
GetGoElementNumber	UniCast
GetGoCBValue	ASCII client/server
SetGoCBValue	ASCII client/server

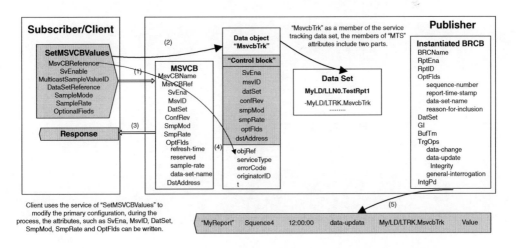

Figure 3.30   Tracking service for MSV (Multicast Sampled Value) control block

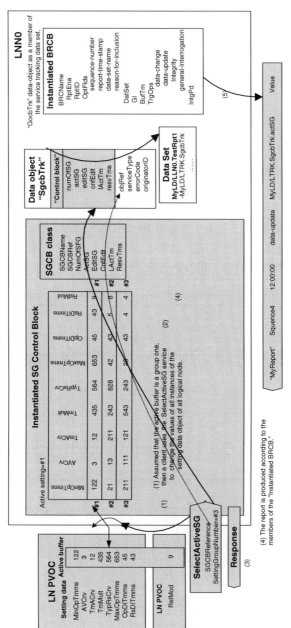

Figure 3.31   Tracking service for setting group control block

3.6.2.3   Service Tracking for Command-Related Services

The control services can be also tracked, as shown in Figure 3.32. The base concept consists in defining within the data model where to store the values of the service parameters used by a service (for any control objects). For all controllable data objects of a given server there shall be a single data object instance (tracking data object) available in the object directory that will mirror the value of the service parameters of the control services. Therewith, the control service can be logged or reported to any client as soon as the tracking data object is a data set member of the data set associated to a LCB or to a BRCB/URCB (Unbuffered Report Control Block).

3.7   Analysis of Relationships among Related Standards

3.7.1   IEC 61970 and IEC 61968

As with IEC 61970, the IEC 61968 series of standards also build on the CIM by extending the CIM Base contained in IEC 61970-301 wherever possible by extending it to include additional specializations of existing classes, but also adding entirely new sets of classes to model objects found in the distribution problem domain.

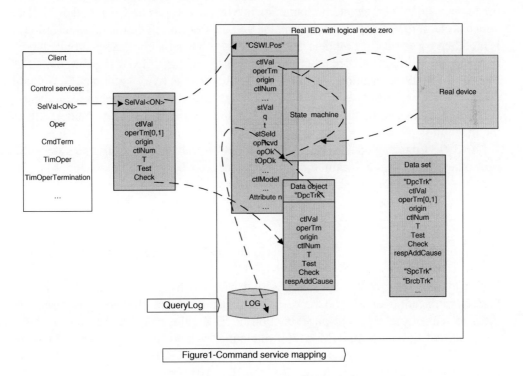

Figure 3.32   Tracking of control service concept

Therefore, to understand the entire scope of the CIM, it is necessary to consider the CIM as described in both the IEC 61970 and IEC 61968 series of standards.

3.7.2   IEC 61850 and IEC 61970

On the basis of IEC 61850, the running state of the communication networks in substation automation can be monitored dynamically using Scalable Vector Graphics (SVG) [56]. SVG provides a real-time curve, which can present dynamically and intuitively the work state of the substation. Theoretically, SVG can define the format of a graphic element. On the basis of the concept of standardization, the CIM graphic element in IEC 61970 can be used.

In addition, both IEC 61850 and IEC 61970 define the access mechanism to the substation electric utility's data and data architecture. But they use a different approach in the access mechanism. IEC 61850 is defined from the internal communication standpoint, whereas IEC 61970 is from the substation management standpoint. Thus, harmonization is required to integrate both sides.

In order to integrate IEC 61850's Logical Node, SCL, and IEC 61970's CIM, the single line diagram must be expressed in IEC 619710 CIM. The next three types of information model unity process are musts [57].

The first is the Component Mapping. Using the CIM-defined model of each utility, modeling is done. The second is Containment Mapping. By using CIM models Fields which is classified by voltage, changing the voltage (the core function of a substation) is modified based on CIM models. By using CIM models Fields classified by voltage to change the voltage (the core function of a substation) is modified using CIM models. The third is Interconnection Mapping. Using the Connectivity Node and Terminal class, connect each utility and allow the Terminal class to take the real measurement.

These methods can be used to modify a single line diagram in CIM model. After the diagram is modified, specific information must be unified. Each data defined by IEC 61850's Logical Node can be redefined by a CIM class attribute. To integrate with each other, Field Device Communication Interface must be done in order to coordinate the CIM class attribute and Logical Node attribute.

3.7.3   IEC 61850 and IEC 60870

The information model of SCADA based on IEC 61850 can be mapped to IEC 60870. Here the Feeder Terminal Unit (FTU) is taken as an example for analysis [58]. FTU is a kind of remote terminal unit along distribution feeders that allows the communication of a local process to a master or central system. In [58], in order to realize the plug-and-play function of FTUs, the FTU information models are proposed by using the IEC 61850 modeling method. Further, the information model and communication services model are mapped to telecontrol protocols IEC60870-5-104. Extensions to the IEC 60870-5-104 are also proposed to support the exchange of the model information. A prototype system is developed. The system demonstrates that a seamless communication between the master station and FTU can be realized using the proposed modeling and mapping methods. In this work, in order to normalize the transform between IEC 61850 and 60870-5-104, IEC TC57 proposes the information-exchange rules IEC 61850-80-1. It describes standardized mapping of device-oriented data models (IEC 61850) with already-defined attributes of a common data class and services onto the already-defined ASDU and services of IEC 60870-5-104 or IEC 60870-5-101. The configuration can be done by using the configurationSCL or other configuration methods. Some mapping rulers are shown as follows.

- The logical device references shall map to the Common Address of Application Service Data Unit (CASDU). The number of logical devices in an FTU is small (less than five). So the same CASDU is adopted in the same FTU.
- The logical node instance ID (Identifier) and data-attribute reference shall map to the information object address (IOA).

3.7.4   TASE.2 and MMS

The TASE.2 resides on top of MMS. It describes a standardized application of MMS using the MMS services and protocol. TASE.2 enhances the functionality of MMS by specifying structured data mapped to MMS objects and assigning specific semantics to it. As an example of pure MMS services, MMS allows reading data from a remote system. The data will be responded without any specific condition. If these data are read depending on very specific conditions (e.g., on change only) then TASE.2 provides appropriate services which are not provided by MMS.

Although the specific requirements agreed upon within IEC TC 57 have led to the definition of TASE.2 there are several other application domains (outside the control centers) with less, very limited, or mixed requirements which may use the TASE.2 services. These other areas are outside the scope of this standard but the use of TASE.2 goes far beyond the specific scope of this standard.

TASE.2 provides an independent and scalable set of services to allow efficient implementations optimized for the respective requirements of a control center. It does this by defining several conformance building blocks (CBBs). MMS offers also a scalability of its services specifying MMS CBBs. A simple TASE.2 implementation requires only a simple MMS implementation.

TASE.2 and MMS provide their services to their respective users. MMS provides its services to TASE.2 and TASE.2 provides its services to the control center application. MMS is an independent standard that can provide its services also to users other than TASE.2 it may serve directly to specific control center applications and to any other application. This means that the use of MMS is not restricted to TASE.2.

For requirements outside the scope of this standard or for future requirements, for example, journaling of data, downloading and uploading of mass data like programs, additional MMS models, and services, that is, Journaling and Domain Loading, respectively can be applied by an real system in addition to TASE.2. This is possible because the additional application of MMS objects and services is independent of the use of TASE.2 and the use of MMS by TASE.2.

3.7.5 Latest Progresses of Related Standards

Recently, further progress has been made on IEC 61850. The FDIS (Final Draft International Standard) IEC 61850-9-2 Ed.2: Communication networks and systems for power utility automation – Part 9-2: SCSM – Sampled values over ISO/IEC 8802-3 has been accepted as International Standard. In addition, the IEC 61850-10 Ed.2 (57/1162/CDV (Committee Draft for Vote)): Communication networks and systems for power utility automation – Part 10: Conformance testing has been published for comments and ballot by 2012-01-06. The following contents have been changed and extended:

- update of Server device conformance test procedures;
- new Client system conformance test procedures;
- new Sampled values device conformance test procedures;
- new Engineering Tool related conformance test procedures;
- new GOOSE performance test procedures.

  IEC 61400-25 is the IEC 61850 adaptation for wind turbines. IEC 61400-25 (Communications for monitoring and control of wind power plants, TC 88) provides uniform information exchange for monitoring and control of wind power plants. This addresses the issue of proprietary communication systems utilizing a wide variety of protocols, labels, semantics, and so on, thus enabling one to exchange information with different wind power plants independently of a vendor. It is a subset of IEC 61400, a set of standards for designing wind turbines. The IEC 61400-25 standard is a basis for simplifying the roles that the wind turbine and SCADA systems have to play. The crucial part of the wind power plant information, information exchange methods, and communication stacks are standardized. They build a basis to which procurement specifications and contracts could easily refer. The standard has specified five mappings (IEC 61400-25-4) to communication protocol stacks in order to address the real wind power business needs for communication. The mappings specified in the part of IEC 61400-25 comprise a mapping to SOAP-based (Simple Object Access Protocol) web services, OPC (OLE for Process Control)/XML-DA (eXtensible Markup Language Data Attribute), IEC 61850-8-1 MMS, IEC 60870-5-104, and a mapping to DNP3.

  The XML encoding rules will be covered in a future part of the IEC 61968 series. The XML is a data format for structured document interchange particularly on the Internet. One of its primary uses is information exchange between different and potentially incompatible computer systems. XML is thus well suited to the domain of system interfaces for distribution management. Where applicable, future parts of the IEC 61968 series will define the information required for "message payloads."

Message Payloads will be formatted using XML with the intent that these payloads can be loaded on to messages of various messaging transports, for example, OAG (Open Application Group) and SOAP. As such, the middleware adapter only goes as far as necessary to make the used set of middleware services conformant to the requirements of one or more of the interface specifications in the series from future IEC 61986-3 onward. In this context, the middleware services represent not one single interface but a set of interfaces to a set of corresponding services for components.

3.8 Conclusion

This chapter tries to identify existing standardization and potential gaps in the IEC portfolio which will be relevant for smart power grid implementation. A number of standards will form a core set of standards, which will be valid or necessary for nearly all smart power grid applications. These standards will be considered priority standards. Their further promotion and development will be a key for the IEC to provide support for smart power grid solutions. These standards are at the core of an IEC roadmap for the standardization of the smart power grid.

Care must be taken to concentrate standardization efforts on providing additional value to the smart power grid implementation. This will be especially true for all interoperability standards, which will help to reach the goal of increased observability and controllability of the power system. In this respect, the IEC offers the absolute precondition for a further promotion of smart power grid. On the other hand, the IEC itself refrains from standardization of solutions or applications. This would actually block innovation and the further development of the smart power grid.

Appendix 3.A An SED File Example (Extensible Markup Language)

```
<?xml version="1.0" encoding="UTF-8"?>
<SCL xmlns="http://www.iec.ch/61850/2003/SCL"
xmlns:xsi="http://www.w3.org/2001/XMLSchema-instance"
xsi:schemaLocation="http://www.iec.ch/61850/2003/SCL SCL.xsd">
<Header id="Substation2 to substation1" />
<Substation name="Substation1">
    <VoltageLevel name="D1">
        <Voltage multiple="k" unit="V">220</Voltage>
        <Bay name="Q1">
<LNode iedName="IED1_1" prefix= "F21" lnInst="1" lnClass="PSCH"
    ldInst="LD1"/>
<LNode iedName="IED1_1" prefix= "F21" lnInst="1" lnClass="PDIS"
    ldInst="LD1"/>
<LNode iedName="IED1_1" lnInst="1" lnClass="PTOC" ldInst="LD1"/>
<LNode iedName="IED1_2" lnInst="1" lnClass="CSWI" ldInst="LD1"/>
<LNode iedName="IED1_3" lnInst="1" lnClass="XSWI" ldInst="LD1"/>
<LNode iedName="IED1_3" lnInst="2" lnClass="XSWI" ldInst="LD1"/>
            <LNode iedName="IED1_3" lnInst="1" lnClass="CILO"
    ldInst="LD1"/>
```

```xml
  <ConductingEquipment name="DS1" type="DIS">
      <Terminal connectivityNode="Substation1/D1/Q1/N1"
   substationName="Substation1" voltageLevelName="D1"
   bayName="Q1" cNodeName="N1" />
  </ConductingEquipment>
  <ConductingEquipment name="ES1" type="DIS">
        <Terminal connectivityNode="Substation1/D1/Q1/N1"
   substationName="Substation1" voltageLevelName="D1"
   bayName="Q1" cNodeName="N1" />
           <Terminal connectivityNode="Substation1/D1/Q1/
   grounded" substationName="Substation1"
   voltageLevelName="D1" bayName="Q1" cNodeName="grounded" />
  </ConductingEquipment>
  <ConnectivityNode name="N1" pathName="Substation1/D1/Q1/N1"/>
  </Bay>
 </VoltageLevel>
</Substation>
<Substation name="Substation2">
    <VoltageLevel name="D1">
        <Voltage maltiple="k" unit="V">220</Voltage>
        <Bay name="Q1">
<LNode iedName="IED2_2" prefix= "F21" lnInst="1" lnClass="PSCH"
     ldInst="LD1"/>
<LNode iedName="IED2_2" prefix= "F21" lnInst="1" lnClass="PDIS"
     ldInst="LD1"/>
<LNode iedName="IED2_2" lnInst="1" lnClass="PTOC" ldInst="LD1"/>
<LNode iedName="IED2_3" lnInst="1" lnClass="CSWI" ldInst="LD1"/>
<LNode iedName="IED2_1" lnInst="1" lnClass="XSWI" ldInst="LD1"/>
<LNode iedName="IED2_1" lnInst="2" lnClass="XSWI" ldInst="LD1"/>
          <LNode iedName="IED2_1" lnInst="1" lnClass="CILO"
     ldInst="LD1"/>
  <ConductingEquipment name="DS1" type="DIS">
      <Terminal connectivityNode="Substation1/D1/Q1/N1"
   substationName="Substation1" voltageLevelName="D1"
   bayName="Q1" cNodeName="N1" />
  </ConductingEquipment>
  <ConductingEquipment name="ES1" type="DIS">
        <Terminal connectivityNode="Substation1/D1/Q1/N1"
   substationName="Substation1" voltageLevelName="D1"
   bayName="Q1" cNodeName="N1" />
        <Terminal connectivityNode="Substation1/D1/Q1/grounded"
   substationName="Substation1" voltageLevelName="D1"
   bayName="Q1" cNodeName="grounded" />
  </ConductingEquipment>
  <ConnectivityNode name="N1" pathName="Substation1/D1/Q1/N1"/>
  </Bay>
 </VoltageLevel>
</Substation>
```

```
<Substation name="Substation1-Substation2" desc="Line between
     Substation1 and Substation2">
<VoltageLevel name="D1" desc="Line Voltage Level">
<Bay name="W1" desc="Bay">
 <ConductingEquipment name="WA1" type="Overhead line">
    <Terminal connectivityNode="Substation1/D1/Q1/N1"
     substationName="Substation1" voltageLevelName="D1"
     bayName="Q1" cNodeName="N1" />
    <Terminal connectivityNode="Substation2/D1/Q1/N1"
     substationName="Substation2" voltageLevelName="D1"
     bayName="Q1" cNodeName="N1" />
     </ConductingEquipment>
 </Bay>
 </VoltageLevel>
</Substation>

<Communication>
<SubNetwork name="Substation2Vlan2" desc="IEC61850 through both
     stations" type="8-MMS">
  <ConnectedAP iedName="IED1_3" apName="S1">
   <Address>
       <P type="IP">172.16.121.77</P>
       <P type="IP-SUBNET">255.255.0.0</P>
       <P type="IP-GATEWAY">172.16.121.126</P>
    <P type="OSI-TSEL">0001</P>
       <P type="OSI-PSEL">00000001</P>
    <P type="OSI-SSEL">0001</P>
    </Address>
    <GSE ldInst="LD1" cbName="GoCB_S2">
    <Address>
       <P type="MAC-Address">00-15-58-50-6B-FF</P>
       <P type="APPID">Itl_IED1_3</P>
    </Address>
               <MinTime unit="s">2</MinTime>
           <MaxTime unit="s">1000</MaxTime>
           </GSE>
  </ConnectedAP>
<ConnectedAP iedName="IED2_1" apName="S1">
   <Address>
   <P type="IP">172.17.0.100</P>
    <P type="IP-SUBNET">255.255.0.0</P>
   <P type="IP-GATEWAY">172.17.121.126</P>
   <P type="OSI-TSEL">0001</P>
   <P type="OSI-PSEL">00000001</P>
   <P type="OSI-SSEL">0001</P>
   </Address>
   <GSE ldInst="LD1" cbName="GoCB_S1">
```

```xml
    <Address>
    <P type="MAC-Address">01-0C-CD-01-0B-02</P>
    <P type="APPID">2001</P>
    </Address>
          <MinTime unit="s">2</MinTime>
          <MaxTime unit="s">1000</MaxTime>
             </GSE>
  </ConnectedAP >
<ConnectedAP iedName="S1OPC1" apName="S1">
   <Address>
   <P type="IP">172.16.0.100</P>
    <P type="IP-SUBNET">255.255.0.0</P>
   <P type="IP-GATEWAY">172.16.121.126</P>
   <P type="OSI-TSEL">0001</P>
   <P type="OSI-PSEL">00000001</P>
   <P type="OSI-SSEL">0001</P>
   </Address>
</ConnectedAP >
</SubNetwork>
</Communication>

<IED name="S1OPC1" type="OPCServer"
     configVersion="1.0" engRight="fix" owner="S1">
<AccessPoint name="S1">
        <LN inst="1" lnClass="IHMI" lnType="IHMI_OPCServer_
     IEC61850"/>
</AccessPoint>
</IED>
<IED name="IED1_1" type="REL316-4" desc="Protecting IED"
     configVersion="1.0" engRight="fix" owner="S1">
<Services>
 <DynAssociation />
 <SettingGroups>
       <SGEdit />
       </SettingGroups >
        <GetDirectory />
  <GetDataObjectDefinition />
  <GetObjectDirectory />
 <GetDataSetValue />
 <ConfDataSet max="50" maxAttributes="240" />
       <ConfReportControl max="100" />
 <ReadWrite />
 <GetCBValues />
 <GOOSE max="20" />
        <ReportSetting cbName="Conf" datSet="Conf" rptID="Dyn"
     optFields="Dyn" bufTime="Dyn" trgOps="Dyn" intgPd="Dyn" />
 <GSESettings cbName="Conf" datSet="Conf" appID="Conf" />
 </Services>
```

```
  <AccessPoint name="S1" >
  <Server>
  <Authentication none="true" />
   <LDevice inst="LD1">
   <LN0 inst="" lnClass="LLN0" lnType="myLLN0">
  <DataSet name="reporttoIHMI">
          <FCDA ldInst="LD1" prefix="F21" lnClass="PSCH" lnInst="1"
     doName="Op" fc="ST" />
<FCDA ldInst="LD1" prefix="F21" lnClass="PSCH" lnInst="1"
     doName="TxTr" fc="ST"/>
 <FCDA ldInst="LD1" prefix="F21" lnClass="PDIS" lnInst="1"
     doName="Str" fc="ST"/>
  </DataSet>
              <DataSet name="statecontrol">
                     <FCDA ldInst="LD1" prefix="F21"
     lnClass="XSWI" lnInst="1"
     doName="Pos" daName="stVal" fc="ST" />
              </DataSet>

<GSEControl name="GOOSE1_1" datSet="statecontrol"
     confRev="1" appID="ctl">
<IEDName>IED1_3</IEDName>
</GSEControl>

<ReportControl name="URCB_controlreport"
     datSet="reporttoIHMI" rptID="report" bufTime="0"
     confRev="1">
 <TrgOps dupd="true" dchg="true"/>
<OptFields seqNum="true" timestamp="true" dataset="true"
     dataRef="true"/>
     <RptEnabled    max      ="5">
             <ClientLN iedName="S1OPC1" lnClass="IHMI" lnInst="1"
     ldInst="LD1">
</ClientLN >
 </RptEnabled>
</ReportControl>
</LN0>
<LN inst="1" lnClass="PDIS" lnType="myPDIS" />
<LN inst="1" lnClass="PTOC" lnType="myPTOC" />
<LN inst="1" lnClass="PSCH" lnType="myPSCH" />
<LN inst="1" lnClass="LPHD" lnType="myLPHD" />
</LDevice>
</Server>
</AccessPoint>
</IED>
<IED name="IED1_2" type="REL316-4" desc="control IED"
     configVersion="1.0" engRight="fix" owner="S1">
```

```
<Services>
 <DynAssociation />
<SettingGroups>
<SGEdit />
</SettingGroups >
<GetDirectory />
 <GetDataObjectDefinition />
  <GetObjectDirectory />
 <GetDataSetValue />
 <ConfDataSet max="50" maxAttributes="240" />
 <ConfReportControl max="100" />
        <ReadWrite />
        <ConfReportControl max="12" />
 <GetCBValues />
        <ReportSettings cbName="Conf" datSet="Conf" rptID="Dyn"
     optFields="Conf" bufTime="Dyn" intgPd="Dyn" />
        <GSESettings cbName="Conf" datSet="Conf" appID="Conf" />
<GOOSE max="20" />
</Services>
 <AccessPoint name="S1" >
 <Server>
 <Authentication none="true" />
  <LDevice inst="LD1" desc="">
  <LN0 inst="" lnClass="LLN0" lnType="myLLN0">

 <DataSet name="controltoXSWI_DS1">
              <FCDA ldInst="LD1" prefix="" lnClass="XSWI"
     lnInst="1" doName="Pos" daName="ctlVal"
     fc="ST"/>
</DataSet>
        <DataSet name="CSWItoIHMI">
              <FCDA ldInst="LD1" prefix="" lnClass="XSWI"
     lnInst="1" doName="Pos" daName="stVal" fc="ST"/>
</DataSet>

<GSEControl name="GoCB_CSWI" desc="to XSWI_DS1"
     datSet="controltoXSWI_DS1" confRev="1" appID="ctl">
<IEDName>IED1_3</IEDName>
</GSEControl>
<ReportControl name="URCB_statereport" datSet="CSWItoIHMI"
     rptID="report" bufTime="0" confRev="1">
 <TrgOps dupd="true" dchg="true"/>
<OptFields seqNum="true" times-
tamp="true" dataset="true" dataRef="true"/>
     <RptEnabled    max       ="5">
           <ClientLN ied-
Name="S1OPC1" lnClass="IHMI" lnInst="1" ldInst="LD1">
</ClientLN >
 </RptEnabled>
</ReportControl>
</LN0>
```

```
<LN inst="1" lnClass="LPHD" lnType="myLPHD" />
<LN inst="1" lnClass="CSWI" lnType="myCSWI" />
</LDevice>
</Server>
</AccessPoint>
</IED>
<IED name="IED1_3" type="REL316-4" desc="interlocking IED"
     configVersion="1.0" engRight="dataflow" owner="S1">
<Services>
 <DynAssociation />
<SettingGroups>
<SGEdit />
</SettingGroups >
<GetDirectory />
 <GetDataObjectDefinition />
  <GetObjectDirectory />
 <GetDataSetValue />
 <ConfDataSet max="4" />
 <ReadWrite />
    <ConfReportControl max="10" />
 <GetCBValues />
 <ReportSettings cbName="Conf" datSet="Conf" rptID="Dyn"
     optFields="Conf" bufTime="Dyn" intgPd="Dyn" />
        <GSESettings cbName="Conf" datSet="Conf" appID="Conf" />
        <GOOSE max="20" />
</Services>
  <AccessPoint name="S1" >
  <Server>
  <Authentication none="true" />
        <LDevice inst="LD1" desc="">
        <LN0 inst="" lnClass="LLN0" lnType="myLLN0">
        <DataSet name="XSWItoCILO">
<FCDA ldInst="LD1" prefix="" lnClass="XSWI" lnInst="1"
     doName="Pos" daName="stVal" fc="ST"/>
<FCDA ldInst="LD1" prefix="" lnClass="XSWI" lnInst="2"
     doName="Pos" daName="stVal" fc="ST"/>
   </DataSet>
<DateSet name="XSWItoCSWI">
<FCDA ldInst="LD1" prefix="" lnClass="XSWI" lnInst="1"
     doName="Pos" daName="stVal" fc="ST"/>
</DateSet>

  <DataSet name="XSWItoS2">
<FCDA ldInst="LD1" prefix="" lnClass="XSWI" lnInst="1"
     doName="Pos" daName="stVal" fc="ST"/>
<FCDA ldInst="LD1" prefix="" lnClass="XSWI" lnInst="2"
     doName="Pos" daName="stVal" fc="ST"/>
  </DataSet>
 <DateSet name="XSWItoPSCH">
<FCDA ldInst="LD1" prefix="" lnClass="XSWI" lnInst="1"
     doName="Pos" daName="stVal" fc="ST"/>
```

```
</DateSet>
 <GSEControl name="GoCB_PSCH" desc="to PSCH"
     datSet="XSWItoPSCH" confRev="1" appID="reporttoPSCH">
<IEDName>IED1_1</IEDName>
</GSEControl>
 <GSEControl name="GoCB_XSWI" desc="to CILO"
     datSet="XSWItoCILO" confRev="1" appID="Itl_IED1_3">
<IEDName>IED1_3</IEDName>
</GSEControl>
<GSEControl name="GoCB_S2" desc="to S2" datSet="XSWItoS2"
     confRev="1" appID="Itl_S2">
<IEDName>IED2_1</IEDName>
</GSEControl>
<ReportControl name="BRCB_XSWI" datSet="XSWItoCSWI"
     rptID="3000" bufTime="0" confRev="1">
     <OptFields seqNum="true" times-
tamp="true" dataset="true" dataRef="true"/>
<TrgOps dupd="true" dchg="true"/>
       <RptEnabled max="5">
 <ClientLN iedName="IED1_2" ldInst="LD1" lnClass="CSWI" lnInst="1" >
</ClientLN >
 </RptEnabled>
</ReportControl>
</LN0>
<LN inst="1" lnClass="LPHD" lnType="myLPHD" />
<LN inst="1" lnClass="CILO" lnType="myCILO" />
<LN inst="2" lnClass="XSWI" lnType="myXSWI" />
</LDevice>
</Server>
</AccessPoint>
</IED>
<IED name="IED2_1" type="REL316-4" desc="interlocking IED"
     configVersion="1.0" engRight="fix" owner="S1">
<Services>
 <DynAssociation />
<SettingGroups>
<SGEdit />
</SettingGroups >
<GetDirectory />
 <GetDataObjectDefinition />
 <GetObjectDirectory />
 <GetDataSetValue />
 <ConfDataSet max="4" />
 <ReadWrite />
    <ConfReportControl max="10" />
 <GetCBValues />
```

```
  <ReportSettings cbName="Conf" datSet="Conf" rptID="Dyn"
      optFields="Conf" bufTime="Dyn" intgPd="Dyn" />
        <GSESettings cbName="Conf" datSet="Conf" appID="Conf" />
      <GOOSE max="20" />
</Services>
  <AccessPoint name="S1" >
  <Server>
  <Authentication none="true" />
   <LDevice inst="LD1" desc="">
   <LN0 inst="" lnClass="LLN0" lnType="myLLN0">
   <DataSet name="XSWItoCILO">
<FCDA ldInst="LD1" prefix="" lnClass="XSWI" lnInst="1"
     doName="Pos" daName="stVal" fc="ST"/>
<FCDA ldInst="LD1" prefix="" lnClass="XSWI" lnInst="2"
     doName="Pos" daName="stVal" fc="ST"/>
   </DataSet>
<DateSet name="XSWItoCSWI">
<FCDA ldInst="LD1" prefix="" lnClass="XSWI" lnInst="1"
     doName="Pos" daName="stVal" fc="ST"/>
</DateSet>
  <DataSet name="XSWItoS1">
<FCDA ldInst="LD1" prefix="" lnClass="XSWI" lnInst="1"
     doName="Pos" daName="stVal" fc="ST"/>
   </DataSet>
<DateSet name="XSWItoPSCH">
<FCDA ldInst="LD1" prefix="" lnClass="XSWI" lnInst="1"
     doName="Pos" daName="stVal" fc="ST"/>
</DateSet>
<GSEControl name="GoCB_PSCH" desc="to PSCH" datSet="XSWItoPSCH"
     confRev="1" appID="reporttoPSCH">
<IEDName>IED2_3</IEDName>
</GSEControl>
<GSEControl name="GoCB_XSWI" desc="to CILO" datSet="XSWItoCILO"
     confRev="1" appID="Itl_IED2_1">
<IEDName>IED2_1</IEDName>
</GSEControl>
 <GSEControl name="GoCB_S1" desc="to S1" datSet="XSWItoS1"
     confRev="1" appID="Itl_S1">
<IEDName>IED1_3</IEDName>
</GSEControl>
<ReportControl name="BRCB_XSWI" dat-
Set="XSWItoCSWI" rptID="3000" bufTime="0" confRev="1">
     <OptFields seqNum="true" times-
tamp="true" dataset="true" dataRef="true"/>
<TrgOps dupd="true" dchg="true"/>
```

```
        <RptEnabled max="5">
 <ClientLN iedName="IED2_2" ldInst="LD1" lnClass="CSWI" lnInst="1" >
</ClientLN >
 </RptEnabled>
</ReportControl>
</LN0>
<LN inst="1" lnClass="LPHD" lnType="myLPHD" />
<LN inst="1" lnClass="CILO" lnType="myCILO" />
<LN inst="2" lnClass="XSWI" lnType="myXSWI" />
</LDevice>
</Server>
</AccessPoint>
</IED>

<IED name="IED2_2" type="REL316-4" desc="control IED"
     configVersion="1.0" engRight="fix" owner="S1">
<Services>
 <DynAssociation />
<SettingGroups>
<SGEdit />
</SettingGroups >
<GetDirectory />
 <GetDataObjectDefinition />
  <GetObjectDirectory />
 <GetDataSetValue />
 <ConfDataSet max="40" />
 <ConfReportControl max="100" />
        <ReadWrite />
        <ConfReportControl max="12" />
 <GetCBValues />
        <ReportSettings cbName="Conf" datSet="Conf" rptID="Dyn"
     optFields="Conf" bufTime="Dyn" intgPd="Dyn" />
        <GSESettings cbName="Conf" datSet="Conf" appID="Conf" />
 <GOOSE max="20" />
 </Services>
  <AccessPoint name="S1" >
  <Server>
  <Authentication none="true" />
   <LDevice inst="LD1" desc="">
   <LN0 inst="" lnClass="LLN0" lnType="myLLN0">
 <DataSet name="controltoXSWI">
            <FCDA ldInst="LD1" prefix="" lnClass="XSWI"
     lnInst="1" doName="Pos" daName="ctlVal" fc="ST"/>
</DataSet>
<GSEControl name="GoCB_CSWI" desc="to XSWI"
     datSet="controltoXSWI" confRev="1" appID="ctl">
```

```
<IEDName>IED2_1</IEDName>
</GSEControl>
</LN0>
<LN inst="1" lnClass="LPHD" lnType="myLPHD" />
<LN inst="1" lnClass="CSWI" lnType="myCSWI" />
</LDevice>
</Server>
</AccessPoint>
</IED>
<IED name="IED2_3" type="REL316-4" desc="Protecting IED"
     configVersion="1.0" engRight="fix" owner="S1">
<Services>
 <DynAssociation />
 <SettingGroups>
       <SGEdit />
       </SettingGroups >
        <GetDirectory />
  <GetDataObjectDefinition />
  <GetObjectDirectory />
 <GetDataSetValue />
 <ConfDataSet max="4" />
       <ConfReportControl max="100" />
 <ReadWrite />
 <GetCBValues />
 <GOOSE max="20" />
         <ReportSetting cbName="Conf" datSet="Conf" rptID="Dyn"
     optFields="Dyn" bufTime="Dyn" trgOps="Dyn" intgPd="Dyn" />
 <GSESettings cbName="Conf" datSet="Conf" appID="Conf" />
 </Services>
  <AccessPoint name="S1" >
  <Server>
  <Authentication none="true" />
   <LDevice inst="LD1">
   <LN0 inst="" lnClass="LLN0" lnType="myLLN0">
                 <DataSet name="statecontrol">
                      <FCDA ldInst="LD1" prefix="F21"
     lnClass="XSWI" lnInst="1" doName="Pos"
     daName="stVal" fc="ST" />
                 </DataSet>
<GSEControl name="GOOSE1_1" datSet="statecontrol" confRev="1"
     appID="ctl">
<IEDName>IED2_1</IEDName>
</GSEControl>
</LN0>
<LN inst="1" lnClass="PDIS" lnType="myPDIS" />
<LN inst="1" lnClass="PTOC" lnType="myPTOC" />
```

```
<LN inst="1" lnClass="PSCH" lnType="myPSCH" />
<LN inst="1" lnClass="LPHD" lnType="myLPHD" />
</LDevice>
</Server>
</AccessPoint>
</IED>
<DataTypeTemplates>
...... ........
</DataTypeTemplates>
</SCL>
```

References

[1] IEC (2003) IEC61850-1. *Communication Network and Systems in Substations-Part 1: Introduction and Overview*, International Electrotechnical Commission.

[2] IEC (2003) IEC61850-2. *Communication Network and Systems in Substations-Part 2: Glossary*, International Electrotechnical Commission.

[3] IEC (2003) IEC 61850-3. *Communication Network and Systems in Substations-Part 3: General Requirements*, International Electrotechnical Commission.

[4] IEC (2003) IEC 61850-4. *Communication Network and Systems in Substations-Part 4: System and Project Management*, International Electrotechnical Commission.

[5] IEC (2003) IEC 61850-5. *Communication Networks and Systems in Substation-Part 5: Communication Requirement for Functions and Device Models*, International Electrotechnical Commission.

[6] IEC (2003) IEC 61850-6. *Communication Networks and Systems in Substation-Part 6: Configuration Description Language for Communication in Electrical Substations Related to IEDs*, International Electrotechnical Commission.

[7] IEC (2003) IEC 61850-7-1. *Communication Networks and Systems in Substation-Part 7-1: Basic Communication Structure for Substation and Feeder Equipment-Principles and Models*, International Electrotechnical Commission.

[8] IEC (2003) IEC 61850-7-2. *Communication Networks and Systems in Substation Part 7-2: Basic Communication Structure for Substation and Feeder Equipment-Abstract Communication Service Interface (ACSI)*, International Electrotechnical Commission.

[9] IEC (2003) IEC 61850-7-3. *Communication Networks and Systems in Substation Part 7-3: Basic Communication Structure for Substation and Feeder Equipment-Common Data Classes*, International Electrotechnical Commission.

[10] IEC (2003) IEC 61850-7-4. *Communication Networks and Systems in Substation Part 7-4: Basic Communication Structure for Substation and Feeder Equipment-Compatible Logical Node Classes and Data Classes*, International Electrotechnical Commission.

[11] IEC (2003) IEC 61850-8-1. *Communication Networks and Systems in Substation- Part 8-1: Specific Communication Service Mapping (SCSM)-Mapping to MMS (ISO/IEC 9506-1 and ISO/IEC 9506-2) and to ISO/IEC 8802-3*, International Electrotechnical Commission.

[12] IEC (2003) IEC 61850-9-1. *Communication Networks and Systems in Substation- Part 9-1: Specific Communication Service Mapping (SCSM)-Sampled Values Over Serial Unidirectional Multidrop Point to Point Link*, International Electrotechnical Commission.

[13] IEC (2003) IEC 61850-9-2 *Communication Networks and Systems in Substation- Part 9-2: Specific Communication Service Mapping (SCSM)-Sampled Values Over ISO/IEC 8802-3*, International Electrotechnical Commission.

[14] Wang, Z. and Ren, Y. (2005) Application of a DATA-SET Model in IEC 61850. *Automation of Electric Power Systems (Chinese)*, **29** (2), 61–63.

[15] IEC (1990) IEC 60870-5-1. *Telecontrol Equipment and Systems – Part 5-1: Transmission Protocols. Transmission Frame Formats*, International Electrotechnical Commission.

[16] IEC (1992) IEC 60870-5-2. *Telecontrol Equipment and Systems – Part 5-2: Transmission Protocols. Data Link Transmission Services*, International Electrotechnical Commission.

[17] IEC (1992) IEC 60870-5-3. *Telecontrol Equipment and Systems – Part 5-3: Transmission Protocols. General Structure of Application Data*, International Electrotechnical Commission.

[18] IEC (1993) IEC 60870-5-4. *Telecontrol Equipment and Systems – Part 5-4: Transmission Protocols. Definition and Coding of Information Elements*, International Electrotechnical Commission.

[19] IEC (1995) IEC 60870-5-5. *Telecontrol Equipment and Systems – Part 5-5: Transmission Protocols. Basic Application Functions*, International Electrotechnical Commission.

[20] IEC (2006) IEC 60870-5-6. *Telecontrol Equipment and Systems – Part 5-6: Guidelines for Conformance Testing for the IEC 60870-5 Companion Standards*, International Electrotechnical Commission.

[21] IEC (1995) IEC 60870-5-101. *Telecontrol Equipment and Systems – Part 5-101: Transmission Protocols. Companion Standard for Basic Telecontrol Tasks*, International Electrotechnical Commission.

[22] IEC (2000) IEC 60870-5-101. *Telecontrol Equipment and Systems – Part 5-101: Transmission Protocols. Companion Standard for Basic Telecontrol Tasks*, Ed. 2, International Electrotechnical Commission.

[23] IEC (1996) IEC 60870-5-102. *Telecontrol Equipment and Systems – Part 5-102: Companion Standard for the Transmission of Integrated Totals in Electric Power Systems*, International Electrotechnical Commission.

[24] IEC (1997) IEC 60870-5-103. *Telecontrol Equipment and Systems – Part 5-103: Transmission Protocols. Companion Standard for Basic Telecontrol Tasks*, International Electrotechnical Commission.

[25] IEC (2006) IEC 60870-5-104. *Telecontrol Equipment and Systems – Part 5-104: Transmission Protocols. Companion Standard for the Informative Interface of Protection Equipment*, International Electrotechnical Commission.

[26] IEC (2003) IEC 60870-6-1. *Application Context and Organization of Standards*, International Electrotechnical Commission.

[27] IEC (2003) IEC 60870-6-2. *Use of Basic Standards (OSI Layers 1–4)*, International Electrotechnical Commission.

[28] IEC (1995) IEC 60870-6-501. *TASE.1 Service Definitions*, International Electrotechnical Commission.

[29] IEC (1995) IEC 60870-6-502. *TASE.1 Protocol Definitions*, International Electrotechnical Commission.

[30] IEC (2002) IEC 60870-6-503. *TASE.2 Services and Protocol*, International Electrotechnical Commission.

[31] IEC (1995) IEC 60870-6-504. *TASE.1 User Conventions*, International Electrotechnical Commission.

[32] IEC (1998) IEC 60870-6-601. *Functional Profile for Providing the Connection-oriented Transport Service in an End System Connected Via Permanent Access to a Packet Switched Data Network*, International Electrotechnical Commission.

[33] IEC (1998) IEC 60870-6-602. *TASE Transport Profiles*, International Electrotechnical Commission.

[34] IEC IEC 60870-6-701. *Functional Profile for Providing the TASE.1 Application Service in End Systems*, International Electrotechnical Commission.

[35] IEC IEC 60870-6-702. *Functional Profile for Providing the TASE.2 Application Service in End Systems*, International Electrotechnical Commission.

[36] IEC (2005) IEC 61970-1 *Ed.: Energy Management System Application Program Interface (EMS-API) Part 1: Guidelines and General Requirement*, International Electrotechnical Commission.

[37] IEC (2004) IEC 61970-2. *Energy Management System Application Program Interface (EMSAPI), Part 2: Glossary*, International Electrotechnical Commission.

[38] IEC (2005) IEC 61970-301. *Energy Management System Application Program Interface (EMSAPI), Part 301: Common Information Model (CIM) Base*, International Electrotechnical Commission.

[39] IEC (1999) IEC 61970-302. *Energy Management System Application Program Interface (EMSAPI), Part 302: Common Information Model (CIM) Financial, Energy Scheduling and Reservations*, International Electrotechnical Commission.

[40] IEC (2005) IEC TS 61970-401. *Energy Management System Application Program Interface (EMSAPI), Part 401: Component Interface Specification (CIS) Framework*, International Electrotechnical Commission.

[41] IEC (2009) IEC TS 61970-402. *Energy Management System Application Program Interface (EMSAPI), Part 402: Component Interface Specification (CIS) – Common Services*, International Electrotechnical Commission.

[42] IEC (2009) IEC TS 61970-402. *Energy Management System Application Program Interface (EMSAPI), Part 403: Component Interface Specification (CIS) – Generic Data Access*, International Electrotechnical Commission.

[43] IEC (2007) IEC 61970-404. *Energy Management System Application Program Interface (EMS-API), Part 404: High Speed Data Access (HSDA)*, International Electrotechnical Commission.

[44] IEC (2007) IEC 61970-407. *Energy Management System Application Program Interface (EMS-API), Part 407: Time Series Data Access (TSDA)*, International Electrotechnical Commission.

[45] IEC (2009) IEC 61970-407. *Energy Management System Application Program Interface (EMS-API), Part 453: Exchange of Graphics Schematics Definitions (Common Graphics Exchange)*, International Electrotechnical Commission.

[46] IEC (2006) IEC 61970-501. *Energy Management System Application Program Interface (EMS-API), Part 501: Specific Communication Service Mapping (SCSM) Common Information Model (CIM) XML Codification for Programmable Reference and Model Data Exchange*, International Electrotechnical Commission.

[47] IEC (2003) IEC61968-1. *Application Integration at Electric Ultilities-System Interfaces for Distribution Management-Part 1: Introduction and General Requirements*, International Electrotechnical Commission.

[48] IEC (2003) IEC61968-2. *Application Integration at Electric Ultilities-System Interfaces for Distribution Management-Part 2: Glossary*, International Electrotechnical Commission.

[49] IEC (2004) IEC61968-3. *Application Integration at Electric Ultilities-System Interfaces for Distribution Management-Part 3: Interface for Network Operations*, International Electrotechnical Commission.

[50] IEC (2007) IEC61968-4. *Application Integration at Electric Ultilities-System Interfaces for Distribution Management-Part 4: Interfaces for Records and Asset Management*, International Electrotechnical Commission.

[51] IEC (2008) IEC61968-9. *Application Integration at Electric Ultilities-System Interfaces for Distribution Management-Part 9: Interface Standard for Meter Reading and Control*, International Electrotechnical Commission.

[52] IEC (2008) IEC61968-11. *Application Integration at Electric Ultilities-System Interfaces for Distribution Management-Part 11: Common Information Model (CIM) Extensions for Distribution*, International Electrotechnical Commission.

[53] IEC (2008) IEC61968-13. *Application Integration at Electric Ultilities-System Interfaces for Distribution Management-Part 13: Common Information Model (CIM) RDF Model Exchange Format for Distribution*, International Electrotechnical Commission.

[54] IEC (1999) IEC 60834-1. *Teleprotection Equipment of Power Systems – Performance and Testing – Part 1: Command Systems*, International Electrotechnical Commission.

[55] IEC (1993) IEC 60834-2. *Teleprotection Equipment of Power Systems – Performance and Testing – Part 2: Analogue Comparison System*, International Electrotechnical Commission.

[56] Gao, H., Xiang, H. and Liu, G. (2006) Application analysis of IEC61850 protocol based on the ethernet technology. *Relay*, **34** (14), 46–57.

[57] Choi, J., Jang, H., and Vice, S.K. (2009) Integration method analysis between IEC 61850 and IEC 61970 for substation automation system. *Transmission and Distribution Conference and Exposition: Asia and Pacific (IEEE T&D Asia 2009)*.

[58] Han, G., Bingyin, X. and Suonan, J. (2012) IEC 61850-based feeder terminal unit modeling and mapping to IEC 60870-5-104. *IEEE Transactions on Power Delivery*, **27** (4), 2046–2053.

4

Smart Storage and Electric Vehicles

4.1 Introduction

Electric Storage (ES), Distributed Energy Resources (DERs), and Electric Vehicles (EVs) are expected to play increasingly important roles in the evolution toward the Smart Grid. ES can enhance grid reliability and enable a more efficient use of base load generation by capturing cheap power during low demand hours and producing power during peak times to meet high loads. As discussed in Chapter 9, *Integration of Variable Renewable Resources* of this book, ES could lead to an increased penetration of intermittent renewable energy resources. DERs minimize the transmission and distribution losses and offer economical and ecological benefits by generating energy close to where it is needed. EVs can mitigate some of the negative consequences of oil dependency on economy, national security, and the environment. EVs can also be integrated with the grid to enable many applications, such as smoothing variable generation from renewable sources, peak load shifting, voltage control, frequency regulation, and providing DERs. However, despite the significant benefits brought by ES, DERs, and EVs, they do face challenges to widespread adoption. Few standards exist related to ES systems which are used in electric power infrastructure and have taken the location-specific requirements into consideration. Standards are also needed to allow new ways of operating DERs, such as the Virtual Power Plant (VPP) and microgrid to achieve further benefits. For a wider use of EVs, barriers such as limited range, inadequate infrastructure, lack of regulations and standards, high cost, inconvenience, and lack of safety should be overcome. Standard gaps related to safety, performance, and interoperability issues need to be filled through the cooperation of a broad set of stakeholders. In general, many of the challenges faced by ES, DERs, and EVs are highly interrelated and interdependent. Therefore, it is essential to understand the interrelationships between the challenges faced by ES, DERs, and EVs. This chapter provides an overview of available technology options and applications and ongoing R&D projects and standardization efforts made by different stakeholders

Smart Grid Standards: Specifications, Requirements, and Technologies, First Edition. Takuro Sato,
Daniel M. Kammen, Bin Duan, Martin Macuha, Zhenyu Zhou, Jun Wu, Muhammad Tariq and Solomon Abebe Asfaw.
© 2015 John Wiley & Sons, Ltd. Published 2015 by John Wiley & Sons, Ltd.

to fill the gaps between standards and technology requirements. The organization of this chapter is as follows: Section 4.2 introduces ES technologies, applications, and standardization efforts made by Standard Development Organizations (SDOs) such as the National Institute of Standards and Technology (NIST), Institute of Electrical and Electronics Engineers (IEEE), and International Electrotechnical Commission (IEC); Section 4.3 introduces various DER technologies, applications, and R&D and standardization efforts made by EPRI (Electric Power Research Institute), NIST, CENELEC (European Committee for Electrotechnical Standardization), DERLab, Consortium for Electric Reliability Technology Solutions (CERT), IEC, and so on. Section 4.4 provides an overview of EV history, battery technologies, Grid-to-Vehicle (G2V) and Vehicle-to-Grid (V2G) technologies, and various R&D and standardization efforts made by IEC, ANSI (American National Standards Institute), NIST, CEN-CENELEC (European Committee for Standardization-European Committee for Electrotechnical Standardization), ChAdeMO, SAE International, EPRI, and so on.

4.2   Electric Storage

4.2.1   An Overview of Electric Storage

ES is a technology used for storing electric energy directly or in other forms of energy. Electrical energy is stored when the production exceeds consumption and the stored energy is used when consumption exceeds production. In contrast, the conventional electric grid is built with little storage of electrical energy, and the power is generated almost at the time it is consumed.

Generally, electricity demand is high during daytime and early evening, and is low in the late evenings and the early mornings when people are sleeping. However, the conventional electric grid must be built to accommodate the highest power demand, resulting in underutilization of assets. ES can enhance grid reliability, and enable a more efficient use of baseload generation by capturing cheap power during low demand hours and producing power during peak times to meet high loads.

In particular, the renewable energy resources, such as solar energy and wind energy, can benefit from ES. Intermittent energy sources have some sort of seasonal and diurnal profile (this is discussed in Chapter 9), which can sometimes generate excess energy depending on the renewable system size and the grid flexibility. With ES, this surplus energy can be stored at the time of excess generation and used at a later time. Thus, ES can be used to adapt energy production to energy consumption, to increase efficiency, and lower the cost of energy production.

ES can be categorized as large, medium,- and small scale. Centralized bulk ES facilities are connected directly into the generation plant or distribution grid to store large amounts of electricity. The common forms of centralized bulk ES are Hydroelectric Pumped Storage or Pumped Hydro Storage (PHS), Compressed Air Energy Storage (CAES), and so on. On the other hand, medium- and small-scale ES can be designed as distributed ES. Distributed ES can be located at generating plants, connected into the transmission or distribution grids, or integrated into smart home and building automation systems. The common forms of distributed ES are batteries (including plug-in hybrid electric vehicles (PHEVs)), flywheels, solar thermal storage, super-conducting magnetic energy storage (SMES), and so on. The primary forms of both centralized bulk ES and distributed ES are introduced and compared in this subsection.

Given the close relationship between ES and the Smart Grid, various international and national Standard Developing Organizations (SDOs) have been working on the standardization of ES. This subsection also introduces the up-to-date standardization processes of ES. The organization of this subsection is as follows: Section 4.2.2 gives an overview of various forms of ES; Section 4.2.3 introduces the standardization projects and efforts of various international/national standardization organizations.

4.2.2 Electric Storage Technologies and Applications

ES already exists in many electrical power systems. The storage of electricity is not as easy as that of gas or fuel oil. Electricity is usually converted to another form of energy. For instance, electric energy is converted to mechanical energy in storage technologies, such as flywheel, CAES, and PHS, while it is converted to electrochemical energy in batteries. This energy is converted back to electricity when needed. This section introduces various storage technologies and their applications in the electric power grid. Table 4.1 presents a comparison of various ES technologies in terms of application, capacity, and cost.

4.2.2.1 Pumped Hydro Storage

PHS uses cheap electricity from coal or nuclear sources to pump water to a reservoir at a higher elevation during off-peak times. During peak periods, this water is released from the higher elevation reservoir to produce electricity. PHS is currently the most cost-effective form of massive energy storage technologies. On the basis of the present technologies, about 75% of the energy used to pump the water to the upper reservoir can be recovered. In 2010, the world installed capacity was more than 90 GW. As compared to less than 80 GW in 2006, the installed capacities increased by more than 12% [1].

In the future Renewable Energy Sources (RES), wind energy or solar energy, for example, can be used to pump water. However, despite the ability to store enormous quantities of energy, PHS requires a huge initial investment, and two nearby reservoirs at considerably different heights. PHS could also have environmental impacts by affecting the ecology if the reservoir creation includes flooding of valleys which could damage wildlife habitats. Actually, there are some PHS capacities which have never been used commercially for environmental reasons.

Table 4.1 A comparison of several electric storage technologies

Technology	Representative applications	Capacity	Cost per kilo watts of capacity
Pumped hydro storage	Peak shaving, load shifting, energy arbitrage, smoothing weather effects	Tens to hundreds of megawatts, discharged over long periods of time	High
Compressed air energy storage	Peak shaving, load shifting, energy arbitrage, smoothing weather effects	Tens to hundreds of megawatts, discharged over long periods of time	Low
Batteries	Peak shaving and load shifting, backup power supply (islanding), PHEV, voltage control, and frequency regulation	Up to 1 MW per unit and multiple units can be combined for larger capacity	Medium
Flywheel	Line or local faults, voltage control, and frequency regulation, spinning reserve	About 25 kW per unit and multiple units can be combined for larger capacity	High
Thermal energy storage	Smoothing weather variations, peak load shifting	Tens to hundreds of megawatts	Low
Ultra capacitor electric storage	Voltage control and frequency regulation	Up to multi-megawatts	Low
Superconducting magnetic energy storage	Line or local faults, voltage control, and frequency regulation	Up to multi-megawatts per unit, and multiple units can be combined for larger capacity	High

4.2.2.2 Compressed Air Energy Storage

CAES uses surplus cheap electricity to compress air. The air is usually stored in a salt dome or some other kind of geological form. During peak periods, when the demand is high and electricity is expensive, the compressed air is heated by natural gas and then used to help drive a turbine generator.

4.2.2.3 Batteries

There are various existing battery storage technologies and applications. Depending on the technology, battery storage can provide several different services to the Smart Grid. The following applications and services can be used for the Smart Grid:

- to provide a local source for time-varying real-time demand;
- to displace or defer the need for additional generating, transmission, or distribution capacity;
- to mitigate the impact of variability of intermittent energy sources such as wind and solar energy;

- to prove a backup source of power to a local area during power-islanding conditions;
- to increase grid penetration of intermittent renewable energy sources.

The three potential services of battery storage and the interfaces to other Smart Grid systems are summarized as follows:

- Battery storage can be directly implemented in the electricity-generating plant or transmission grid for electricity control and management. This kind of battery storage is usually controlled by Energy Management Systems (EMSs) and interfaced with a Wide Area Situational Awareness (WASA) systems.
- Battery storage can be integrated into the distribution grid for distribution voltage control. This kind of battery storage is usually controlled by Supervisory Control and Data Acquisition (SCADA) and interfaced with a Distribution Management System (DMS).
- Battery storage can also be integrated directly into a smart home or building automation system for demand response and load control, power source backup, and so on. This kind of battery storage is usually controlled by the Home Energy Management System (HEMS) or Building Energy Management System (BEMS) and interfaced with the Energy Service Interface (ESI) systems.

Although the use of battery storage can be dated back to the last century, batteries are generally expensive, hard to maintain, and have limited life spans. Nevertheless, interest in battery storage has been growing with technological advancements which can make battery storage a more practical means of ES for the Smart Grid. Lead-acid batteries have been in use for backup power in power plants for a number of decades. Sodium–sulfur and vanadium redox flow batteries become more effective in large-scale ES. Rechargeable batteries such as nickel cadmium (NiCd), nickel metal hydride (NiMH), lithium ion (Li-ion), and lithium ion polymer (Li-ion polymer) are applicable for PHEV. The technical characteristics and applications of various batters can be referred to in Section 4.4 E-mobility/PHEV.

4.2.2.4   Flywheel

A flywheel stores electricity in the form of kinetic energy in a spinning wheel or disc. In the storage mode, a spinning wheel is accelerated by an electric motor to a higher speed, which acts like a load and draws power from the grid. The flywheel is usually placed in a vacuum and advanced magnetic bearings are used to keep the friction at the minimum possible. To recover the electricity, the flywheel is slowed down to drive the generator, which in turn creates electricity. With current technology, individual flywheel units can store and deliver 25 kWh of extractable energy. Multiple flywheels can be connected together and deployed in integrated arrays to provide various megawatt-level power capacities. Flywheels are usually considered as a short-duration storage technology, and used to address power-quality disturbances and frequency regulation.

4.2.2.5 Thermal Energy Storage

Thermal Energy Storage (TES) comprises a number of technologies which can be generally divided into two categories: heat thermal storage and cool thermal storage [2]. The basic types of heat thermal storage can be described as sensible heat storage, in which the temperature of the storage material varies with the amount of energy stored; and latent heat storage, in which a substance changes from one phase to another by melting (e.g., from ice to water). For sensible heat storage, both liquid media storage (e.g., water, oil-based fluids, molten salts) and solid media storage (e.g., rocks, metals, building fabrics) can be used. For latent heat storage, the energy is stored as latent heat of fusion at a constant temperature corresponding to the phase transition temperature of the Phase Change Materials (PCMs). PCMs can undergo solid–solid, liquid–gas, and solid–liquid phase transformations. A typical application of heat thermal storage is solar thermal storage. In such a system, the concentrated solar energy is used to melt salts to store solar heat. The solar heat is used to drive a steam generator to produce steam, which then drives a power turbine to generate electricity.

Thermal storage can also be used to change heating or cooling electricity load from network stress time to other times. Heat thermal storage can be applied where heating needs are significant. On the other hand, cool thermal storage is applied where cooling demands significantly contribute to high demand charges. For example, a refrigeration system could be operated during the off-peak nighttime hours and then supply the daytime cooling to reduce the peak demand electricity charges. Chilled water, ice, and eutectic salt are the three most used types of cool storage systems. By using a cool thermal storage system, the load of cooling is not eliminated, but shifted from the peak hours to the non-peak hours as indicated earlier.

4.2.2.6 Ultra Capacitor Electric Storage

Conventional capacitors have not been able to match a battery's ES capacity because of their structures and materials. Ultra capacitors are new types of capacitors which have multiple advantages over conventional capacitors, including longer lifetime, larger ES capacity, and higher discharging efficiency. Unlike batteries, in which electricity is stored as chemical potential, ultra capacitors store energy in the form of electric fields established by charge accumulated on the bearing surfaces. Unlike batteries in which internal electrochemical reactions occur and the cycle life is limited, ultra capacitors have the ability to be cycled thousands of times. The latest technology advancement enables ultra capacitors to perform even better than Li-ion batteries by using nanocomposite materials.

4.2.2.7   Superconducting Magnetic Energy Storage

SMES systems store energy in the magnetic field created by charging a superconducting coil. To recover the energy, the superconducting coil is discharged. SMES systems cooled by liquid helium and liquid nitrogen have been developed. However, the high cost of superconductors is constraining the commercial use of SMES systems.

4.2.3   Standardization Projects and Efforts

ES is a key functionality of the Smart Grid. The corresponding standards should be treated as a key priority by SDOs and other stakeholders. With technology advancement, ES is expected to play an increasingly important role in smart gird, particularly to accommodate the increasing penetration of intermittent renewable energy resources, such as wind and solar energy. However, few standards exist to integrate ES systems with the electric power infrastructure. Furthermore, the ES functions and requirements will depend on the grid domain where the storage is used (e.g., transmission, distribution, end-user side) and few standards exist to take that into consideration. Therefore, a broad set of stakeholders and SDOs should cooperate to address the interoperability and evolution problems of ES. In this subsection, various standardization projects and efforts regarding ES is introduced and summarized.

4.2.3.1   NIST Energy Storage Interconnection Guidelines (PAN 07)

NIST has initiated the Energy Storage Interconnection Guidelines project and listed it as the Priority Action Plan 07 (PAP 07). The objectives of this guideline are summarized as follows [3]:

- Involve a broad set of stakeholders and SDOs to address the Electric Storage-Distributed Energy Resource (ES-DER) electric interconnection issues.
- Develop a scoping document to identify the ES-DER interconnection and operational interface requirements for application issues, including high penetration of ES-DER and plug-in electric vehicles (PEVs).
- Develop use cases to identify ES-DER interconnection and object modeling requirements, including coordination with PEVs and Wind Use Cases.
- Update or augment the IEEE 1547 standard series to accommodate ES-DER requirements and IEC 61850-7-420 object models.
- Initiate development of transmission-level standards for ES-DER.
- Harmonize the distribution and transmission-level standards.

NIST will cooperate with other key stakeholders such us Underwriters Laboratories (UL), National Electrical Code-(NEC)-National Fire Protection Association (NFPA) 70, Canadian Standards Association (CSA), IEC, IEEE, and SAE to ensure safe and reliable implementation of ES-DER.

4.2.3.2 IEEE 1547 and IEEE 2030 Standardization Projects

IEEE 1547 *Standard for Interconnecting Distributed Resources with Electric Power Systems* was approved in 2003 by IEEE. It defines the interconnection of DERs with electric power systems (EPSs) and provides the requirements relevant to the performance, operation, testing, safety, and maintenance of the interconnection [4]. However, IEEE 1547 was developed for interconnected systems with limited considerations of DERs and renewable energy resources. Some limitations of IEEE 1547 are summarized as follows:

- There is no distinction between energy storage devices and generator has been specified within the DER portfolio.
- It is only applicable to aggregate-capacity-rated 10 MVA and less.
- It does not provide voltage support and does not define maximum capacity.
- There are no ramp rate specifications to enable hybrid generation – storage to mitigate the intermittency of renewable resources.
- It does not specify DER self-protection.

In order to update and revise the IEEE 1547 standard, nine complementary standards have been designed and five of them have been published. A summary of the published standards and ongoing drafts are as follows:

- IEEE 1547.1 2005 *Standard for Conformance Tests Procedures for Equipment Interconnecting Distributed Resources with Electric Power Systems* – This standard was published in 2005. It provides conformance test procedures to confirm the suitability of interconnections of DERs with EPS.
- IEEE 1547.2 2008 *Application Guide for IEEE 1547 Standard for Interconnecting Distributed Resources with Electric Power Systems* – This standard was published in 2008. It provides technical descriptions, application guidance, and interconnection examples to enhance the use of IEEE 1547.
- IEEE 1547.3 2007 *Guide For Monitoring, Information Exchange, and Control of Distributed Resources Interconnected with Electric Power Systems.*
- IEEE 1547.4 2011 *Guide for Design, Operation, and Integration of Distributed Resource Island Systems with Electric Power Systems* – This standard was published in 2011. It provides alternative approaches for the design, operation, and integration of Demand Response (DR) island systems with EPS.
- IEEE P1547.5 *Draft Technical Guidelines for Interconnection of Electric Power Sources Greater than 10 MVA to the Power Transmission Grid* – This standard was withdrawn in 2011.
- IEEE 1547.6 2011 *Draft Recommended Practice for Interconnecting Distributed Resources With Electric Power Systems Distribution Secondary Networks* – This standard was published in 2011. It specifies the criteria, requirements, and test for the interconnection of DER with EPS distribution secondary networks.
- IEEE P1547.7 *Draft Guide to Conducting Distribution Impact Studies for Distributed Resource Interconnection* – This standard is designed to describe criteria,

scope, and extent for studies of the potential impact of a DER interconnected to an area EPS system.

- IEEE P1547.8 *Recommended Practice for Establishing Methods and Procedures that Provide Supplemental Support for Implementation Strategies for Expanded Use of IEEE Standard 1547* – This standard is designed to expand the usefulness and uniqueness of IEEE 1547 through the identification of innovative designs, processes, and operational procedures.
- IEEE P1547a *Standard for Interconnecting Distributed Resources with Electric Power Systems – Amendment 1* – This standard is designed to establish updates to voltage regulation and other necessary changes to IEEE 1547.

4.2.3.3   IEEE P2030 Standard Series

The IEEE P2030 standard series are developed by IEEE to provide guidelines for Smart Grid interoperability. IEEE P2030.2 *Draft Guide for the Interoperability of Energy Storage Systems Integrated with the Electric Power Infrastructure* is developed to provide guidelines for discrete and hybrid storage systems that are integrated with the electric power infrastructure, including end-use applications and loads [5]. IEEE 2030.3 *Standard for Test Procedures for Electric Energy Storage Equipment and Systems for Electric Power Systems Applications* defines test procedures for electric energy storage equipment and systems for EPSs applications. Storage equipment and systems that are connected to EPS need to be verified to meet the requirements specified in related IEEE standards. Both IEEE 2030.2 and 2030.3 standards are still in the development stage.

4.2.3.4   IEC Smart Grid Standardization Roadmap

The IEC *Smart Grid Standardization Roadmap* presents an inventory of existing standards and analyses the gaps between existing standards and future requirements. On the basis of this analysis, recommendations for future evolution are presented. The requirements for ES specified by IEC are summarized as follows [6]:

- Safe operation and handling are major requirements.
- Information about capacity of the storage units and forecasts of pricing information should be available.
- Optimal scheduling of the storage units is required.
- Data models and semantic information models must be available.
- Communications must be available for the whole chain, including power grid, battery management system and battery modules.
- The key parameters including cell type, rating, start-up rate, accumulated kilowatt hour, charging condition, temperature, load history, availability, and manufacturer are necessary to be communicated.

The existing standard IEEE 61850-7-410, *Communication Networks and Systems for Power Utility Automation – part 7-410: Hydroelectric Power Plants –*

Communication for Monitoring and Control defines the connection of hydroelectric power plants to the power automation. It specifies the additional common data classes, logical nodes, and data objects required for the application of IEC 61850 in a hydroelectric power plant.

IEC 61850-420 *Communication Networks and Systems for Power Utility Automation – part7-420: Basic Communication Structure – DERs Logical Nodes* addresses the IEC 61850 information modeling for DERs. The IEC 61850 information models for DER are to be used in the exchange of information with DER including reciprocating engines, fuel cells, microturbines, photovoltaics (PVs), combined heat and power (CHP), and energy storage. These information models are fully compatible with the Common Information Model (CIM) concepts developed by IEC TC 57.

At present, there is no standard for other bulk ES systems other than for hydroelectric storage. Equivalent standards for connection of large and distributed storage systems should be developed by IEC TC 57. Profiles needed to be developed to specify the amount and kind of data which are necessary to be exchanged. Testing and verification procedures for immobile and mobile batteries and battery stacks are required and should be developed by TC 21 and TC 35.

4.3   Distributed Energy Resources

4.3.1   An Overview of Distributed Energy Resources

DERs offers economic and ecological benefits by generating energy close to the consumers. DERs could minimize transmission and distribution losses and offer continuous and reliable supply of electricity to customers when a power outage occurs. DER has significant impact on the design of the Smart Grid and important issues such as reliability, power quality, and cost should be well analyzed by various stakeholders. Many standards organizations or industry stakeholders have proposed slightly different definitions for DER. In the United States, DER defined by EPRI and its industry stakeholders refers to "small or modular" electricity generation or energy storage resources which are usually less than 60 MW [7]. In Australia, the word "distributed energy" is replaced by "decentralized energy" [8] to "communicate better and more intuitively the paradigm shift from the current model of large-scale centralized energy generation and delivery." IEC has directed the interest to VPPs and specifies the technical requirements for a successful operation of VPP in [6]. In contrast, NIST has focused on the ES-DER interconnection issues [9]. IEEE has revised and updated the IEEE 1547 standard for issues related to the interconnection of DERs with EPSs. In [10], CEN/CENELEC/ETSI (European Telecommunications Standards Institute) recommends related technical committees to work on unified standards for electrical connection and installation rules of DERs.

In this subsection, the current landscape of DERs including applications, technologies, costs and benefits, and challenges are introduced in Section 4.3.2. Then, the efforts made by various stakeholders including utilities, SDOs, and government agencies to develop new DER standards and technologies are explained in Section 4.3.3.

4.3.2   Technologies and Applications

In order to successfully deploy the DER system, it is important to understand the DER technology options with their applications and the specific energy solutions they enable. This section describes some of the technologies for DER, their applications for end users and utilities. Available technological options, and their advantages, disadvantages, and possible applications are summarized in Table 4.2.

Conventional DER technologies, such as the combustion turbine and reciprocating engine are evolving in the direction of decreasing cost, increasing efficiency, lower emissions, and higher reliability. They will likely remain competitive in many applications and dominate the landscape for the foreseeable future. However, the lack of power electronics, which provide the control flexibility required by DER, minimizes their role in the Smart Grid. Renewable generation, such as PV and wind turbine, provides a significant environmental benefit. PV cells convert sunlight directly into electricity and can be installed on rooftops or other sunny areas. Wind turbines are usually placed at the top of a tall tower to harness the wind at a greater velocity and free of turbulence caused by obstacles such as trees, hills, and buildings. Microturbines are small combustion turbines that produce both heat and electricity. Waste heat recovery technologies can be used together with microturbines to achieve higher energy conversion efficiency. The initial installation costs, operation costs, and maintenance costs are low due to their small size.

Fuel cells convert the chemical energy of a fuel into electricity by electrochemical reactions and offer high efficiency and low emissions. The Proton Exchange Membrane Fuel Cell (PEMFC) which operates on pure hydrogen fuel has attracted interest from several major automotive companies. High-temperature fuel cells, including Solid Oxide Fuel Cell (SOFC) and Molten Carbonate Fuel Cell (MCFC), which run on natural gas have been demonstrated by several energy companies such as FuelCell Energy, Bloomenergy, Fuji Electric, ClearEdge Power. Compared to low-temperature fuel cells, high-temperature fuel cells are particularly suitable for CHP systems. However, both types of fuel cells still face challenges such as expensive catalysts, high-temperature corrosion, and breakdown.

Figure 4.1 summarizes the total cost of energy for several DER technologies, sorted from lowest to highest cost [7]. The costs are shown within a sensitivity range, which is a driven by a combination of capital cost, financing cost, fuel costs, maintenance costs and waste heat recovery [11]. The results have demonstrated that, with technology advancement, some DERs can be cost effective enough in comparison to the retail rates (the cost of delivering energy to end users). For more information on the installation costs, operation and maintenance costs, and cost of electricity for DER technologies, please refer to [12].

Table 4.2   A comparison of available DER technology options

Technology options	Advantages	Disadvantages	Applications
Microturbine	Low capital cost	Insufficient thermal output	Peak shaving
	Compact size	Low efficiency	CHP
	Low emissions		Standby power
			Power quality and reliability
Fuel cell	High conversion efficiency	Expensive	Backup power
	Low emissions	High-temperature corrosion and breakdown (SOFC, MCFC)	Electric vehicles
	High power density	No mature technology	CHP
	Quiet operation		Distributed generation
Photovoltaic	No emissions	High installation costs	Distributed generation
	No moving parts and low maintenance cost	Variable energy output	Peak shaving
	No operation noise	Low conversion efficiency	
Wind turbine	Mature technology	Variable energy output	Distributed generation
	No emissions	Strict site requirements	Peak shaving
	No variable costs for fuel	Low conversion efficiency	
Combustion turbine	Mature technology	Environmental issues – emission and noise	CHP
	Relatively high efficiency	Consumes more fuel than reciprocating engine when idle	Peak shaving
	Low operation cost	Lack of power electronic	Backup power
	Great reliability		Power quality and reliability
Reciprocating engine	Low initial installation cost	Environmental issues – emission and noise	Peak shaving
	Short start-up times	High maintenance	Backup power
	Mature technology	Low efficiency	Power quality and reliability
	High reliability	Lack of power electronics	CHP

DER technologies can be deployed either by customers to achieve energy cost savings and higher energy reliability, or by utilities to lower infrastructure investment and to improve asset utilization. The end-user applications include backup power systems, heat recovery/cooling applications such as CHP, demand response, and load management. The utility grid applications include peak shifting, self-healing, microgrid, VPP, and power quality and reliability. Therefore, it is necessary to capture both end user and utility benefits for DERs to be widely deployed.

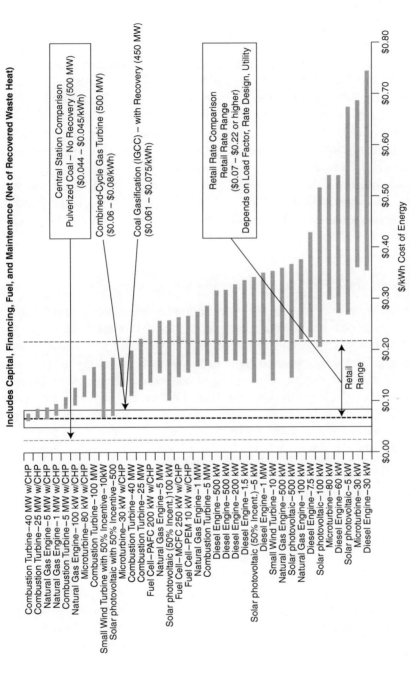

Figure 4.1 The total cost of energy for several DER technologies. Reproduced from "Distributed energy resources: current landscape and a roadmap for the future", November, 2004, Copyright © 2004 EPRI [7], with permissions from Electric Power Research Institute (EPRI)

4.3.3   Various Standardization Processes and Projects

4.3.3.1   EPRI Efforts

EPRI conducts research, development, and demonstration projects which span nearly every area of electricity generation, delivery and use, management and environmental responsibility. EPRI has conducted a number of projects related to energy storage and DER. EPRI has launched the seven-year Smart Grid Demonstration Initiative, which includes core Smart Grid research and several large-scale Smart Grid projects. The Hawaiian Electric Company Smart Grid demonstration project, which is part of the EPRI Smart Grid Demonstration initiative, was founded to identify approaches for interoperability and widespread integration of DER, including demand response, storage, Distributed Generation (DG), and distributed renewable generation. In [7], several pathways to the 2015 vision of DER have been proposed. These pathways include three individual pathways: "Grid Support Pathways," "End-Use Pathways," and "Energy Supply Pathways"; and two joint pathways: "Joint End-Use/Distribution Pathways," and "Joint Supply/Transmission Pathways."

4.3.3.2   DERlab and CENELEC Collaboration Efforts

DERlab is the association of leading DER laboratories and research institutes with the aim to develop joint requirements and quality criteria for the connection and operation of DERs. DERlab has actively engaged in activities about DER technology research and development, prestandardization, and national/international collaborations. It has published numerous technical reports on grid-connected storage, electromagnetic compatibility (EMC), DER protection, and so on. Furthermore, DERlab has cooperated with the European standards organization CENELEC TC8X/WG3 for developing requirements about the connection of generators above 16 Å per phase to the Low Voltage (LV) distribution system or to the Medium Voltage (MV) distribution system.

4.3.3.3   CERTS MicroGrid Concept

The Consortium for Electric Reliability Technology Solutions (CERTS) has proposed its MicroGrid concept for the US Department of Energy, which appears to the grid as a single self-controlled entity providing both power and heat. Compared to conventional DER systems which might have unexpected impacts on the grid, the advantage of a MicroGrid is that it can be regarded as a controlled entity within the power system and complies with grid rules as an existing customer. Advanced power electronics are used to provide the required control and communication capabilities. The MicroGrid can provide several benefits such as congestion relief, local voltage support, and response to load changes.

4.3.3.4   European DER Projects

The European Union (EU) has initiated the Framework Programme (FP) to provide funding for research, technological development, and demonstration. This section introduces several DER projects from FP5 (1998–2002), FP6 (2002–2006), and FP7 (2007–2013).

The FP5 DISPOWER project proposed a new DG structure for power supply in regional, local and island grids, and investigated basic solutions for DG technical problems [13]. The FP5 MICROGRIDS project addressed the issues related to large-scale integration of micro-generation to low-voltage grids [14].

The FP6 EU-DEEP project was initiated by eight European utilities to address a number of technical and nontechnical barriers to the massive deployment of DER in Europe [15]. The main topics of DER include market integration, regulation adaptation, and connection technologies to the grid, grid impact, and DER systems. The IRED [16] project was initiated to facilitate the integration of RESs and DG into the future European electricity network. IRED is a large research cluster which was formed by seven FP5 projects dealing with RES-DG integration issues in 2002. The FP6 FENIX project was founded to boost integration of DERs by maximizing their contribution to the EPS, through aggregation into Large-Scale Virtual Power Plants (LSVPPs) and decentralized management [17].

The FP7 iGREENGrid project is a four-year project from January 2013 to December 2015. Its objective is to increase the hosting capacity for distributed RES in grids while ensuring the reliability and the quality of supply [18]. The FP7 ADDRESS project was founded to design interactive distribution energy networks integrating DER and demand response [19].

4.3.3.5   IEC Smart Grid Standardization Roadmap

In the IEC *Smart Grid Standardization Roadmap* [6], the interest of the DER is directed to VPP, which is a collection of small and very small decentralized generation units providing both power and heat. The technical equipment requirements for a successful operation of VPP include EMS, forecasting system, energy data management system, and powerful front end for the communication of the EMS with the DER units. The components/units of VPP can be modeled in the DER based on the IEC 61850-420 standard, which addresses the IEC 61850 information modeling for DERs. The introductions of IEC 61850-7-420 and its relationship with IEC 61850 can be found in Section 4.2.3. IEC 61850-7-410, IEC 61400-25, and IEC 61727 are the equivalent standards to IEC 61850-7-420 for hydro power plants, wind turbines, and PV systems respectively. IEC will seek close cooperation with relevant national and regional standards bodies such as IEEE, CENELEC, and so on, to develop DER standards.

4.3.3.6 NIST Priority Action Plans and IEEE DER Interconnection Standards

NIST has listed the Energy Storage Interconnection Guidelines project as the PAP 07. The objective of this guideline is to address the ES-DER electric interconnection issues. In PAP 07, NIST has recommended to update or augment the IEEE 1547 standard series to accommodate ES-DER requirements and IEC 61850-7-420 object models. IEEE 1547 defines the interconnection of DER with EPS and provides the requirements relevant to the performance, operation, testing, safety, and maintenance of the interconnection. In order to update and revise the IEEE 1547 standard, nine complementary standards have been designed and five of them have been published. The details of NIST PAP 07 and IEEE 1547 can be found in Section 4.2.3.

4.3.3.7 Efforts in Australia, Japan, Germany, and China

In Australia, the Commonwealth Scientific and Industrial Research Organization (CSIRO) has published the *Decentralized Energy Roadmap for Australia*. The objective of the roadmap is to provide an overview of the current status of decentralized energy in Australia, and to outline policy actions for increasing the penetration rate of decentralized energy. The word "distributed" is replaced by "decentralized" to communicate the paradigm shift from the current large-scale centralized energy generation and delivery to future decentralized energy generation [8].

In Japan, the New Energy and Industrial Technology Development Organization (NEDO) has initiated numerous research projects about DER and demonstration projects in Sendai, Ota City, Hachinohe, and so on. In Germany, the E-Energy program has been initiated to demonstrate how the Information and Communication technologies (ICT) can best be used to enhance the efficiency, reliability, and environmental compatibility of the power supply. ICT has been used in network integration of DER and electricity demand in order to better balance supply and demand. In China, the State Grid Corporation of China (SGCC) has launched a two-year project (from January 2013 to December 2014) which studies the integration of DER with electrical grid.

4.4 E-Mobility/Electric Vehicles

4.4.1 Introduction of E-Mobility/Electric Vehicles

According to [20, 21], 29% of CO_2 emissions in the 27 EU Member States and 33.1% in the United States are from transportation. The conventional transportation systems' reliance on oil as a source of energy has created serious economic and security vulnerabilities for nations. Electrification of the transportation systems with EVs has the potential to mitigate some of the negative consequences of oil dependency on the economy, national security, and the environment. EVs can also be integrated with electrical grid for smoothing variable generation from renewable sources, peak load shifting, voltage control, and frequency regulation, and providing DERs. However, the relatively high cost of EVs compared to conventional vehicles, limited range, long battery charging time, short battery life span, and security concerns are barriers to massive adoption of EVs. In this subsection, EV technologies and related specifications and standards are introduced, with a special focus on integration with the electrical grid.

The organization of this subsection is as follows: Section 4.4.2 gives an introduction of the history of the PEV and issues related to vehicle deployment; Section 4.4.3 introduces the various types of EVs; Section 4.4.4 introduces various kinds of EV batteries; Section 4.4.5 introduces the issues related to the integration of EVs with electric grid. Section 4.4.6 introduces the standardization projects and efforts of various international/national SDOs.

4.4.2   The Rise and Fall of Electric Vehicles

Several inventers are being credited as the first EV inventors. In 1828, Hungarian engineer, Ányos Jedlik, invented the EV model car. Between 1834 and 1835, Thomas Davenport, an American inventor, built a battery-powered EV, a small locomotive that was operated on a short section of tracks. Other electric car inventors around this early period include Robert Anderson of Scotland and Sibrandus Stratingh of the Netherlands.

By the twentieth century, EVs were commonplace and took the majority of the market. At that time, the light, powerful Internal Combustion Engines (ICEs) had not been developed yet. By 1912, there were around 30 000 EVs on the roads in the United States, while a third of these EVs were actually commercial vehicles [22, 23]. The advantages of EVs over other types of vehicles are that the electric car was quicker to start up, cleaner, and doing better in the snow. However, by the late 1920s, EVs had nearly gone from the market and were mostly used for specialist roles, for example, platform trucks, forklift trucks, tow tractors, and urban delivery vehicles. Compared to ICE, the EV was very expensive for consumers. Henry Ford sold the popular Model T in 1908 for $850, while the price of EVs at the same time was around $2000. The Model T was later sold for as little as $260, due to the production savings of assembly lines. Another primary reason for the downfall of personal EVs was the limited range. Electric cars in the early 1900s would last about 35 miles per charge, while Model T could get 13–21 miles per gallon of gasoline and could hold 9–10 gallons of gasoline [24].

The barriers to massive adoption of EVs are limited range, inadequate infrastructure, lack of regulatory framework and standards, high cost, inconvenience, and lack of safety, and so on. These barriers are briefly introduced here for a better understanding of the challenges faced by researchers and engineers.

- **Range**: Range anxiety is a common consumer concern about EVs, due to the fact that battery-powered EVs have a range less than 100 miles. In a survey, the insufficient battery range of EVs is regarded as the top reason to choose non-battery-powered EVs [25]. The "expected range" of 63% of respondents was around 300 miles on a single charge, which was not available at that time due to the technology limitations (see [26]). Range concerns could be reduced by technological advancement of batteries and the installation of public charging infrastructure, which is described in Sections 4.4.4 and 4.4.5.

- **Infrastructure**: The current scarcity of public charging infrastructure is one of the major reasons lack of massive adoption of EVs. However, the paradox lies in the fact that without the mass adoption of EVs, there is not enough incentive to build public charging infrastructures.

- **Regulatory**: Regulatory issues of integrating EVs into the electrical power grid should be addressed. Utilities should be encouraged to make investments in infrastructures and provide services for EV users. Consumers should be encouraged to buy EVs by offering different forms of incentives such as government subsidies, lower costs, free parking rights.

- **Standards**: Standards regarding the integration of the EV with the electrical grid needs to be developed by SDOs for the reassurance of standardized, safe, convenient charging points everywhere. Open standards needs to be developed by various stakeholders for roaming between EV charging sites and communication of metering data.

- **Cost**: The estimated cost per kilowatt-hour of EV lithium-ion batteries is around $600 and prices for mass production of batteries are expected to reach $500/kWh by 2015 [27, 28]. However, EV may not become cost competitive with conventional ICE vehicles until the price is as low as $300/kWh [29].

- **Convenience**: Consumers are used to buying a car from a car dealer and driving it home directly. However, for EVs, permits are required for the installation of EV chargers at home and consumers need to wait at home for the licensed electrician. It may take one or two months or even longer before the charger can be installed, depending on the business model and collaboration efficiency between government, utilities, charger operation companies, and so on. Furthermore, it may take from half an hour to even a day to charge EVs, depending on the battery capacity, state of charge, and the type of chargers installed. While a fast charger is technologically available, consumers still facing problems such as the load limitations of local legacy distribution grid, the high cost of fast chargers, safety concerns, and so on. Besides having chargers at home, charging infrastructure needs to be available in public places. Consumers need to have access to on-time information such as where to charge, price, how long it will take, how far they can go on a charge, and so on.

- **Safety**: Consumers are concerned about many EV safety issues, such as battery safety, charging safety, operation safety, driving safety, and fire and electric safety.

4.4.3 Types of Electric Vehicles

There are various ways of classifying EVs. One classification method is based on how EVs are charged. EVs can be plugged into a charging station and recharged from

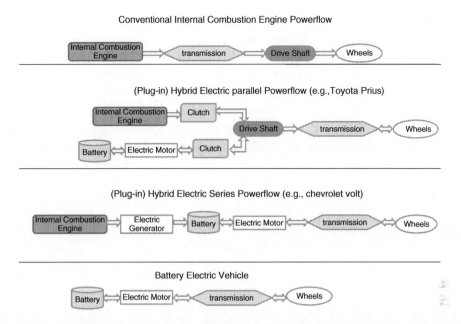

Figure 4.2   A comparison of the conventional ICE vehicle and various types of potential EV solutions. Reproduced with permission of the Center for Climate and Energy Solutions [30]

an external source of electricity. In contrast, EVs can also be untouchably charged by inductive charging, in which energy is transferred between electromagnetic coils attached to the ground and the bottom of the car. The former type is usually called a Plug-in Electric Vehicle (PEV) and can also be integrated with the functionality of inductive charging. In this book, PEV refers to PHEVs and Battery Electric Vehicles (BEVs). On the other hand, EVs can also be classified into two types based on the power sources, that is, pure EVs and Hybrid Electric Vehicles (HEVs). HEVs employ more than one power source, usually combing an ICE propulsion system with an electric propulsion system. The advantage of HEVs is that it can achieve better fuel economy than conventional ICE vehicles. PHEV is one type of HEVs which can be recharged by connecting a plug to an external electric power source. The conventional ICE vehicle and various types of potential EV solutions are compared in Figure 4.2.

The power flows for different vehicle types are summarized as follows:

- **Conventional ICE vehicles**: In conventional ICE vehicles, the combustion of a fossil fuel occurs with an oxidizer (usually air) to produce the expansion of high-temperature and high-pressure gases. This expansion of gases produces the power that is transmitted to wheels. With technology advancement, hydrogen internal combustion engine vehicle (HICEV) has been developed by replacing fossil fuel with hydrogen because the combustion of hydrogen with oxygen produces water as its only product.
- **PHEV**: In a PHEV, power is simultaneously transmitted to the wheels from the ICE engine as well as the battery-powered electric motor (e.g., Toyota Prius).

- **BEV**: The BEV is also called a pure electric vehicle or all-electric vehicle. It uses only battery-powered electric systems for propulsion and contains no ICE systems.
- **Extended range electric vehicle** (EREV): The EREV is similar to the PHEV except that the ICE does not power the wheels directly. In contrast, the ICE is used to charge the battery system (e.g., Chevrolet Volt).

Although the ICE system integrated into a PHEV adds extra cost, PHEVs require a lower energy capacity than BEVs. Thus, the addition of the ICE system lowers the total cost of the battery pack and does not necessarily make PHEVs less economically competitive than BEVs.

4.4.4   Electric Vehicle Batteries

Different from batteries used for starting, lighting, and ignition, Electric Vehicle Batteries (EVB) are characterized by higher power-to-weight ratio, energy-to-weight ratio, and energy density. Power-to-weight ratio determines a vehicle's ability to reach a given distance at a specified time, that is, a vehicle with a high power-to-weight ratio will accelerate faster than one with a lower value. Energy-to-weight ratio, usually called specific energy, determines the battery weight required to achieve a given range, while energy density determines the battery size required to achieve a given range. A vehicle with high specific energy and energy density will have a longer range than an automobile with lower values under the constraints of weight and space. EVBs benefit from technology advancements which have been driven by demands for larger, brighter displays and longer battery life in the field of consumer electronics. However, the cost of batteries remains quite high – it is usually half the retail cost of an electric car. Another major barrier to the mass adoption of EVs is the much lower energy density of batteries compared to gasoline, for example, the energy density of today's lithium-ion batteries is only 1% that of gasoline [30]. This lower energy density limits the maximum range EVs can achieve. However, as the system efficiency of EV is much higher than the conventional ICE system, an EV might achieve comparable range with a much lower energy density.

Table 4.3 shows the requirements for the advanced EVB, which are set by the US Advanced Battery Consortium (USABC) [31], an organization whose members are Chrysler LLC, Ford Motor Company, and General Motors Corporation. The storage system performance goals for power-assisted HEVs and the requirements of end-of-life energy storage system for PHEVs can also be found in [32] and [33]. In order to compare the USABC's requirements with currently available EVBs, the power-to-weight ratio and energy-to-weight ratio for various batteries, electrochemical capacitors, and fuel cells are illustrated in Figure 4.3 in the form of a Ragone plot [34]. A superposition of the USABC requirements onto Figure 4.3 shows that while Li-ion batteries can easily satisfy HEV requirements, the energy is much smaller than the requirements for EVs. The energy-to-weight ratio of lead-acid batteries is too low

Table 4.3 USABC goals for advanced batteries for EVs

Parameters (units) of fully burdened system	Minimum goals for long-term commercialization	Long-term goal
Power density (W/l)	460	600
Specific power – discharge 80% DOD/30 s (W/kg)	300	400
Specific power – C/3 discharge rate (Wh/l)	150	200
Energy density – C/3 discharge rate (Wh/kg)	230	300
Specific energy – C/3 discharge rate (Wh/kg)	150	200
Specific power/specific energy ratio	2 : 1	2 : 1
Total pack size (kWh)	40	40
Life (years)	10	10
Cycle life – 80% DOD (cycles)	1000	1000
Power and capacity degradation (% of rated spec)	20	20
Selling price – 2500 units @ 40 kWh ($/kWh)	< 150	100
Operating environment (°C)	−40 to +50 20% performance loss (10% desired)	−40 to +85
Normal recharge time	6 h (4 h desired)	3–6 h
High rate charge	20–70% SOC in < 30 min @ 150 W/kg (< 20 min @270 W/kg desired)	40–80% SOC in 15 min
Continuous discharge in 1 h – no failure (% of rated energy capacity)	75	75

Source: Reproduced from United States Council for Automotive Research LLC

to meet the requirements of USABC for HEVs, PHEVs, and EVs. The nickel–metal hydride battery is relatively mature and widely used for HEVs. However, its energy cannot meet the USABC's requirements for PHEVs and EVs. The discharge time of the capacitors is too short compared to that of the batteries.

Although it appears possible that Li-ion batteries may meet the requirements of PHEVs, other important parameters such as cost, life span, and safety, and so on, need to be examined. One of the major concerns for the Li-ion battery is the aging problem, that is, the maximum storage capacity appears to diminish over time. Therefore, criteria such as cost, life span, and safety remain challenges in all applications, and research efforts have been directed to address these issues. Nevertheless, Lithium-ion batteries are more suitable for use in vehicular applications owing to their high energy and power capability, and possibly lower cost in the future. Batteries today are very far from achieving the theoretically maximum-possible energy limit, and the energy density of batteries has been increasing slowly over the last 20 years. Therefore, there is a

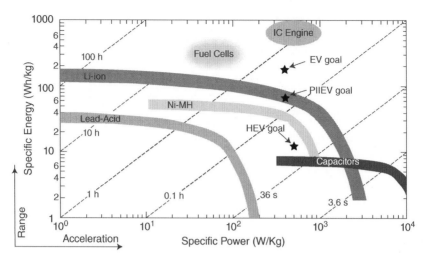

Figure 4.3 Ragone plot (specific power density in watts per kilogram versus specific energy density in watt-hours per kilogram) of various electrochemical energy storage and conversion devices [34]

lot of room for improvement and the future of batteries remains very strong and optimistic. The key factors to commercialize batteries systems for vehicular applications are high energy, long life, good safety characteristics, and low cost.

4.4.5 Grid-to-Vehicle (G2V) and Vehicle-to-Grid (V2G) Opportunities and Challenges

4.4.5.1 Grid-to-Vehicle (G2V) Opportunities and Challenges

One of the barriers to the mass adoption of electric vehicles is the long charging time. Table 4.4 shows the relationship between the required charging time for full charge with the available power at the terminal output, and the relationship between the touring range of one hour's recharging with the available power at the terminal

Table 4.4 The relationship between the required time of full charge with the output power, and the relationship between the touring range (by one hour's recharging) with the output power [35]

Charging Mode	Output Power	Time of full charge (h)	Touring range (km)
Single phase 230 V	16 A/3 kW	8	40
	32 A/7 kW	6	70
Three phase 400 V	16 A/11 kW	4	110
	32 A/22 kW	2	150
Fast charging	500 V/125 A	0.3	150

output [35]. Table 4.4 demonstrates that it takes 6–8 h to fully recharge a BEV by a single-phase 16 A 230 V domestic socket. The slow charging infrastructures are usually installed at consumers' home or at workplaces to provide a full (100%) charge. Charging will be performed mostly at home during off-peak hours when electricity is cheapest. Table 4.4 also shows that 1 hour recharging by the same kind of home-charging infrastructure can only enable a BEV to travel around 40 km, which is hardly enough to meet the daily driving range of most drivers. Although slow charging cannot provide enough driving range, slow-charging infrastructures are cheaper than fast-charging infrastructures and widely installed at homes and public streets. Most EVs can be slow-charged with a standard Blue Commando connector (IEC 60309) at the charging point end or a gun-shaped Society of Automotive Engineers (SAE) J1772 socket for connection to the vehicle.

Table 4.5 shows the charging levels included in the SAE J1722 standard [36]. The Level 1 AC charging takes nearly 17 hours to recharge a BEV, which may be inconvenient for EV consumers. Therefore, in order to fully charge a BEV overnight, a Level 2 charger is needed to be installed, which may require a system upgrade as 240 V outlets. These outlets are not common in some countries or regions. Fast charging can be performed by using a higher voltage, for example, 450 V, which requires not only the upgrading of the local home electric socket but also the upgrading of the distribution grid. The fast-charging infrastructures can be too expensive for installing in households and will likely only be available in public spaces or large charging stations.

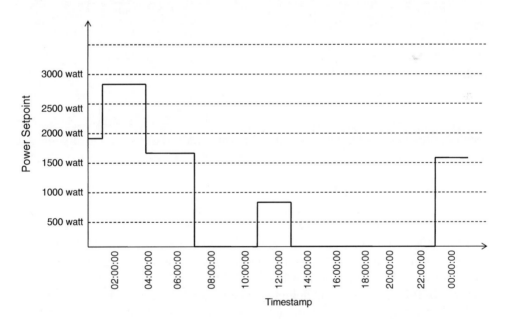

Figure 4.4   Illustration of an EDISON charging schedule. Reproduced from Lars T. Berger, Krzysztof Iniewski, Smart Grid Applications, Communications, and Security, pp. 390. Copyright © 2012 John Wiley and Sons [40], with permissions from John Wiley and Sons.[40]

Table 4.5 Charging levels defined by the Society of Automotive Engineers (SAE) J1772 standard [36]

Level	Electric potential difference (V)	Current (A)	Power (kW)	BEV charging time (h)			
				3.3 kW charger	7 kW charger	20 kW charger	45 kW charger
AC Level 1	120	12/16	1.4/1.92	BEV: 17			
				PHEV: 7			
DC Level 1	200–450	80	Up to 36	—	—	BEV:1.2	—
						PHEV: 0.37	
AC Level 2	240	80	Up to 19.2	BEV: 7	BEV: 3.5	BEV: 1.2	—
				PHEV: 3	PHEV: 1.5	PEV: 0.37	
DC Level 2	200–450	200	Up to 90	—	—	—	BEV: 0.3
							PHEV: 0.17
DC level 3	200–600	400	Up to 240	—	—	—	BEV (only): <0.17

According to Joule's first law, for a specific time and resistance, the heat generated by the current flowing through a conductor is proportional to the square of the current. Therefore, for fast charging with large currents, additional cooling systems and temperature sensors should be included to provide protection for both the vehicle and the charger side. For DC Level 2 chargers, the Japanese auto industry has proposed the CHAdeMO connectors, which allows the charger to communicate with the car regarding how much electricity to be sent in a given period. The CHAdeMO standard is introduced in more detail in Section 4.4.6. There is no official Level 3 charging today because of the concerns regarding cost, safety, and so on.

Although fast charging technologies enable a BEV to be fully charged in 20–30 minutes, the recharging time is still longer than the time it takes to refuel a conventional ICE vehicle. To overcome this challenge, battery swapping technology which takes the same amount of time it takes to refuel at a gas station has been proposed and researched [37, 38]. With better swapping technology, a user can drive into a battery swapping station and have the entire battery pack removed from their car and replaced with a fully charged pack in several minutes, instead of charging for several hours. Consumers no longer have to pay for the price of a battery pack and only lease the battery from a third party to help reduce the total cost of EVs. Despite the promising benefits brought by battery swapping, barriers to commercialization still exist. The pure capital cost of battery swapping infrastructure is incredibly high. For example, the cost of a Better Place battery swapping station is as much as $500 000 [39]. Furthermore, enabling battery swapping technology brings up new issues such as design and safety. One of the design limitations is that the battery pack may be unique to each vehicle model and cannot be used for other models. Batteries can weigh more than 200 kg, and are dangerous to handle and hazardous to store.

Despite conventional fossil-fuel energy resources, charging stations can also be designed to make use of renewable energy resources. One example is the solar-powered automobile charging stations, which employ solar power to charge pure electric or plug-in hybrid vehicles. The power used to charge an EV is derived from a solar PV system and this means the EV is running entirely on clean energy.

The installation of charging infrastructure is one of the several key issues related to the integration of EVs into the electrical grid. Charging infrastructures enable consumers to charge their EVs during off-peak hours, which helps maintain the reliability of the grid. If a large number of EVs are recharged at the same time, this could put stress on the grid by increasing the instantaneous demand for electricity. This stress becomes more severe when consumers expect Level 2 fast charging infrastructures to be used to reduce charge time. The higher the charging level, the more instantaneous demand is put on the electrical grid. Daytime charging of EVs during peak hours may cause the instantaneous demand to exceed the supply provided by the existing electrical grid and large capital investment is needed to provide additional capacity to meet the increase in demand. In order to balance demand and supply, financial incentives for consumers to perform off-peak charging should be created by the government or utilities.

G2V technologies enable the EVs to exchange the necessary information with the grid for smart charging, including the state of charge, charging duration, price, and estimated driving range. With the available information, the charging schedule for each vehicle can be computed remotely by the utilities using some predetermined optimization targets as well as a set of constraints. The charging schedule provides many advantages, including minimizing energy cost, increasing energy efficiency, and reducing instantaneous demand. A charging schedule of the European Electric vehicles in a Distributed and Integrated market using Sustainable energy and Open Networks (EDISON) project is illustrated in Figure 4.4 [40]. More details about the EDISON project are provided in Section 4.4.6.

4.4.5.2   Vehicle-to-Grid (V2G) Opportunities and Challenges

The V2G technology can help regulate imbalances in supply and demand and serve as spinning reserves, that is, meet sudden demands for electricity. In this scenario, EVs are used as energy storage units to store energy during off-peak hours when the electricity price is lowest, and provide energy back to the grid during peak hours. In particular, the renewable energy resources such as solar energy and wind energy can benefit from V2G. Intermittent energy sources are by nature unpredictable and V2G can be used to smooth variable generation from renewables by adapting energy production to energy consumption. However, despite the promising future provided by V2G, it also faces serious barriers such as expensive infrastructures, low energy efficiency (energy is lost during the processes of recharging a battery and then reversing energy from the battery), short battery cycle life, unknown impacts on the grid, security issues, and lack of communication standards. Innovative business models need to be developed to capture the emerging EV market. Strategic partnership among auto manufactures, charging infrastructure operators, utilities, standardization bodies, and government organizations needs to be formed to overcome the obstacles facing integration of EVs into the electrical grid.

4.4.6   Standardization of E-Mobility/Electric Vehicles

4.4.6.1   Standardization Efforts of IEC

In the *IEC Smart Grid Standardization Roadmap*, requirements for E-mobility have been proposed for the following aspects: lifecycle and cyclic stability of batteries; safety; EMC; availability of pricing information; and communication. The Roadmap provides a list of the existing standards and standards currently under development. Gaps between the E-mobility requirements and the existing standards are identified and recommendations have been proposed. In this subsection, we mainly introduce the standard series related to communication, charging system, and conductive charging, as shown in Table 4.6.

The ISO/IEC (International Organization for Standardization) 15118 series, which are still under development, specify the communication between EVs and EVSE (Electric Vehicle Supply Equipment). The ISO/IEC 15118-1 has identified the following use-case elements: (i) plug-in process; (ii) communication setup; (iii) identification and authentication; (iv) payment; (v) charging; (vi) value-added services; and (vii) plug-out process. The ISO/IEC 15118 specifies both the physical layer and higher layers, as opposed to IEC 61850, which only deals with high-level communication protocol. The data models defined in ISO/IEC 15118 should be harmonized with IEC 61850 and IEC 61968.

The IEC 61851 series specify the requirements for charging EVs by on-board and off-board equipment. IEC 61851 also defines the data interface for EV−EVSE

Table 4.6 The IEC standards related to V2G communication, EV charging system, and conductive charging

Sector	Standard
Communication	IEC/ISO 15118 Vehicle to Grid Communication Interface
	Part 1: Definitions and Use-case
	Part 2: Sequence Diagrams and Communication Layers
	Part 3: Physical Communication Layers
Charging system	IEC 61851 Electric Vehicle Conductive Charging System
	Part 1: General Requirements
	Part 21: Electric Vehicle Requirements for Conductive Connection to an AC/DC Supply
	Part 22: AC Electric Vehicle Charging Station
	Part 23: DC Electric Vehicle Charging Station
	Part 31: Data Interface for Recharging of Electric Road Vehicles Supplied from the AC Mains
	Part 32: Data Interface for Recharging of Electric Road Vehicles Supplied from an External DC Charge
Conductive charging	IEC 62196 Plugs, Socket-outlets, Vehicle Couplers, and Vehicle Inlets - Conductive Charging of Electric Vehicles
	Part 1: Charging of Electric Vehicles up to 250 A AC and 400 A DC
	Part 2: Dimensional Interchangeability Requirements for AC Pin and Contact-tube Accessories
	Part 3: Dimensional Interchangeability Requirements for Pin and Contact-tube Coupler with rated Operating Voltage up to 1000 V DC and rated Current up to 400 A for Dedicated DC Charging.

communication via a control pilot wire using a Pulse-Width-Modulated (PWM) signal with a variable voltage level. Four different EV charging modes have been defined in IEC 61851-1, which are summarized as follows:

- **Mode 1 charging**: slow charging utilizing standardized socket-outlets.
- **Mode 2 charging**: slow charging utilizing standardized socket-outlets with additional protection mechanisms.
- **Mode 3 charging**: slow or fast charging utilizing dedicated EVSE.
- **Mode 4 charging**: rapid charging utilizing an off-board charger.

Some charging modes defined in IEC 61851 requires dedicated supply and charging equipment incorporating control and communication circuits. IEC 62196 series specify the mechanical, electrical, and performance requirements for dedicated plugs, socket outlets, vehicle connections, and vehicle inlets for interfacing between such dedicated charging equipment and EVs. IEC 62196-1 defines the requirements for plugs, socket-outlets, connectors, inlets, and cable assemblies for charging of EVs up to 250A AC. and 400A DC. In IEC 62196-2, which was published in 2011, three plug types for AC charging have been specified:

- **Type 1**: single-phase vehicle coupler based on the SAE J1772/2009 automotive plug specifications, which is proposed by the SAE International and specific to North American and Japan as a 120 V option;
- **Type 2**: single-and three-phase vehicle coupler based on the VDE-AR-E 2623-2-2 plug specifications, which is developed by the connector manufacturer Mennekes and standardized by the German DKE/VDE (German Commission for Electronics of the Association for Electrical, Electronic, and Information Technologies).
- **Type 3**: single-and three-phase vehicle coupler with shutters based on the EV Plug Alliance proposal, which uses shutters for safety requirements of European countries.

In the incoming IEC 62196-3, which specifies the plug types of rapid DC charging and is still under development, the following plug types are under consideration:

- the Japanese CHAdeMO connector which is developed by the CHAdeMO Association and uses a CAN-based protocol for communications between EVs and EVSE;
- the US/German Combined Charging System combo connector which is developed as an extension of the SAE J1772 connector and uses the Homeplug GreenPHY power line carrier (PLC) protocol for communications;
- the Chinese connector (currently under development) which requires unique control and communication interface and uses CAN-based protocol for communications similar to CHAdeMO.

4.4.6.2 Standardization Efforts of the ANSI Electric Vehicles Standards Panel (EVSP)

The ANSI Electric Vehicles Standards Panel (EVSP) is formed in response to suggestions that the US standardization community needed a more coordinated approach to keep pace with EV initiatives moving forward in other parts of the world. Its main objective is to foster coordination and collaboration on standardization matters among public and private sector stakeholders to enable the safe, mass deployment of EVs and associated infrastructures in the United States with international coordination, adaptability, and engagement [41]. The EVSP has developed the *Standardization Roadmap for EVs – Version 1.0 (April 2012)*, and *ANSI EVSP Roadmap Standards Compendium – Version 1.0 (April 2012)* with various participants, including representatives of the automotive, electrotechnical, and utilities industries, relevant trade associations, standards development and conformity assessment organizations, and academic and government agencies.

The ANSI EVSP roadmap mainly focus on light duty, on-road PEVS, including BEV, PHEV, and EREV, while conventional hybrid EVs which are recharged by an

ICE are not considered. In order for EVs to be broadly successful, the roadmap has identified several major challenges, including safety, affordability, interoperability, performance, and environmental impact. The goals of the roadmap are as follows:

- Facilitate the development of comprehensive, robust, and streamlined standards and conformance landscape.
- Maximize the coordination and harmonization of the standards and conformance environment domestically and with international partners.

In this roadmap, standards and conformance activities are framed under three broad domains: Vehicles, Infrastructure, and Support Services. Within those three domains, seven broad topical areas related to standards and conformance programs for EVs were identified: Energy Storage, Vehicle Components, Vehicle User Interface, Charging Systems, Communications, Installation and Education, and Training. Important issues within the topical areas are in general highly interrelated and interdependent. Gaps related to safety, performance, and interoperability, and so on, which have not been addressed by any existing standard, code, regulation, or conformance program have been identified. Beside gaps, partial gaps which have been partially addressed by an existing standard, code, regulation, or conformance programs have also been identified. In conclusion, a total of 36 gaps or partial gaps and corresponding recommendations across the three domains and seven topical areas have been identified.

In addition, this roadmap is supplemented by the *ANSI EVSP Roadmap Standards Compendium*, a searchable spreadsheet which was developed to identify and assess the currently existing or developing standards, guidelines, codes, and regulations that relate to the safe, mass deployment of EVs and associated charging infrastructure and support services in the United States.

4.4.6.3   Standardization Efforts of the CEN-CENELEC Focus Group on European Electro-Mobility

The CEN-CENELEC have established a Focus Group on European Electro-Mobility, which comprises representatives of the CEN and CENELEC national members and of European-level associations related to electro-mobility. In October 2011, the Focus Group produced a report, titled *Standardization for Road Vehicles and Associated Infrastructure*, as a reply to the European Commission/EFTA Mandate M/468. In order to support coordination of standardization activities related to eMobility, the Focus Group established the CEN-CENELEC eMobility Co-ordination Group (eM-CG).

The report published by the CEN-CENELEC Focus Group has considered several most important issues related to eMobility, such as electrical vehicle charging modes, connection systems for charging, smart charging, communications, batteries, EMC, regulations, and standards. This report provides the specific standardization requirements for European electro-mobility, and an overview of current activities, necessary fields of action, international cooperation, and strategic recommendations [42]. Furthermore, this report also provides a list of standards developed by SDOs such as CEN, CENELEC, ISO, IEC, SAE, and UL. An example of this list is shown in Table 4.7 [42]. The column "Type" shows the types of standards, that is, A-General Information, B-Test Methods, C-Safety or EMC requirements, D-Miscellaneous. The column "Class" shows the importance of standards, with "1" denoting the standards of utmost importance, and "4" denoting the superseded standards. The "Technical domain/Standardization corpus" column shows the related sector of EV-HEV for which the standards have been proposed. In summary, the Focus Group has recommended the following standardization activities should be given priority: safety of charging installations, plug-in interoperability, EMC provisions for charging station and vehicles, communication protocols for V2G, and quick battery exchange.

In December 2012, a Transatlantic Roundtable was organized by CEN, CENELEC, and ANSI to hold discussions on the four key areas: coupler safety and interoperability of fast charging, V2G communications, wireless charging, and safety of EV infrastructure and batteries. Information sharing efforts have also been underway among standards organizations such as IEC, ISO, and SAE.

4.4.6.4   NIST Common Object Models for Electric Transportation (PAP 11)

NIST has initiated the Common Object Models for Electric Transportation project and listed it as the Priority Action Plan 11 (PAP 11). PAP 11 has developed the information models under Smart Energy 2.0 as a CIM extension. The objectives of PAP 11 are summarized as follows [43]:

- Extract interface requirements from enhanced and polished use cases (based on SAE and NIST workshop USE Cases).
- Draft high-level information model (CIM/IEC 61850) in Unified Modeling Language (UML).
- Improve collaboration with standards bodies for developing the PEV information exchange requirements.
- Cooperate with IEC to harmonize the IEC 61968 and 61850.
- Review the limitations of current regulations/use cases.
- Coordinate standards activities for electrical interconnection and safety standards for chargers and discharging, as well as weights and standards certification and seal for charging/discharging.

PAP 11 also supports energy storage integration with distribution grid as addressed by PAP 07, which is introduced in Section 4.4.1.

4.4.6.5   The Japanese CHAdeMO Standard

CHAdeMO is the DC fast charging standard, which provides EV drivers with an opportunity to charge within 5–10 minutes for a 40–60 km drive, and 80%

Table 4.7 An example of the identified existing standards impacting EV-HEV

Type	Class	Technical domain/ standardization corpus	EN (CEN)	EN (CENELEC)	ISO	IEC	SAE	UL
A	2	Electric road vehicle – vocabulary						
A	2	Electric road vehicle – vocabulary			ISO 8713: 2005 under revision			
A	3	Electrically propelled road vehicles – terminology	EN 13447: 2001		ISO 8713			
A	2	Graphical symbols for use on equipment				IEC 60417		
A	2	Basic and safety principles for man–machine interface, marking, and identification. Identification of conductors by colors or numerals		EN 60446		IEC 60446		
A	3	Degrees of protection provided by enclosures (IP code)		EN 60529		IEC 60529		

IP, Internet Protocol.

charge in less than 30 minutes [44]. The name CHAdeMO originates from Japan, standing for "Let's CHArge and Move" in English and "while having a cup of tea" in Japanese. The CHAdeMO association is comprised of more than 430 organizations in 26 countries around the world, representing international partners from multiple sectors. CHAdeMO is the world's first charging solution and as of September 2012, there were more than 1600 CHAdeMO chargers and more than 57 000 CHAdeMO-compatible EVs in operation around the world. In August 2012, the Japanese Industrial Standards Committee (JISC) decided to issue the CHAdeMO standard as a technical specification of the Japanese Industrial Standard (JIS). IEC has also include CHAdeMo in the drafts of IEC 61851-23 for charging systems, IEC 61851-24 for communication, and IEC 62196-3 for connector, which was published on 19 June, 2014.

4.4.6.6 Efforts of SAE International

SAE International is a global association for engineering professionals and researchers in the aerospace, automotive, and commercial-vehicle industries [45]. SAE International creates and manages more aerospace and ground vehicle standards than any other entity in the world. For EVs, SAE has published standards to provide references for performance rating of EV batteries, battery system safety, determination of the maximum available power from a rechargeable energy storage system, packaging of EV batteries, communications between EVs and utility grid and EVSE, communications between EVs and customers, interoperability with EVSE, and so on. In the NIST Roadmap 2.0, the SAE J2847 series, SAE J1772, and SAE J2836 series have been identified by the NIST Smart Grid Interoperability Panel (SGIP) working group as critical standards for the development of the Smart Grid.

SAE J2847 standards are critical to enable communications between PEVs and the utility grid, the EVSE, customers, and the utility for reverse power flow. Requirements for diagnostics between PEV and EVSE for charge or discharge sessions have been also specified.

The SAE J1772 standard specifies the conductive charge coupler for EV charging in North America, which has been included in the international IEC 62196-2 standard. It covers the general physical, electrical, functional, and performance requirements to facilitate conductive charging using a single-phase SAE J1772-2009 connector [46]. SAE has also developed a combo coupler for DC fast charging, which is proposed as the Combined Charging System to compete with the Japanese CHAdeMO standard. The Combined Charging System uses PLC technology for communication between the EV and utility grid and EVSE, as opposed to CHAdeMO which uses the CAN bus for communication. The Combined Charging System is likely to be included in the international IEC 62196-3 standard.

Some critical SAE standards for V2G and batteries have been summarized in Table 4.8.

Table 4.8 Critical SAE standards for V2G and batteries

Sector	Standard
V2G communication	SAE J2847
	Part 1: Communication between Plug-in Vehicles and the Utility Grid
	Part 2: Communication between Plug-in Vehicles and Off-board DC Chargers
	Part 3: Communication between Plug-in Vehicles and the Utility Grid for Reverse Power Flow
	Part 4: Diagnostic Communication for Plug-in Vehicles
	Part 5: Communication between Plug-in Vehicles and Their Customers
V2G communication use cases	SAE J2836
	Part 1: Use Cases for Communication between Plug-in Vehicles and the Utility Grid
	Part 2: Use Cases for Communication Between Plug-in Vehicles and the Supply Equipment (EVSE)
	Part 3: Use Cases for Communication between Plug-in Vehicles and the utility Grid for Reverse Power Flow
	Part 4: Use Cases for Diagnostic Communication for Plug-in Vehicles
	Part 5: Use Cases for Communication between Plug-in Vehicles and Their Customers
V2G EVSE communication	SAE J2931
	Part 1: Electric Vehicle Supply Equipment Communication Model.
	Part 2: Inband Signaling Communication for Plug-in Electric Vehicles.
	Part 3: PLC Communication for Plug-in Electric Vehicles
	Part 4: Broadband PLC Communication for Plug-in Electric Vehicles
	Part 5: Telematics Smart Grid Communications between Customers, Plug-In Electric Vehicles (PEV), Energy Service Providers (ESP) and Home Area Networks (HAN)
	Part 6: Digital Communication for Wireless Charging Plug-in Electric Vehicles
	Part 7: Security for Plug-in Electric Vehicle Communications
V2G interoperability	SAE J2953 Plug-in Electric Vehicle Interoperability with Electric Vehicle Supply Equipment (EVSE)
V2G energy transfer systems	SAE J2293 Energy Transfer System for Electric Vehicles
	Part 1: Functional Requirements and System Architectures.
	Part 2: Communication Requirements and Network Architectures
V2G rechargeable energy storage system	SAE J2758 Determination of the Maximum Available Power from a Rechargeable Energy Storage System on a Hybrid Electric Vehicle
V2G safety	SAE J2344 Guidelines for Electric Vehicle Safety
V2G emission and fuel economy measurement	SAE J1711 Recommended Practice for Measuring the Exhaust emissions and fuel economy of hybrid electric vehicles

(*continued overleaf*)

Table 4.8 (*continued*)

Sector	Standard
Battery vibration testing	SAE J2380 Vibration Testing of Electric Vehicle Batteries
Battery abuse testing	SAE J2464 Electric Vehicle Battery Abuse Testing
Battery packaging	SAE J1797 Recommended Practice for Packaging of Electric vehicle battery modules
Battery performance rating	SAE J1798 Recommended Practice for Performance Rating of Electric Vehicle battery modules
Battery life cycle testing	SAE J2288 Life Cycle Testing of Electric Vehicle Battery Modules
Battery functional guidelines	SAE J2289 Electric-Drive Battery Pack System: Functional Guidelines
Battery performance levels and methods of measurement	SAE J551/5 Performance Levels and Methods of Measurement of Magnetic and Electric Field Strength from Electric Vehicles, Broadband, 9 kHz to 30 MHz
Battery storage	SAE J537 Storage Batteries
Battery labeling	SAE J2936 Vehicle Battery Labeling Guidelines
Battery system safety	SAE J2929 Electric and Hybrid Vehicle Propulsion Battery System safety standard
Battery control system	Battery Electronic Fuel Gauging Recommended Practices

4.4.6.7 Efforts of Underwriters Laboratories

The UL was founded in 1894 and is the world's largest safety testing and certification organization. UL works with a diverse group of stakeholders and has developed more than 1000 standards for safety. UL has been working on safety standards for EVs and working with manufacturers to develop safer EV equipment. UL has also launched the EV charging equipment installation training program.

4.4.6.8 Other EV Projects

Some projects and their short descriptions have been summarized in Table 4.9.

4.5 Conclusion

This chapter provided an overview of ES, DER, and EV technologies, applications, and standardization status. Some of the ES technology options discussed in this chapter such as PHS and CAES have already been deployed around the world, while some other emerging technologies are still in the research and development stage. However, despite the popularity of PHS and CAES, they require specific geographical conditions which limit their implementations. We also face the same situation when choosing the suitable DER technologies. Conventional DER technologies, such as the reciprocating engine and the combustion turbine, have been widely deployed

Table 4.9 A summary of EV projects

Project name	Description
Open V2G project	The main scope of the **openV2G project** is to provide an open source implementation of the latest draft of "vehicle 2 grid communication interface" (V2G CI) included in the ISO/IEC 15118 standard [47]. The openV2G library allows messages to be exchanged between BEV or PHEV and EVSE followed by the DIN 70121 standard
MERGE project	The main scope of the MERGE project is to evaluate the impacts that EVs will have on the EU electric power systems regarding planning, operation, and market functioning. A management and control concept, that is, the MERGE concept, will be developed [48]
G4V project	The G4V Project aims to evaluate the impacts of the mass introduction of EVs on the grid infrastructure and a visionary "roadmap" for the year 2020 and beyond [49]
EDISON project	The Danish EDISON project has been launched to study how EVs can be used to provide the required balancing power for increasing the production of renewable energy to 50% of total generation by 2020. EDISON is the abbreviation for "Electric Vehicles in a Distributed and Integrated Market Using Sustainable Energy and Open Networks"
Ten cities thousand vehicles project	The Chinese Ten Cities Thousand Vehicles Project was initiated by Chinese government ministries to expand to 10 pilot cities each year over a period of three years, bringing at least 1000 EVs into operation in each pilot city
University of Delaware V2G project	The V2G project led by Professor Willett Kempton from University of Delaware aims to develop technologies, policies, and market strategies to achieve values brought by Grid Integrated Vehicle (GIV) and V2G
MOLECULES project	The MOLECULES project, funded by the European Commission, aims to use ICT services to help achieve a consistent, integrated uptake of Smart Connected Electromobility (SCE) with three pilot cities, that is, Barcelona, Berlin, and Grand Paris [50]
KAIST Online Electric Vehicle (OLEV) project	The Online Electric Vehicle (OLEV) project, conducted by the Korea Advanced Institute of Science and Technology (KAIST), aims to develop a noncontact magnetic charging method for charging EVs

DIN, Deutsches Institut für Normung

around the world and will likely remain competitive in many applications, while some emerging technologies are still in the research and development stage and have not fully demonstrated their potential benefits. Therefore, for real applications, the benefits brought by a technology should be compared against the total cost of installation and maintenance.

We believe that with technology advancement, some emerging ES and DER technologies can be cost-effective enough to be widely deployed by either end consumers or utilities, or both. This chapter also provided an overview of various EV technologies, which can mitigate some of the negative consequences of oil dependency on economy, national security, and the environment. EVs can be integrated with the grid to enable many applications such as smoothing variable generation from renewable sources, peak load shifting, voltage control, and frequency regulation, and DERs. However, despite the benefits brought by EVs, the mass adoption of the EV still faces challenges such as high cost, concern over range, lack of charging infrastructures, safety problems, and gaps between standards and market needs.

ES, DER, and EVs are interconnected and interdependent on each other. ES interconnected with DER, that is, ES-DER, has been identified by the Federal Energy Regulatory Commission (FERC) as a key technology of the Smart Grid [3]. Furthermore, EVs can be treated as ES during charging periods, that is, G2V, or treated as DER when EVs are used to provide electricity, that is, V2G. ES, DER, and EVs make it necessary to address the interconnection and interoperability issues for the design of the Smart Grid system. Therefore, in order to bridge the gap between the Smart Grid requirements and the existing standards, NIST has initiated the Priority Action Plan 07 (PAP 07) to involve a broad set of stakeholders and SDOs to address the ES-DER interconnection issues. IEEE has updated and revised the existing IEEE 1547 interconnection standards and issued the IEEE P2030.2 standard to address the interoperability issues of ES systems. IEC has extended the IEC 61850 information models to the field of PHS and DER. Numerous V2G and G2V projects have been launched to address issues related to the integration of EVs with the electric grid. Efforts have been made by IEC, IEEE, NIST, SAE International, CEN/CENELEC, ANSI, and so on, to promote the development of interoperable EV standards. In the future, continued R&D and standardization efforts and strategic partnership among auto manufactures, charging infrastructure operators, utilities, standardization bodies, customers, and government institutions are needed to realize the full potential of ES, DER, and EVs.

References

[1] U.S. Energy Information Administration (2013) *World Hydroelectricity Installed Capacity from 2006–2010*, www.eia.gov/cfapps/ipdbproject/iedindex3.cfm?tid=2&pid=33&aid=7&cid=ww,& syid=2006& yed=2010&unit=MK (accessed 2 December 2012).

[2] Hasnain, S.M. (1998) Review on sustainable thermal energy storage technologies. *Energy Converts*, **39** (11), 1127–1153.

[3] NIST.(2013) *NIST Energy Storage Interconnection Guidelines (6.2.3)*, www.nist.gov/smartgrid/ upload/7-Energy_Storage_Interconnection.pdf (accessed 21 December 2012).

[4] Institute of Electrical and Electronic Engineering (2003) IEEE 1547. *Standard for Interconnecting Distributed Resources with Electric Power Systems*, IEEE.

[5] Institute of Electrical and Electronic Engineering (2013) IEEE P2030.2. *Draft Guide for the Interoperability of Energy Storage Systems Integrated with the Electric Power Infrastructure*, IEEE.

[6] IEC SMB Smart Grid Strategic Group (2010) *IEC Smart Grid Standardization Roadmap Edition 1.0*, www.iec.ch/smartgrid/downloads/sg3_roadmap.pdf (accessed 27 December 2012).

[7] Rastler, D. (2004) *Distributed Energy Resources: Current Landscape and a Roadmap for the Future*. EPRI Technical Update.

[8] Commonwealth Scientific and Industrial Research Organization (2011) *"THINK SMALL" The Australian Decentralised Energy Roadmap 1st Issue*, http://igrid.net.au/resources/downloads/

project4/Australian_Decentralised%20Energy_Roadmap_December_2011.pdf (accessed 5 December 2012).

[9] National Institute of Standards and Technology (2012) *NIST Framework and Roadmap for Smart Grid Interoperability Standards, Release 2.0*, www.nist.gov/smartgrid/upload/NIST_Framework_Release_2-0_corr.pdf (accessed 2 December 2012).

[10] CEN/CENELEC/ETSI Joint Working Group (2011) *Final Report of the CEN/CENELEC/ETSI Joint Working Group on Standards for Smart Grids*, .ftp://ftp.cen.eu/CEN/Sectors/List/Energy/SmartGrids/SmartGridFinalReport.pdf. (accessed 27 December 2012).

[11] Rastler, D. (2004) *Economic Costs and Benefits of Distributed Energy Resources*. EPRI Technical Update, Energy and Environmental Economics Inc., San Francisco, CA.

[12] Herman, D. (2003) *Installation, Operation, and maintenance Costs for Distributed Generation Technologies*. EPRI Technical Report 1007675. (http://www.epri.com/abstracts/Pages/ProductAbstract.aspx?ProductId=000000000001007675)

[13] European Commission (2002) *Distributed Generation with High Penetration of Renewable Energy Sources Project*, www.dispower.org (accessed 10 December 2012).

[14] European Commission. (2002) *The EU Frame Programme 5 MICROGRIDS Project*, http://microgrids.power.ece.ntua.gr/ (accessed 10 December 2012).

[15] European Commission (2006) *The European Distributed Energy Partnership Project*, http://cordis.europa.eu/search/index.cfm?fuseaction=result.document&RS_LANG=ES&RS_RCN=12477109&q= (accessed 10 December 2012).

[16] European Commission (2006)*The Integration of Renewable Energy Sources and Distributed Generation into European Electricity Grid Project*, www.ired-cluster.org/ (accessed 10 December 2012).

[17] European Commission (2006) *The Flexible Electricity Network to Integrate the Expected "Energy Evolution" Project*, www.fenix-project.org/ (accessed 12 December 2012).

[18] European Commission (2013) *The Integrating Renewables in the European Electricity Grid Project*, www.iberdrola.es/ (accessed 12 December 2012).

[19] European Commission (2012) *The Active Distribution Networks with Full Integration of Demand and Distributed Energy Resources Project*, www.addressfp7.org/ (accessed 13 December 2012).

[20] Transport & Environment. (2012) *CO_2 Emissions from Transport in the EU27*, www.transportenvironment.org/sites/te/files/media/2009%2007_te_ghg_inventory_analysis_2007_data.pdf (accessed 15 December 2012).

[21] U.S. Energy Information Administration (2012) *US Emissions of Greenhouse Cases Report*, www.eia.gov/oiaf/1605/ggrpt/carbon.html#transportation. (accessed 15 December 2012).

[22] Ipakchi, A. and Albuyeh, F. (2009) Grid of the future. *IEEE Power and Energy Magazine*, **7** (2), 52–62.

[23] Vojdani, A.F. (2008) Smart integration. *IEEE Power and Energy Magazine*, **6** (6), 72–79.

[24] Ford (2012) *Model T Facts*, http://media.ford.com/article_display.cfm?article_id=858 (accessed 15 December 2012).

[25] Accenture (2011) *Plug-in Electric Vehicles: Charging Perceptions, Hedging Bets*.

[26] Deloitte Global Services Ltd (2011) *Gaining Traction: Will Consumers Ride the Electric Vehicle Wave?*

[27] Produced for the United States Securities and Exchange Commission (2010) *10-K: Annual Report Pursuant to Section 13 and 15 (d)*, Enerl, Inc., New York.

[28] Boston Consulting Group (2010) *Batteries for Electric Cars: Challenges, Opportunities, and the Outlook to 2020*, Boston Consulting Group,Detroit, MI.

[29] MIT Energy Initiative (2010) *Electrification of the Transportation System*, MIT, Cambridge, MA.

[30] Ralston, M. and Nigro, N. (2011) *Plug-in Electric Vehicles: Literature Review*, www.c2es.org/docUploads/PEV-Literature-Review.pdf (accessed 16 December 2012).

[31] USABC (2012) *USABC Goals for Advanced Batteries for EVs*, www.uscar.org/commands/files_download.php?files_id=27 (accessed 17 December 2012).

[32] USABC (2012) *FreedomCAR Energy Storge System Performance Goals for Power Assist Hybrid Electric Vehicles*, www.uscar.org/commands/files_download.php?files_id=83 (accessed 17 December 2012).

[33] USABC. (2012) *USABC Requirements of End of Life Energy Storage Systems for PHEVs*, www.uscar.org/commands/files_download.php?files_id=156 (accessed 17 December 2012).

[34] Srinivasan, V. (2012) *Battery Choices for Different Plug-in HEV Configurations*, .www.nrel.gov/ vehiclesandfuels/energystorage/pdfs/40378.pdf. (accessed 17 December 2012).

[35] Periyaswamy, P. and Vollet, P. (2011) *The Electric Vehicle: Plugging in to Smarter Energy Management. Schneider Electric White Paper*, www2.schneider-electric.com/documents/ support/white-papers/electric-vehicle-smarter-energy-management.pdf (accessed 18 December 2012).

[36] SAE International (2011) *SAE Charging Configurations and Ratings Terminology*, www.sae .org/smartgrid/chargingspeeds.pdf (accessed 18 December 2012).

[37] Becker, T.A., Ikhlaq, S., and Burghardt, T. (2009) *Electric Vehicles in the United States: A New Model with Forecasts to 2030.*

[38] Kohchi, A. (1996) System apparatus for battery swapping. US Patent US005585205A.

[39] Yarow, J. (2009) *The Cost of a Better Place Swapping Station: $500,000. Business Insider*, www.businessinsider.com/the-cost-of-a-better-place-battery-swapping-station-500000-2009-4 (accessed 20 December 2012).

[40] Andersen, P.B., Hauksson, E.B., Pedersen, A.B. *et al.* (2012) Smart charging the electric vehicle fleet, in *Smart Grid – Applications, Communications, and Security* (eds L.T. Berger and K. Iniewski), John Wiley & Sons, Ltd, Chichester, pp. 381–408.

[41] American National Standards Institute (2012) *ANSI Electric Vehicles Standards Panel (EVSP)*, www.ansi.org/standards_activities/standards_boards_panels/evsp/overview.aspx?menuid=3 (accessed 20 December 2012).

[42] CEN/CENELEC Focus Group on European Electro-Mobility (2011) *Standardization for Road Vehicles and Associated Infrastructure Version 2.0.*

[43] National Institute of Standards and Technology (2010) *NIST PAP11 Common Object Models for Electric Transportation*, http://collaborate.nist.gov/twiki-sggrid/bin/view/SmartGrid/PAP11PEV (accessed 12 December 2012).

[44] CHAdeMo (2012) *CHAdeMO Long Brochures* www.chademo.com/wp/wp-content/uploads/2012/ 12/Brolong.pdf (accessed 10 January 2013).

[45] SAE International www.sae.org/ (accessed 10 January 2013).

[46] SAE International (2012) SAE J1772. *SAE Electric Vehicle and Plug in Hybrid Electric Vehicle Conductive Charge Coupler*, SAE International.

[47] OpenV2G Project http://openv2g.sourceforge.net/ (accessed 14 January 2013).

[48] European Commission (2012) *Mobile Energy Resources in Grids of Electricity (MERGE) Project Homepage*, www.ev-merge.eu/ (accessed 14 January 2013).

[49] European Commission (2012) *Grid for Vehicle (G4V)*, www.g4v.eu/ (accessed 15 January 2013).

[50] European Commission (2012) *Mobility based on Electric Connected Vehicles in Urban and Interurban Smart Clean Environments (MOLECULES)*, www.molecules-project.eu/ (accessed 15 January 2013).

5

Smart Energy Consumption

5.1 Introduction

Demand Response (DR), Advanced Metering Infrastructure (AMI), and Smart Home and Building Automation are important systems to enable smart energy consumption by consumers. These three systems are closely interconnected and interdependent on each other. DR provides customers with the ability to manage load demand in response to supply conditions, such as the price of electricity. AMI communicates the record of energy consumption data of utilities such as electricity, gas, and water to both users and utilities through various transmission media. These data collected by AMI can be displayed in a useful form to make consumers more aware of their energy usage. Smart home and building automation systems automatically control the operation of customers' devices in response to supply conditions such as electricity price. This chapter provides a brief introduction into the various DR, AMI, and smart home and building automation technologies, specifications, and standards developed by different organizations.

This chapter also introduces the major barriers for fully realizing the potential benefits of DR, AMI, smart home and building automation, and efforts made by National Institute of Standards and Technology (NIST), ZigBee, Open Automated Demand Response (OpenADR), Organization for the Advancement of Structured Information Standards (OASIS), European Committee for Standardization (CEN), European Committee for Electrotechnical Standardization (CENELEC), and European Telecommunications Standards Institute (ETSI) to address the interoperability issues. The organization of this chapter is as follows: Section 5.2 introduces DR technologies, implementation barriers, and standardization efforts; Section 5.2.3.5 introduces the two major AMI standards: the IEC 62056 and ANSI C12 standards (American National Standards Institute), and the various metering standardization projects and efforts initiated by CEN, CENELEC, ETSI, NIST, International Electrotechnical Commission (IEC), International Telecommunication Union (ITU),

Smart Grid Standards: Specifications, Requirements, and Technologies, First Edition. Takuro Sato,
Daniel M. Kammen, Bin Duan, Martin Macuha, Zhenyu Zhou, Jun Wu, Muhammad Tariq and Solomon Abebe Asfaw.
© 2015 John Wiley & Sons, Ltd. Published 2015 by John Wiley & Sons, Ltd.

UtilityAMI, and so on. Section 5.3.3.5 provides an introduction into numerous smart home and building automation standards, including ISO (International Organization for Standardization)/IEC Information Technology-Home Electronic System (HES), ZigBee/HomePlug Smart Energy Profile (SEP) 2.0, Z-Wave, Energy Conservation and HOmecare NETwork (ECHONET), ZigBee Home Automation (ZHA) Public Application Profile, Building Automation and Control Network (BACnet), LONWORKS, INSTEON, KNX, and ONE-NET.

5.2 Demand Response

5.2.1 An Overview of Demand Response Technologies

DR is an important feature of the Smart Grid, which is closely connected to other Smart Grid systems such as AMI, smart home and building automation systems, Distributed Energy Resource (DER), and electric storage. DR includes Direct Load Control (DLC) programs that are implemented by utilities to adjust or control the load to balance supply and demand. DLC brings many benefits such as shifting peak demand, reducing electricity cost, and eliminating the need for large capital expenditures. However, it has a limited influence on the control of individual loads due to user privacy. In comparison, indirect load-control programs provide customers with the ability to manage load demand in response to supply conditions, such as the price of electricity. In DLC programs implemented by utilities, the operation of a home appliance is remotely controlled by the utility or an aggregator and responds passively to control signals. In comparison, in indirect load-control programs implemented by customers, the operation of a home appliance is not directly controlled by utilities. Instead, the decision of operation is made by the users themselves and is usually based upon cost concerns, that is, to reduce electricity cost by turning off some devices when the electricity price is high. In conclusion, DR can significantly decrease the peak demand for electricity, thus reducing overall plant and capital cost requirements.

DR can be performed in two aspects:

- **Fast DR**: energy demands need to be balanced at near real time or real time, for example, frequency stabilizing applications require a response time of several seconds.
- **Slow DR**: slow response such as the day-ahead DR where the signals are sent significantly before the events are called and it has a response time of several days.

Compared to slow DR, fast DR requires seamless communications between consumer devices and the power grid, fully automated load control activities, and consistent signals for DR. Furthermore, the advancement of other Smart Grid technologies such as DER, Electric Vehicle (EV), electric storage, and so on, make DR systems more complex and difficult to realize. This section provides a brief overview of the DR technologies, major barriers for fully realizing the potential benefits of DR, and efforts to address the interoperability issues. The organization of this section is as follows: Section 5.2.2 gives a brief introduction of DER technologies and barriers; Section 5.2.3 introduces the standardization projects and efforts initiated by various Standards Developing Organizations (SDOs) to address the DR interoperability issues.

5.2.2 Demand Response Technology and Barriers

A broad set of pricing models have been developed to engage customers in DR: Real-Time Pricing (RTP), Critical Peak Pricing (CPP), Variable Peak Pricing, (VPP), Time-Of-Use Pricing (TOUP), Day-Ahead Pricing (DAP), and so on. In all of these pricing models, the key idea is twofold: firstly, allowing electricity prices to fluctuate to reflect changes in electricity supply and demand; secondly, encouraging customers to shift the usage of high energy consumption home appliances to offpeak hours [1]. The pricing-incentive-based DR decisions are not easily predictable because they also depend on whether a consumer will change his/her own electricity consumption habit according to the set price incentives or not. The pricing incentive itself is not sufficient in the long run to enable mass adoption of DR devices. Furthermore, recent studies have shown that the lack of knowledge among users about how to respond to complex time-varying electricity prices and the lack of effective automation control systems are the two major barriers for realizing the maximum benefits of DR programs [2, 3]. Therefore, it is important to integrate DR systems with smart home and building automation systems, AMI systems, and DER systems to make full use of the flexibility and energy storage and generation options for consumers.

The integration of DR systems with these Smart Grid systems is illustrated in Figure 5.1. DR systems are closely connected with DER systems, AMI systems, and smart home and building automation systems. DER offers economical and ecological

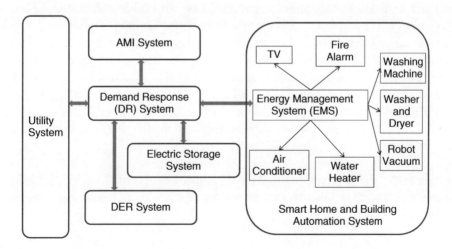

Figure 5.1 Systems related to the demand response (DR) system

benefits by generating energy close to the consumers to minimize the transmission and distribution losses and providing continuous and reliable supply of electricity to customers when a power outage occurs. AMI communicates the energy consumption data record of utilities such as electricity, gas, and water to both users and utilities through various transmission media. These data collected by AMI can be displayed in useful forms to make consumers more aware of their energy usage. Smart home and building automation systems automatically control the operation of customers' devices in response to supply conditions such as electricity price. The Energy Management System (EMS) within a home is responsible for carefully scheduling the energy consumption for each home device to minimize the total electricity bill. The Energy Services Interface (ESI) in a home device enables DR applications through bidirectional communications with EMS and the automatic control and coordination functions of the utilities. The DR messages are sent to an ESI to identify the device that will participate in the DR event. The identified home device receives the control signal initiated either by the customer EMS or by remote utilities and adjusts the load accordingly. Therefore, the bidirectional communication between the DR service providers and customers is one of the most important factors to realize both pricing-related DR programs and load-management-related DR programs. The standardization of open, consistent, and transparent DR signals has been initiated by several SDOs and is still underway, as introduced in the next section.

5.2.3   Standardization Efforts Related to Demand Response

5.2.3.1   OpenADR 2.0 Profile Specification

The Open Automated Demand Response Communications Specification (OpenADR) was developed by Lawrence Berkeley National Laboratory and California Energy Commission to define the communication protocols for enabling DR applications. The OpenADR 1.0 specification was accepted as part of the OASIS Energy Interoperation (EI) standard, which was developed to define an information and communication model to enable DR and energy transactions. OpenADR 2.0 was developed to define profiles that are specific to DR and DER applications. In particular, OpenADR 2.0 supports the following eight services to enable DR and DER applications: registration (EiRegisterParty), enrollment (EiEnroll), market contexts (EiMarketContext), event (EiEvent), quote or dynamic prices (EiQuote), reporting or feedback (EiReport), availability (EiAvail), and opt or override (EiOpt). OpenADR 2.0 is comprised of the following three different profile subsets:

- **OpenADR 2.0a**: OpenADR 2.0a Feature Set Profile was developed for resource-constrained or low-end embedded devices, which only need to support limited EiEvent services.

- **OpenADR 2.0b**: OpenADR 2.0b Feature Set Profile was developed for advanced DR devices, which need to support most services including EiEvent, EiReport, EiRegisterParty, and EiOpt.
- **OpenADR 2.0c**: OpenAFR 2.0c Feature Set Profile was developed for the most sophisticated demand respond devices, which support all DR services. OpenAFR 2.0c is still under development.

In OpenADR, nodes or devices are classified into two types: Virtual Top Node (VTN) and Virtual End Node (VEN). VTNs are responsible for announcing DR events and VENs control the electrical energy demand in response to the DR events. Bidirectional communications between VTN and VEN are necessary to enable DR services. A VEN may serve as a VEN in one interaction and take the role of a VTN in another interaction, depending on the deployment scenario.

5.2.3.2   ZigBee/HomePlug Smart Energy Profile (SEP) 2.0

The ZigBee/HomePlug SEP 2.0 is an application-layer profile to enable open, standardized, and interoperable information flow between meters, smart appliances, plug-in electric vehicle (PEV), EMS, electric storage systems, and DER. It has specified the application-layer business objective technical requirements for demand response and load control (DRLC). More details of SEP 2.0 are introduced in Section 5.4 Smart Home and Building Automation.

5.2.3.3   OASIS Energy Interoperation 1.0 and Energy Market Information Exchange (EMIX) 1.0 Specifications

The OASIS is a consortium to drive the development, convergence, and adoption of open standards for areas including security, cloud computing, the Smart Grid, and emergency management. The OASIS EI 1.0 standard was developed to define an information and communication model to enable DR and energy transactions. The OASIS EI 1.0 has specified requirements for exchange of Smart Grid signals, including dynamic pricing signals, reliability signals, emergency signals, and load predictability signals.

The OASIS Energy Market Information eXchange (EMIX) 1.0 has been developed to standardize messages communicating market information including energy price, delivery time, characteristics, availability, and schedules. All of these various types of information are necessary to enable the full automation of DR decision making. The OASIS EMIX 1.0 has been included as part of the NIST Priority Action Plan (PAP), and is now facilitated by the NIST Smart Grid Interoperability Panel (SGIP).

5.2.3.4   CEN/CENELEC/ETSI Joint Efforts

In the final report released by the CEN/CENELEC/ETSI Joint Working Group [4], the gaps between existing DR standards and Smart Grid requirements have been identified. The standards that define consistent signals and process interfaces as part of the Common Information Model (CIM), Companion Specification for Energy Metering (COSEM), and IEC 61850 for successfully deploying DR applications are missing. Different standardization bodies are working on integrating EVs with DR applications in parallel, that is, ISO/IEC joint WG, Working Group, V2G is in charge of defining the communication interface between EVs and grid, and the European CEN/CENELEC/ETSI is working on EV charging systems for European Union (EU). In order to fill these gaps, CEN/CENELEC/ETSI has recommend related SDOs to focus on defining subfunctions of DR including main system level DR use cases in a first stage rather than focus on broad and business-model-dependent specifications.

5.2.3.5   NIST SGIP Priority Action Plan 09

The NIST SGIP PAP 09 Standard DR and DER Signals has been initiated to define consistent signals for DR and DER, including pricing signals, grid safety or integrity signals, and DER support signals. NIST has cooperated with various SDOs including the ZigBee/HomePlug Alliance, OASIS, OpenADR Alliance, North American Energy Standards Board (NAESB), and IEC to develop a common semantic model for standard DR signals. Consistent DR signals are necessary to improve the responsiveness of the entire power generation and delivery systems to take advantage of renewable and other intermittent resources. The PAP 09 WG has conducted a survey of existing DR standards and efforts to identify their overlap and gaps related to DR signaling. The standardization work is done in alignment with IEC 61968 and incorporates work from OpenADR and OASIS, that is, OpenADR 2.0 and OASIS EI 1.0.

5.3   Advanced Metering Infrastructure Standards

AMI integrates Smart Grid infrastructure with smart metering, which is not a single technology implementation but rather a fully configured infrastructure. Different standardization organizations or groups have similar definitions of AMI. For example, in [5], AMI refers to systems that measure, collect, analyze, and control energy distribution and usage, with the help of advanced energy distribution automation devices such as distribution network monitoring and controlling devices, network switching devices, load/source-shedding devices, and electricity/gas/water meters through various communication media on request or on a predefined schedule. In [6], AMI refers to the full measurement and collection system that includes meters at the customer site, communication networks between the customer and a service provider, such as an electric, gas, or water utility, and data reception and management systems that make the information available to the service provider.

In this section, we introduce key standards proposed by different organizations for AMI. The problem of how to describe a common set of requirements within these standards to facilitate exchange of confidential and authentic information across standards must be solved for the realization of the Smart Grid. Therefore, the metering standardization projects and efforts initiated by various organizations and groups, including IEC, ITU, IEEE, NIST, Utility AMI, and European Standards' Organizations (ESOs) such as CEN, CENELEC, and ETST will also be covered.

Table 5.1 contains all these standards, which are categorized according to their function fields. Each standard is then explained in detail. Section 5.3 describes the AMI system. Section 5.3.1 describes the two most popular standard suites: IEC 62056 and ANSI C12. Section 5.3.2.2 describes the metering standardization projects and efforts initiated by various organizations and groups.

5.3.1 The AMI System

The AMI system is made up of the smart meter, communication module, Data Concentrator (DC), and Meter Data Management System (MDMS). The AMI system diagram is shown in Figure 5.2. At the consumer level, the energy consumption data are communicated to both the user and the utility by smart meters. Smart meters have the ability to transmit the collected data through different media. The meter data are received by the DC and sent to the MDMS. MDMS manages data storage and analyzes the consumption data to provide the information in useful form to the service provider. Detailed and timely meter information enables the service provider to support better outage detection, to rapidly address grid deficiencies, and to improve management of utility assets and asset maintenance. Smart meters can also communicate with in-home displays (IHDs) through the Home Area Network (HAN) to make consumers more aware of their energy usage. Detailed and timely electric pricing information can also be provided by the service provider to enable users to modify their energy usage to reduce cost and environmental impacts. The HAN gateway provides a capability for the service provider to connect with home appliances in the customer's premises. It provides added functionality and facilitates a broader range of DR benefits. For example, the service provider can remotely turn off or turn on the power of some loads to optimize energy usage.

5.3.2 The IEC 62056 and ANSI C12 Standards

5.3.2.1 IEC 62056 Standards

IEC 61107 is a communication protocol widely used for smart meters in the European Union and is superseded by IEC 62056. IEC 61107 is a half-duplex protocol

Table 5.1 Standard list of advanced metering infrastructure (AMI)

Function field	Standard name	Short introduction
Product	IEC 62051	Electricity metering – glossary of terms
Product	IEC 62052-11, 62052-21, 62052-31	General requirements, test and test conditions for electricity metering equipment (AC)
Product	IEC 62053 series	Particular test requirements and test methods for electricity metering equipment
Product	IEC 62054-11, 62054-21	Tariff and load control requirements for electricity metering
Product	IEC 62058-11, 62058-21, 62058-31	Acceptance inspection requirements for electricity metering equipment
Product	IEC 61968-9	Interfaces for meter reading and control
Transmission	IEC 61334	Metering automation using narrowband PLC
Transmission	EN 13757	Communication systems for remote reading of meters based on M-bus
Transmission	PRIME	Iberdrola specs-based PLC modem standard for smart meters
Transmission	ITU G3-PLC	ERDF specs-based PLC modem communication standard for smart meters
Transmission	HomePlug Netricity PLC	HomePlug PLC standards targeting smart meter to grid applications
Transmission	IEEE 802.15.4	Wireless WPANs PHY and MAC specification
Transmission	IEEE 802.11	Wireless LAN PHY and MAC specification
AMI	UtilityAMI high level requirements	High-level requirements for AMI
AMI	OPEN meter deliverables	A comprehensive set of open and public standards for AMI
Payment	IEC 62055 series	Payment systems for electricity metering
Reliability	IEC 62059	Dependability prediction and assessment methods for electricity metering equipment
Data exchange	IEC 62056 series	Data exchange for meter reading, tariff, and load control
Data exchange	ANSI C12 series	Standard suite of data formats, data structures, and communication protocols specified by ANSI for smart meters
Data exchange	EN 1434-3	Data exchange and interfaces for heat meters
Data exchange	AEIC Guideline V2.0	Guideline for vendors and utilities desiring to implement ANSI C12 standards
Data exchange	NEMA SG-AMI	Requirements for smart meter upgradability
Security	AMI-SEC (security) AMI system security requirement	Security requirements developed by the AMI-SEC Task Force for AMI

ERDF, Électricite Réseau Distribution France and LAN, local area network.

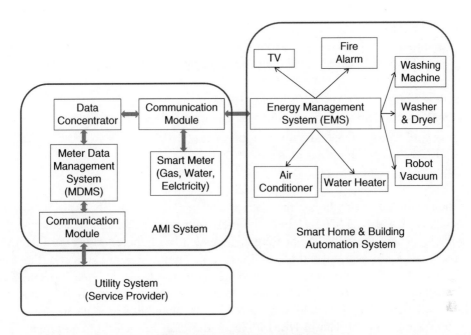

Figure 5.2   AMI system diagram

that sends ASCII data using a serial port such as a twisted pair EIA-485 or an optical port. The IEC 62056 series of standards is a more modern meter-reading protocol widely used in Europe. IEC 62056 is based on Device Language Message Specification (DLMS), which has been developed and maintained by the DLMS User Association. DLMS is comparable to a set of rules or a common language, which standardizes the communication profile, the data objects, and the object identification codes. COSEM defines an application layer protocol which specifies the procedures for the transfer of information for application association control, authentication, and for data exchange between COSEM clients and servers. The DLMS/COSEM specification defines data model, messaging, and communication protocol standards for data exchange over various communication media, for example, Public Switched Telephone Network (PSTN), Global System for Mobile communication (GSM), General Packet Radio Service (GPRS), Internet, and Power Line Carrier, (PLC). The DLMS User Association defines the DLMS/COSEM specification as four technical reports, that is, the Green Book, Yellow Book, Blue Book, and White Book:

- **Green Book**: COSEM Architecture and Protocols
  - **IEC 62056-53**: COSEM Application layer
  - **IEC 62056-47**: COSEM transport layers for Internet Protocol v4 (IPv4) networks
  - **IEC 62056-46**: Data link layer using HDLC, High-Level Data Link Control, protocol

- **IEC 62056-42**: Physical layer (PHY) services and procedures for connection-oriented asynchronous data exchange
- **IEC 62056-21**: Direct local data exchange describes how to use COSEM over a local port (optical or current loop)
- **Yellow Book**: COSEM Conformance Test Process
- **Blue Book**: COSEM Identification System and interface Objects
  - **IEC 62056-61**: Object identification system (OBIS)
  - **IEC 62056-62**: Interface classes
- **White Book**: COSEM Glossary of Terms.

Among all these standards, the data model and data identification standards, IEC62056-61 and IEC 62056-62 are the most important standards. They define the modeling of any metering application independently of the energy type (electricity/gas/water), messaging method, and communication media. The same data can be accessed in the same way from any meter using DLMS services specified in IEC 62056-53. This ensures the interoperability among meters of different vendors. The OBIS defined in IEC 62056-61 provides an unambiguous data identification system for all kinds of data in DLMS/COSEM-compliant metering equipment. The OBIS codes are used to identify COSEM object instances and data on the display of metering equipment. The object is a collection of attributes and methods. Attributes represent the characteristics of an object by means of attribute values, and methods specify the ways to either examine or modify the values of the attributes. Objects that share common characteristics are generalized as an interface class. IEC 62056-62 defines 19 interface classes for modeling the various functions of the meter, including demand registration, tariff and activity scheduling, handling time synchronization and power failures, power quality metering, and secure access to selected portions of the information at the metering equipment. In addition to the IEC 62056 for electricity metering, the DLMS/COSEM is also used for gas, water, and heat metering in the EN 13757 standard series. IEC has been working on an update of the IEC 62056 standards, and has issued the following new standards for IEC 62056 in 2013:

- **IEC 62056-76**: The three-layer, Connection-oriented HDLC-based Communication Profile
- **IEC 62056-83**: Communication Profile for PLC Spread-Frequency Shift Keying (S-FSK) Neighborhood Networks.

5.3.2.2 ANSI C12 Standards

The ANSI C12 Standard Suite is used for metering protocols in North America instead of the IEC 62056 standards used in Europe. The ANSI C12 standards include the following:

- **ANSI C12.18**: Protocol Specification for ANSI Type 2 Optical Port
- **ANSI C12.19**: Utility Industry End Device Data Tables

- **ANSI C12.21**: Protocol Specification for Telephone Modem Communication
- **ANSI C12.22**: Protocol Specification for Interfacing to Data Communication Networks.

The ANSI C12.19 standard specifies the Data Table Elements, which supports gas, water, and electric sensors and related appliances. The purpose of the tables is to define common structures for transport data between end devices by reading from or writing to a particular table or portion of a table. The tables are grouped together into sections called decades, which pertains to a particular feature-set and related function such as time-of-use and load profile. Compared to the previous ANSI C12.19-1997, the ANSI C12.19-2008 includes new tables, decades, syntax, XML-based (eXtensible Markup Language) table description language (TDL/TDL), and documentation of series supporting the needs of AMI.

ANSI C12.18 details the criteria required for transporting the tables defined in ANSI C12.19 between a C12.18 device and a C12.18 client via an optical port. The C12.18 client may be a handheld reader, a portable computer, a master station system, or some other electronic communication device that implements an ANSI Type 2 Optical Port for communication. In comparison, ANSI C12.21 provides detail of the criteria required for transporting the tables defined in ANSI C12.19 between a C12.21 device and a C12.21 client via a modem connected to the switched telephone network. The reason for adapting C12.18 to C12.21 is to be able to send and receive ANSI tables remotely through the telephone network. Unlike C12.18, which describes every detail of the physical attributes of optical ports (dimensions, wavelength, etc.), C12.21 omits many lower-layer details to achieve interoperability with existing telecommunication modems.

ANSI C12.22 defines the process of transporting ANSI C12.19 table data over a variety of networks. It uses Advanced Encryption Standard (AES) encryption to enable strong, secure communications, and it is also extensible to support additional security mechanisms. ANSI C12.22 provides both session and sessionless communications. Unlike C12.18 or C12.21 protocols, which only support session-oriented communications, the sessionless communication has the advantage of requiring less complex handling on both sides of the communication links and reduces the number of signaling overhead. ANSI C12.22 has a common application layer (layer 7 in the OSI, Open System Interconnection, reference model), which provides a minimal set of services and data structures required to support C12.22 nodes for the purposes of configuration, programming, and information retrieval in a networked environment. The application layer is independent of the underlying network technologies. This enables interoperability between C12.22 with already existing communication systems. C12.22 also defines a number of application layer services, which are combined to realize the various functions of the C12.22 protocols. The application layer services provided in C12.22 are

- **Identification Service**: This service is used to obtain information about C12.19 device functionality, including the reference standard, the version and revision of the reference standard implemented, and an optional feature list.

- **Read Service**: It is used to cause a transfer of table data to the requesting device. It allows both complete and partial table transfers, with the inclusion of additional error response codes and the addition of the capability to receive table data in excess of 65 535 bytes.
- **Write Service**: It is used to transfer table data to the target device. It allows both complete and partial table transfers.
- **Logon Service**: It is used to establish a session without establishing access permissions.
- **Security Service**: It is used to establish access permissions by a simple unencrypted password.
- **Logoff Service**: It is used to terminate the session that was established by the Logon Service.
- **Terminate Service**: It provides for an orderly abortion of the session that was established by the Logon Service.
- **Wait Service**: It is used to maintain an established session during idle periods to prevent automatic termination.
- **Disconnect Service**: It is used to remove a C12.22 node from the C12.22 network segment.
- **Registration Service**: It is used to add and keep routing-table entries of C12.22 Relays active.
- **Deregistration Service**: It is used to remove routing-table entries of C12.22 Relays.
- **Resolve Service**: It is used to retrieve the native network address of a C12.22 node.
- **Trace Service**: It is used to retrieve the list of C12.22 relays that have forwarded the specified C12.22 message to a target C12.22 node.

5.3.3 Metering Standardization Projects and Efforts

In order to promote the development and adoption of AMI, the problem of how to describe a common set of requirements within these standards to facilitate exchange of confidential and authentic information across standards must be solved for the realization of the Smart Grid. Therefore, the main focus of this section is on the metering standardization projects and efforts initiated by various organizations and groups, including CEN, CENELEC, ETSI, NIST, IEC, IEEE, ITU, and UtilityAMI.

5.3.3.1 European Commission Mandate M/441

The European Commission has issued the Mandate M/441 to the ESO CEN, CENELEC, and ETSI for the standardization of an open architecture for utility meters involving communication protocols enabling interoperability. The general objective is to enable interoperability of utility meters (water, gas, electricity, heat). The interoperability will enable mass production and full competition on the scale of the EU market to reduce the price of meters. CEN, CENELEC, and ETSI are requested to develop a European standard comprising a software and hardware open architecture for utility meters. The architecture must be scalable to support various applications and must be adaptable for future communication media. It should support secure bidirectional communication and allow advanced information and management and control systems for consumers and service suppliers. The standards should provide harmonized solutions within an interoperable framework, which should be based on communication protocols within an open architecture. CEN, CENELEC, and ETSI are advised to consider other international, European, and national standards, and any overlaps should be indicated. They have set up a Smart Meters Coordination Group (SM-CG) to respond to this request. In the final report of the CEN/CENELEC/ETSI Joint Working Group on Standards for Smart Grids [4], the following recommendations have been proposed:

- Various standards including EN 62056, EN 13757-1, and IEC 61968-9 are being developed to cover the exchange of metering data. However, some standardization initiatives go beyond the scope of M/441. It is necessary to prevent further development of different (competing) standards for smart metering.
- CEN/CENELEC/ETSI should take into consideration the use cases involving smart metering, building/home automation and EVs, and the standardization work in these areas.
- CEN/CENELEC/ETSI should jointly undertake an investigation of the interfaces related to the Smart Grid and e-Mobility to ensure harmonization with existing metering models and other relevant standardization initiatives.

5.3.3.2 OPEN Meter European Project

The OPEN meter project is a European collaborative project to develop a set of open and public standards for AMI in order to support metering based on the agreement of all the relevant stakeholders. In addition, all the relevant aspects such as regulatory environments, smart metering functions, communication media, protocols, and data formats are studied in this project. The result of the project will be a set of new standards complemented with the already existing and accepted standards such as IEC 62056 DLMS/COSEM Standards and EN 13757 to form the new body of AMI standards. The project should remove the barrier by developing a comprehensive set of open standards covering all utility commodities, all AMI requirements, and all communication media. The OPEN meter project can be divided into the following seven work packages (WPs):

- **WP1**: Functional Requirements and Regulatory Issues
- **WP2**: Identification of Knowledge and Technology Gaps
- **WP3**: Prenormative Research Activities
- **WP4**: Testing
- **WP5**: Specification and Proposal of a Standard
- **WP6**: Dissemination
- **WP7**: Coordination.

5.3.3.3 The UtilityAMI Working Group

The UtilityAMI Working Group was formed by the Utility Communications and Architecture International Users Group (UCAIUG) to address utility-specific issues related to AMI systems. UtilityAMI will develop high-level policy statements that define serviceability, security, and interoperability guidelines for AMI from a utility perspective. The specification development should use a common language that minimizes confusion and misunderstanding between utilities and vendors. The UtilityAMI Working Group has issued the following high-level requirements for AMI systems:

- **Standard Communications Board Interface**: the ability to adapt to various communication protocols and meters;
- **Standard Data Model**: the ability to exchange data between multiple vendors' equipment using the same system;
- **Security**: the ability to protect customers' data and information;
- **Two-Way Communications**: the ability to reliably send and receive data to and from the customer site;
- **Remote Download**: the ability to remotely update the metering settings and configurations;
- **Time-of-Use Metering**: the ability to record the usage information of the meter;
- **Bidirectional and Net Metering**: the ability to record energy flow in either direction and calculate net usage;
- **Long-Term Data Storage**: the ability to store all data within the meter for at least 45 days;
- **Remote Disconnect**: the ability to remotely disconnect or reconnect a customer's electrical service;
- **Network Management**: the ability to remotely manage the AMI communication network;
- **Self-healing Network**: the ability to detect and repair network problems automatically;
- **HAN Gateway**: the ability to act as a gateway for HAN devices;
- **Multiple Clients**: the ability to permit multiple clients to access the metering data;
- **Power-Quality Measurement**: the ability to measure and report the power-quality information;
- **Tamper and Theft Detection**: the ability to detect and report tampering or theft;
- **Outage Detection**: the ability to detect and report failures caused by power outages;
- **Scalability**: the ability to be scalable to any particular component;
- **Self-locating**: the ability to geographically locate the meter.

5.3.3.4 IEC Smart Grid Standards Development

The current IEC standards DLMS/COSEM are mainly concentrated on meter data exchange with metering units and have not fulfilled other requirements of the Smart

Grid such as power quality support, fraud detection, and load/source-shedding. TC 13 and TC 57 of IEC will work on the definition of the meter functions and the communication functions for the Smart Grid and smart meters. The IEC 61850 standards will be expanded to include the DLMS/COSEM objects, which would promote the coexistence of smart meter and Smart Grid applications. TC 8, TC 13, and TC 57 will jointly develop the set of objects and profiles of the common interfaces for different domains such as energy market, transmission and distribution, DERs, smart home, and E-mobility. IEC will also operate with relevant ISO/CEN TCs to extend the DLMS/COSEM object model and architecture to accommodate new requirements and new communication technologies. The coexistence with IEC 61334 PLC standards will be maintained and extended by taking the latest developments in smart metering into consideration.

5.3.3.5 NIST Priority Action Plans

NIST has established PAPs to address the standard-related gaps and issues. The AMI-related PAPs are Meter Upgradability Standard (PAP 00) and Standard Meter Data Profiles (PAP 05). PAP 00 has already been completed. The goal of PAP 00 was to define the requirements for smart meter firmware upgradability in an AMI system for regulators, utilities, and vendors. The standard defined by PAP 00 was completed by the National Electrical Manufacturers Association (NEMA) and is entitled NEMA Smart Grid Standards Publication SG-AMI 1-2009 – Requirements for Smart Meter Upgradability. The final version of the standard can be accessed freely at [7].

The Standard Meter Data Profiles (PAP 05) action plan is to define meter data in standard profiles, which will benefit both utility and customers. Meter information that can be made available in common data tables defined in ANSI C12.19 can be easily accessed and this reduces the latency of implementing Smart Grid functions such as DR and real-time usage information. The common data tables created under PAP 05 will be used for PAP 06 "Translate ANSI C12.9 to the Common Semantic Model of CIM." NIST will collaborate with the Association of Edison Illuminating Companies (AEIC), ANSI WGs, IEC TC 13 and TC 57, IEEE Standards Coordinating Committee (SCC 31), NEMA, and other standardization organizations and user groups together to facilitate the PAP 05.

5.4 Smart Home and Building Automation Standards

With the advancement in Information Technology (IT) applications, the deployment of smart home and building automation systems is becoming more and more important for the Smart Grid. With timely information about electrical usage and price, end-use consumers will manage their electricity usage, promote energy efficiency, and lower overall energy costs. Smart home and building automation systems will be a core part of the Smart Grid in achieving these goals.

Different organizations offer standards for interoperable products enabling smart home and building automation systems that can control appliances, lighting, energy management, and security environment, as well as the expandability to connect with different networks. All these standards have been developed in parallel by different organizations. It is therefore necessary to arrange these standards in such a way that it is easier for potential readers to easily understand and select a particular standard that they are interested in without going into the depths of each standard, which often span from hundreds to thousands of pages.

In this section, we introduce key standards proposed by different organizations for smart home and building automation. This section provides an overview of smart home and building automation standards. The main focus is on standards that have been identified by NIST in the *NIST Framework and Roadmap for Smart Grid Interoperability Standards, Release 2.0* [8]. These standards include ISO/IEC Information Technology-HES, ZigBee/HomePlug SEP 2.0, Open Home Area Network (OpenHAN) V2.0, BACnet, LONWORKS, and Z-Wave. These standards are identified by NIST because they support interoperability of Smart Grid devices and systems. However, those standards that are not identified by NIST but have been already widely deployed by industry and consumers such as KNX, INSTEON, ONE-NET, ZHA Public Application Profile, and Japanese standard ECHONET are also covered and explained. The corresponding communication and security standards for this section will be introduced in detail in *Chapter 6. Communications in Smart Grid* and *Chapter 7. Security and Safety for Smart Grid*.

Table 5.2 contains all these standards, which are categorized according to their function fields. Each standard is then explained in detail. Section 5.4.1 explains about ISO/IEC HES. Section 5.4.2 explains the popular ZigBee/HomePlug SEP 2.0; Section 5.4.3 explains the OpenHAN 2.0; Section 5.4.4 explains the Z-Wave; Section 5.4.5 explains the Japanese newly developed standard ECHONET; Section 5.4.6 explains ZHA; Section 5.4.7 explains BACnet; Section 5.4.8 explains LONWORKS; Section 5.4.9 explains INSTEON; Section 5.4.10 explains KNX; and Section 5.4.11 explains ONE-NET.

5.4.1   ISO/IEC Information Technology – Home Electronic System (HES)

The HES is a standard developed by ISO/IEC JTC, Joint Technical Committee, 1 SC25/WG 1 to support applications such as entertainment, lighting, comfort control, life safety, health, and energy management. ISO/IEC JTC 1 SC25/WG 1 is the international standards body writing IT standards for interconnecting home electrical and electronic equipment and consumer products since 1983 [9].

HES allows consumer electronic products, networks, and services to interoperate or to operate, where feasible, as a single coherent system. HES benefits all stakeholders including manufacturers, developers, service providers, installers, utilities, and consumers. HES includes standards for (i) product interoperability; (ii) residential gateway (RG); and (iii) energy management [10].

Table 5.2 Standard list of smart home and building automation

ISO/IEC Information Technology-Home Electronic System (HES)

Function field	Standard name	Short introduction
HES architecture	ISO/IEC 14543-2 series, 14543-3 series, 14543-4 series, 14543-5 series	Standards specify the HES architecture, including the communication layer, user process, system management, media and media-dependent layers, and intelligent grouping and resource sharing
HES gateway	ISO/IEC 15045-1, 15045-2	Standards specify the architecture and requirements for the HES residential gateway
HES application model	ISO/IEC 15067-1, 15067-2, 15067-3, 15067-4	Standards specify the HES application services and protocol, lighting and security models, and energy-management model
HES overview	ISO/IEC JTC 1/SC 25/WG 1 N 1516	An overview of standards developed by WG 1 related to the Smart Grid
HES interoperability	ISO/IEC 18012-1, 18012-2	Standards specify requirements for interoperability and application model
HES WiBEEM standard	ISO/IEC 29145 series	Standards specify the WiBEEM standard for HES, including physical layer, MAC layer, and network layer specifications
ZigBee/HomePlug Smart Energy Profile 2.0		
HAN	ZigBee/HomePlug SEP 2.0	Technical requirements for implementing SEP 2.0 on ZigBee, HomePlug, Wi-Fi, Ethernet, and other IP capable platforms
Link layer	GSM/CDMA	The second-generation telecommunication standards (2G)
Link layer	IEEE 802.3 series	Defines the PHY/MAC layers of wired Ethernet
Link layer	IEEE 802.11 series	Standards for implementing WLAN
Link layer	IEEE 802.15.4	MAC/PHY specifications for WPANs
Link layer	IEEE P1901 series	Standard for broadband over power line networks
Link layer	IEEE P1775	Standard for power line communication equipment
Link layer	IEEE P1905	Standards for interoperable hybrid HAN
Link layer	ITU G.9960/9961 (G.hn)	Defines networking over power lines, phone lines, and coaxial cables with data rates up to 1 Gbit/s

(*continued overleaf*)

Table 5.2 *(continued)*

ISO/IEC Information Technology-Home Electronic System (HES)		
Function field	Standard name	Short introduction
Link layer	ITU G.9954 (HomePNA)	Describes the generic transport architecture for HAN and interfaces to a provider's access network
Link layer	HomePlug series	PLC specifications for HAN
Link layer	LTE	The OFDM-based next-generation telecommunication standard
Link layer	WiMAX/WCDMA/CDMA2000/ TD-SCDMA, Time Division Synchronous Code Division Multiple Access	The third-generation (3G) telecommunication standard
Adaptation layer	ID-6ND	6LoWPAN Neighbor Discovery Standard
Adaptation layer	IEEE 802.2	Local and metropolitan area networks part 2: Logical Link Control
Adaptation layer	RFC 2464	Transmission of IPv6 packets over Ethernet networks
Adaptation layer	RFC 4919	IPv6 over low-power wireless personal area networks (6LoWPANs)
Adaptation layer	RFC 4944	Transmission of IPv6 packets over IEEE 802.15.4 networks
Network layer	RFC 1042	IP datagrams
Network layer	RFC 4291	IETF IPv6 addressing architecture
Network layer	RFC 2460	Internet protocol version 6
Network layer	RFC 4443	IETF ICMPv6 services
Network layer	RFC 4861	IPv6 Neighbor Discovery
Network layer	RFC 4862	IPv6 stateless address autoconfiguration
Application architecture	REST	Representational state transfer
Application requirements	ZigBee/HomePlug MRD	Marketing requirements document (MRD) for SEP and the next-generation smart energy use cases
Application protocol	EXI 1.0	Efficient XML interchange (EXI) Format 1.0
Application protocol	RFC 2616	Hypertext transfer protocol-HTTP/1.1
Data model	IEC 61850 series	Specifies the design of electricity substation automation
Data model	IEC 61970-301	CIM for energy management system application program interface (API)

Table 5.2 *(continued)*

ISO/IEC Information Technology-Home Electronic System (HES)

Function field	Standard name	Short introduction
Data model	IEC 61968 series	Defines the system interface for distribution management
PEV application requirements	SAE J2836 series	Defines use cases for communication between PEV and utility grid, supply equipment, customers, and so on
PEV application requirements	SAE J2847 series	Defines communications between PEV and utility grid, supply equipment, and a utility grid for reverse power flow
Prepayment	IEC 62055	Electricity metering – payment systems
Security	RFC 2409	Provides security associations between network nodes
Security	RFC 4279	Preshared key cipher suites for transport layer security (TLS)
Security	RFC 4302	Provides data integrity, data origin authentication, and protection against replay attacks
Security	RFC 4303	Provides confidentiality, data integrity, data-origin authentication, and protection against replay attacks
Security	RFC 4347	Datagram transport layer security (DTLS)
Security	RFC 4492	Elliptic curve cryptography (ECC) cipher suites for TLS
Security	RFC 5238	DTLS over the datagram congestion control protocol (DCCP)
Security	RFC 5246	TLS protocol version 1.2
Security	RFC 5247	Extensible authentication protocol (EAP) key management framework (KMF)
Security	RFC 5288	AES Galois counter mode (GCM) cipher suites for TLS
Security	ANSI series	Public key cryptography for financial services industry
Security	FIPS, Federal Information Processing Standards, series	NIST standards that define cryptographic modules, hash, advanced encryption, Hash-based Message Authentication Code (HMAC), and so on

(continued overleaf)

Table 5.2   (*continued*)

ISO/IEC Information Technology-Home Electronic System (HES)		
Function field	Standard name	Short introduction
Security	SEC-1, SEC-4	Standards define efficient cryptography group and ECQV, Elliptic Curve Qu-Vanstone, scheme
OpenHAN 2.0		
HAN	UCAIug HAN SRS V2.0	A specification developed by UCAIug that provides common architecture, language, and requirements for HAN
Z-Wave		
HAN	Z-Wave	A wireless mesh networking protocol developed by the Z-Wave Alliance for HAN
ECHONET		
HAN	ECHONET	A Japanese standard suite for HAN
ZigBee Home Automation Public Application Profile		
Home automation	ZigBee home automation public application profile	ZigBee standard for controlling of home appliances, lighting, environment, energy use, and security
BACnet		
Building automation	ANSI/ASHRAE standard 135-2008	A building automation and control networking protocol developed by ASHRAE
Building automation	ISO 16484 series	Standards for building automation and control systems
LONWORKS		
Building automation	ANSI/EIA-852	Enhanced IP-tunneling channel specification
Building automation	ANSI/CEA-709.1	The standard accepted by ANSI for control networking based on LONTALK
Building automation	ISO/IEC DIS 14908 series	Standards for interconnection of information technology equipment
Building automation	IEEE 1473-L	The control network protocol for rail car network
Building automation	LONWORKS	Building automation standards developed by Echelon Corporation
Building automation	LONTALK	An open control protocol for networking devices over various media
Building automation	LONMAKER	A software package for development of local control networks

Table 5.2   (*continued*)

ISO/IEC Information Technology-Home Electronic System (HES)		
Function field	Standard name	Short introduction
INSTEON		
Building automation	INSTEON compared 2006	A white paper compares INSTEON technology with X10, UPB, Universal Power line Bus, LONWORKS, HomePlug, INTELLEON, CEBus, ZigBee, Wi-Fi, Bluetooth, and so on
Building automation	INSTEON the details 2005	A white paper explains INSTEON overview, messaging, signaling details, network usage, and application development
KNX		
Building automation	KNXVol1	A general overview of the whole KNX system
Building automation	KNXVol2	Provides sufficient detail about how to develop the products based on the KNX technology
Building automation	KNXVol3	Provides the information about hardware and software development of products
Building automation	KNXVol4	Requirements for the KNX devices
Building automation	KNXVol5	Provides the requirements, steps, and procedures of products or services for obtaining the KNX trademark
Building automation	KNXVol6	Profiles that define a set of minimum requirements for each of the system specification categories
Building automation	KNXVol7	Function block specifications for different application fields
Building automation	KNXVol8	Specifies system conformance test
Building automation	KNXVol9	Specifies standardized components, devices, and tests
Building automation	KNXVol10	Provides the application-domain-specific standards that are primarily the HVAC easy extension (HEE) parts
Building automation	GB/Z 20965	Chinese standard for building automation based on KNX
Building automation	En 500090	European norm for Home and Building Electronic Systems (HBESs)

(*continued overleaf*)

Table 5.2   (*continued*)

ISO/IEC Information Technology-Home Electronic System (HES)		
Function field	Standard name	Short introduction
ONE-NET		
Building automation	ONE-NET specification V1.6.2 2011	Open specification for designing a low-cost and low-bandwidth wireless control network
Building automation	ONE-NET device payload format V1.6.2 2010	Provides various formats of ONE-NET device payloads

WLAN, Wireless Local Area Network, OFDM, Orthogonal Frequency Division Multiplexing, WCDMA, Wideband Code Division Multiple Access.

ISO/IEC 18012-1 *HES-Guidelines for Product Interoperability-Part 1: Introduction* provides the standard enabling products from multiple manufacturers to interoperate seamlessly to present a single, uniform network, and hence to deliver a variety of applications. Examples of such applications include lighting control, environmental control, audio/video equipment control, and home security. If there are two or more dissimilar networks that conform to this standard and are linked by some physical means, they are expected to behave as if both networks were logically the same. This is shown in Figure 5.3 [11].

ISO/IEC 18012-1 specifies interoperability requirements including safety, addressing, applications, transport of information, setup of devices/elements within home networks, and management. ISO/IEC 18012-2 *Part 2: Taxonomy and Application*

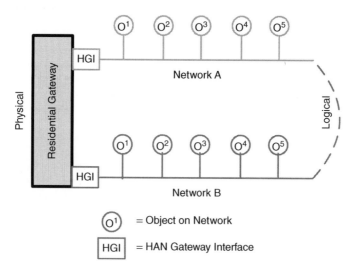

Figure 5.3   Two interoperating networks. Reproduced with permission from International Electrotechnical Commission (IEC) (ISO/IEC 18012-1 ed.1.0 Copyright © 2004 IEC Geneva, Switzerland. www.iec.ch [11])

Interoperability Model specifies the requirements of the application model to enable a common way of describing applications to allow for transparent interoperability. Without being able to describe applications, application-level interoperability cannot be achieved.

ISO/IEC 15045-1 *Information Technology-HES Gateway-Part 1: A RG Model for HES* provides an introduction to the fundamental architecture of the RG and its functions.

The RG is a device of the HES that connects HANs to network domains outside the house, such as wide area networks (WANs), through various communication interfaces. The diagram of possible RG connections and interfaces is shown in Figure 5.4 [12].

The specification of physical connections and interfaces is beyond the scope of HES standards. ISO/IEC 15045-1 applies to the transmission of information between the external environment of the premises and the internal environment of the premises. RG ensures that information can be transmitted between networks in a secure, safe, and transparent manner. There are two basic categories of RG architectures:

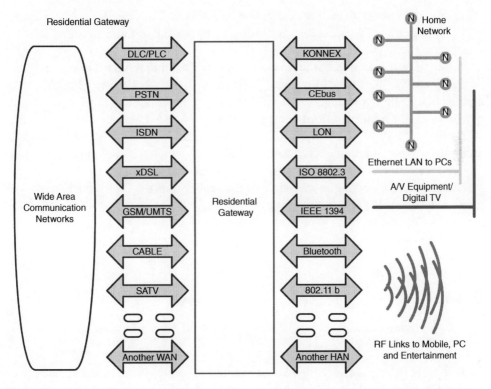

Figure 5.4 Diagram of possible RG connections and interfaces. Reproduced with permission from International Electrotechnical Commission (IEC) (ISO/IEC 15045-1 ed.1.0 Copyright © 2004 IEC Geneva, Switzerland. www.iec.ch [12])

Figure 5.5   Unit architecture. Reproduced with permission from International Electrotechnical Commission (IEC) (ISO/IEC 15045-1 ed.1.0 Copyright © 2004 IEC Geneva, Switzerland. www.iec.ch [12])

the unit and the modular architectures. The unit architecture has no internal RGIP (Residential Gateway Internet Protocol) interfaces and is based upon fixed interfaces between WANs and HANs, as shown in Figure 5.5. The protocols are converted directly, which makes unit architecture more cost efficient than modular architecture. Therefore, the unit architecture is also called a black box approach, and is found in satellite antenna boosters, Data Service Units (DSUs), for Integrated Services Digital Networks (ISDN), Asymmetric Digital Subscriber Line (ADSL) splitters, and so on. The Specific WAN Interface (SWI) provides an interface that conforms to the standards and requirements for connection to a WAN. The Specific HAN Interface (SHI) provides an interface that conforms to the standards and requirements for connection to the HAN.

On the other hand, modular architecture is more flexible to adapt to user requirements by combining a WAN Gateway Interface with a HAN Gateway Interface. The architecture is shown in Figure 5.6. RGIP in the modular architecture requires high performance for multiple information processing.

ISO/IEC 15045-2 *Modularity and Protocol* provides three specific modular architectures for alternative gateway implementations [13]: (i) Simple gateway – interconnecting one-to-one networks, and is nonexpandable; (ii) Multinetwork gateway – interconnecting more than two networks; and (iii) Distributed gateway – interconnecting multiple gateway units.

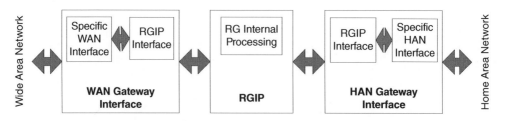

Figure 5.6   Modular architecture. Reproduced with permission from International Electrotechnical Commission (IEC) (ISO/IEC 15045-1 ed.1.0 Copyright © 2004 IEC Geneva, Switzerland. www.iec.ch [12])

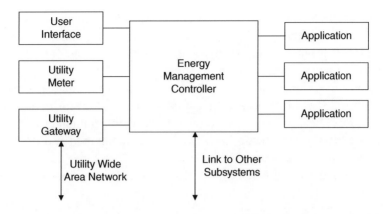

Figure 5.7 Logical model of HES energy management. Reproduced with permission from [14]

ISO/IEC 15067 *HES-Application Model* consists of four parts:

Part 1: Application services and protocol
Part 2: Lighting models for HES
Part 3: Model of an EMS for HES
Part 4: Model of a security system for HES.

ISO/IEC 15067-3 presents a high-level model of EMS for residences, which extends the set of HES application models [14]. These models include the already accepted lighting and security models developed in ISO/IEC 15067-2 and ISO/IEC 15067-4, respectively. These models should validate the language specified for HES in ISO/IEC 16057-1 to foster interoperability among products from competing or complementary manufacturers. Figure 5.7 shows the logical model of the HES energy management model. Electricity price data are sent to all houses in real time over a WAN, such as radio, telephone, or cable television. An energy-management controller in the house receives the electricity rate information and combines it with stored data about appliance power requirements and customer information. Having processed this information, the controller issues control signals to the relevant appliances. The controller may be linked to other home control systems or to a home control coordinator, which is responsible for providing common scheduling and subsystem interaction. A user interface is included to allow the user to override the control signal for some implementations but at the cost of a penalty for the override.

5.4.2 ZigBee/HomePlug Smart Energy Profile 2.0

ZigBee and HomePlug Alliances have jointly proposed a smart energy standard, called the ZigBee/HomePlug SEP 2.0 Technical Requirements Document (TRD) [15].

The purpose of the standard is to incorporate the requirements of smart energy through input from the IEC 61968 standard, UCAIug, OpenSG, OpenHAN, and OpenADE [16]. It promotes the use of widely available and interoperable standards, combined with industry models and best practices. The ZigBee Alliance, Wi-Fi Alliance, and the HomePlug Alliance have formed the Consortium for Smart Energy Profile 2 Interoperability (CSEP) in October, 2011. CSEP provides a platform for the Consortium members to cooperate closely with each other to develop common testing documents and processes to ensure interoperability of products manufactured by different vendors. CSEP issued its first interoperability plugfest for SEP 2.0 on 31st July 2012 [17]. More details about the Wi-Fi Alliance's works can be found in Chapter 6.

5.4.2.1 Link Layer

Smart energy management for the consumer side of the distribution grid is the key for sustainable energy supply. Both wired and wireless communications can be used in smart home communications among appliances, user interfaces, controllers, sensors, and a gateway.

The ZigBee/HomePlug Marketing Requirement Document (MRD) lists various requirements related to different deployment scenarios [18]. For example, it supports all feasible PHYs such as IEEE 802.3, 802.11, 802.15.4, Broadband HomePlug, HomePlug Green PHY (GP), and IEEE P1901. In addition, it supports both wireless and PLC at the Energy Services Interface (ESI), with the ability to configure them in order to operate in combination or independently. Similarly, it supports the bridging between wireless and PLC HANs, while satisfying compatibility requirements.

In SEP 2.0, some technical requirements are established, which are consistent with MRD requirements. It is mentioned in MRD that ESI shall support an IEEE 802.15.4 or a HomePlug physical interface. Furthermore, the ESI may support other alternate Medium Access Control (MAC)/PHYs including ITU G.9960/9961 Gigabit Home Networking (G.hn), ITU G.9954 (HomePNA), G3, IEEE 802.3, IEEE 802.11, IEEE P1901, MoCA, Multimedia over Coax Alliance, LTE (Long term Evolution), World-wide Interoperability for Microwave Access (WiMAX), GSM/Code Division Multiple Access (CDMA), Prime, ISO 14908, and Bluetooth. Consistent with MRD, some requirements are established for IEEE 802.15 in SEP 2.0. It is specified that the 802.15.4 networks shall use IEEE 802.15.4-2006 in the 2.4-GHz band, which is one of the core requirements for interoperability. The 802.15.4 MAC/PHY implementations may also be used with the SEP 2.0 in the 900-MHz band [19]. However, they shall not be labeled or marketed in any way that will allow retail consumers to mistakenly purchase them instead of interoperable MAC/PHY products.

For SEP 2.0, the HomePlug Alliance defines several MAC/PHY combinations. There must be a full interoperation-based implementation that is supported to facilitate the retail market place. HomePlug Alliance is currently working on this matter. HomePlug AV and HomePlug GreenPHY are currently considered as suitable candidates, provided that they offer a fully interoperable mode of operation.

In SEP 2.0, it is mentioned that smart energy systems shall support one specific set of interoperable PLC MAC/PHY implementations, provided by the HomePlug Alliance. Similar to the IEEE 802.15.4 networks, the HomePlug MAC/PHY implementations may also be used with SEP, however, they shall not be branded or marketed in any way that misleads the customers to purchase the wrong interoperable MAC/PHY products. In addition, the interoperable set of HomePlug MAC/PHYs shall be standardized through an external SDO just like the IEEE. Furthermore, the smart energy systems shall support HAN device requests to join the HAN using its local MAC/PHY.

Routing may be required for network admission when multiple MAC/PHYs coexist. Network admission requests may be routed from one network segment to another. The smart energy management shall support network readmission where a HAN device actively leaves the network with a network leave command or passively leaves the network. A device shall be readmitted only if it is capable of retaining security materials after initial authentication, with a state authenticated and authorized by the service provider authorization and security key establishment. The system may support device requests to join a network using an optional MAC/PHY such as IEEE 802.3, 802.11, or any other supported optional MAC/PHY. The SEP specification shall provide guidance for devices to recover when ESI leaves or when it is reset.

Some requirements are established to support SEP 2.0 application deployments in different network topologies. For example, most Personal Area Networks (PANs) devices are expected to utilize a single PHY standard in an instance of a product. For better performance, PAN device installation procedures shall support networks with multiple different PHYs and routers. Similarly, PAN devices shall provide a facility to discover available networks, for example, IEEE 802.15.4 or IEEE P1901 according to their physical interface support, and select a network to join autonomously that shall be the default mode of operation. According to its physical interface support, PAN devices may provide a facility to discover available networks and allow the user to manually choose a network to join.

In certain installations, devices that need to be part of a single HAN may not be colocated closely enough to facilitate reliable communications. Generally, such networks should be connected together with alternate network substrates linked through routers. However, it may also be desirable to use specialized extension devices. For example, a multifamily dwelling may require an extension device located at the ESI, which enables connectivity to the premises of the customer. The IEEE 802.15.4 and Home-Plug networks may be interconnected through a Network Layer, neither a bridging nor a Link Layer. For some installations, it may be necessary to provide an extension device between the ESI and the premises. Provided that these devices do nothing other than link the ESI to premises, they can be operated as either bridges or routers.

The network or architecture limitations for a Neighbor Area Network (NAN) are not yet very clear. In existing IEEE 802.15.4 addressing, there is a maximum limit of 65 000 devices on a network. In reality, networks consisting of several thousand devices are very rare, as larger networks are split into smaller subnetworks. Typically, these smaller networks consist of hundreds to thousands of devices. The limit on these networks is generally the data throughput to the gateway of the network.

Networks with smaller amounts of data to transmit can scale larger with a single gateway. However, as the local link level addressing is expected to be 2 bytes to follow IEEE 802.15.4, there is no practical reason to impose a maximum limit on the size of these networks.

5.4.2.2 Adaptation Layer

Figure 5.8 shows the overview of the Adaptation Layer. Most of the currently used and standardized Link Layer technologies, which are developed for communication in HAN, are not IPv6 compatible. In order to enable IPv6. Network Layer communication in such networks, the Adaptation Layer needs to be implemented. The Adaptation Layer works as an interface between Link Layer technology and IPv6 Network Layer. Therefore, the actual Adaptation Layer is dependent on the Link Layer technology. In this part, an overview of Adaptation Layer for the most common Link Layer technologies is provided.

6LoWPAN Adaptation Layer for IEEE 802.15.4 Networks
Internet Engineering Task Force (IETF) IPv6 over Low Power Wireless Personal Area Networks (6LoWPANs) Adaptation Layer [20] defines a frame format for the transmissions of IPv6 packets and header compression scheme for packet delivery in Low power Wireless Personal Area Networks (LoWPANs), defined in IEEE 802.15.4 PHY/MAC standard. It also describes the formation of IPv6 link-local addresses.

6LoWPAN Adaptation Layer is also responsible for the support of Neighbor Discovery, which is an important mechanism of IPv6 Network Layer and is described in the IETF Internet Draft [20].

HomePlug Adaptation Layer
HomePlug network allows communication over existing electrical wiring at home. HomePlug MAC layer is compatible with Ethernet, IEEE 802.3 frame formats, and utilizes Logical Link Control (LLC) sublayer, IEEE 802.2 [21], which represents an

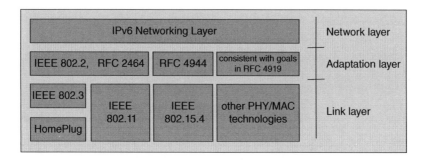

Figure 5.8 Overview of an adaptation layer

interface to Internet Protocol (IP) Network Layer. Specification of IPv6 packet transmission over Ethernet is provided in Request for Comments (RFC), 2464 [22].

Other Networks

The Adaptation Layer of other networks actually depends on the used PHY/MAC technology. In general, it is important to provide IP connectivity. Thus, it should be consistent with goals for LoWPAN networks transmitting over IP given in [23]. A large number of devices connected in a network creates the need for a large address space and IPv6 address can meet such requirements. However, given the limited packet size of many current home automation communication solutions, headers for IPv6 and layers above should be compressed whenever possible.

Many of the current HAN technologies (e.g., INSTEON, X10) are not IP compatible. In order to provide interworking between those networks and IPv6 compatible devices, a home gateway has to support bridging functionality, address mapping, protocol translation, and other features (e.g., INSTEON-to-IP Bridge).

5.4.2.3   Network Layer

In order to allow smooth communication, ensure high security requirements and network flexibility, IPv6 has been adopted in the Network Layer. IPv6 is defined in RFC 2460 [24], which is the Internet standard document developed by IETF. It is the successor of the widely used IPv4 Network Layer protocol. It eliminates the problem of IPv4 address exhaustion by using a 128-bit length size of the address. In addition, IPv6 has been improved to support extensions and options in order to provide greater flexibility and a simplified header format. IPv6 provides extensions to support authentication and implement network security with IPSec [25] as fundamental interoperability requirements. All devices in the HAN should have autoconfigurable addressing system and support RFC 4862, which defines IPv6 stateless address autoconfiguration [26]. The IPv6 stateless autoconfiguration requires no manual configuration of hosts. Hosts can generate their own address as a combination of locally available information and subnet information that is advertised by routers. If there is no router, the host will generate an address for communication restricted to the host on the same link. IPv6 addresses are leased for a fixed length of time. The stateless autoconfiguration mechanism may be used with Dynamic Host Configuration Protocol for IPv6 networks (DHCPv6) simultaneously [27].

As for the addressing architecture, three types of addresses are defined in RFC 4291: unicast, anycast, and multicast [28]. In the case of unicast, packets sent to a unicast address are delivered to the interface that is identified by that address. In the case of anycast, the packets are delivered to one of the interfaces identified by that address. Usually, an anycast address is identified as a set of interfaces. A multicast address also identifies a set of interfaces. However, packets sent to a multicast address are delivered to all interfaces that are identified by the multicast address.

All devices in the HAN should support multicast, either at the Network Layer (IPv6) or at the Application Layer. It is necessary for functions that require multicast transactions (e.g., service discovery). If multicast is implemented at the Application Layer, it is realized by a series of unicast messages to all members of the multicast group.

Another requirement for devices in the HAN is to support IPv6 Neighbor Discovery, defined in RFC 4861 [29]. Neighbor Discovery allows determination of the link-layer addresses of the hosts in the neighborhood. Neighbor Discovery can also detect the reachability of the hosts in the neighborhood and change of link-layer address.

In HAN, various different network configurations and topologies will be likely combined. Therefore, it is necessary to support different routing strategies and routers. Edge routers should be capable to exchange control information between HAN and the routers outside. This is achieved by using Internet Control Message Protocol (ICMP) [30] and Route Advertisements. Routers should support IPv6 netmask routing by using of route metrics from Router Advertisements [18]. A sample scenario is shown in Figure 5.9. As for the communication within the HAN, LoWPAN routers should be supported and IETF RPL 5548 [31] routing protocol should be used for a single 6LoWPAN interface.

The LoWPAN network MAC standard [19] does not define any restriction or recommendations to the Network Layer, and does not provide a native meshing solution. Therefore, routing of the information within PAN depends on the selected

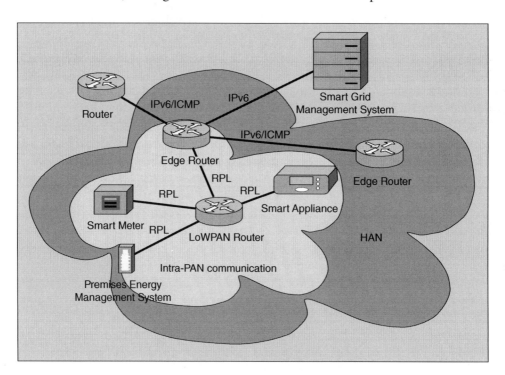

Figure 5.9   IPv6 routing at the network layer – a sample scenario

routing protocol. SEP 2.0 requires support for Low Power and Lossy Networks (LLNs) defined as an IETF Internet draft [18]. The IPv6 Routing Protocol for Low Power and Lossy Networks (RPL) [31] supports various traffic flows, including point-to-point, point-to-multipoint, and multipoint-to-point. Routers typically operate with constraints on energy, memory, and processing power. Routers are usually interconnected by lossy links and support only low data rates. These networks may consist of hundreds or possibly thousands of nodes.

Security on IP Layer

Security on IP layer is provided by IPSec, which is an open standard protocol suite for the protection of data flow between two nodes in the network. It provides mutual authentication and encryption of all IPv6 packets within the communication session. One of the key components of the IPSec suite is the Authentication Header (AH) [32], which provides connectionless integrity and data origin authentication for IPv6 packets and protection against replay attacks. Another key component of the IPSec suite is the Encapsulating Security Payload (ESP) [25], which provides confidentiality, data origin authentication, data integrity, and protection against replay attacks. The ESP header is inserted after the IP header and before the next layer protocol header (Transport mode) or before an encapsulated IP header (Tunnel mode). In Transport mode, the IPv6 headers are not encrypted, while the whole IPv6 packet is encrypted and/or authenticated in Tunnel mode.

5.4.2.4 Transport Layer

Similar to the IP layer, Transport layer should also be compliant with Transmission Control Protocol (TCP)/IP stack. Thus, all devices in the network should support User Datagram Protocol (UDP) [33] and TCP [34]. UDP is a core protocol in IP networks and does not require connection establishment before communication. The datagrams are sent without prior handshake and there is no guarantee that a datagram will be delivered to the destination and also the order of arriving datagrams might not be the same as the order of sent datagrams. Thus, the UDP protocol is unreliable.

On the other hand, the TCP protocol is reliable and provides ordered delivery of information. TCP is realized by three phases: connection establishment, data transfer, and connection termination. The data flow is controlled by windowing, thus after a certain amount of data, acknowledgment of successful delivery should be sent by the receiving host.

Security on the Transport Layer

Transport Layer Security (TLS) [35] is a communication protocol offering security in the IP network. It is a successor of the widely used Secure Sockets Layer (SSL) protocol defined in RFC 6101. TLS encapsulates the application specific protocols such as HyperText Transfer Protocol (HTTP) and File Transfer Protocol (FTP). It encrypts the segments of network connections and consists of TLS Record Protocol and TLS Handshake Protocol. For private information, symmetric cryptography, for example, AES is used and the keys are generated per connection as a result of a TLS Handshake Protocol negotiation between two communicating points. Negotiation is reliable and negotiated secrets cannot be detected by any attacker.

5.4.2.5 Application Layer

Application Layer Data Model
The Application Layer data model is based on both CIM, which is defined in the IEC 61970-301 and IEC 61968-11, and the communication networks and systems in substations, which is defined in IEC 61850.

The IEC 61970 series specify the Application Program Interface (API) for an EMS. The objective of API is to facilitate the integration of EMS applications developed independently by different vendors. CIM specifies the semantics for API. It is an abstract model that represents all the major objects in an electricity enterprise by providing a standard way of representing power system resources as object classes and attributes, along with their relationships. CIM specification uses the Unified Modeling Language (UML) notation, which divides CIM into a set of packages.

IEC 61968 is intended to facilitate interapplication integration as opposed to intra-application integration. Intraapplication integration aims at programs in the same application system. By contrast, IEC 61968 is intended to support the interapplication integration of dissimilar application systems. It provides for interoperability among different computer systems, platforms, and programming languages. IEC 61968-9 specifies the information content of a set of message types that can be used to support many of the business functions related to meter reading and control. Typical uses of the message types include meter reading, meter control, meter events, customer data synchronization, and customer switching.

According to [18], IEC 61850 data models are used where no CIM data model exists, and where harmonization is planned between IEC 61968 and 61850. IEC 61850 is a standard for the design of Substation Automation Systems (SAS). The abstract data models defined in IEC 61850 can be mapped to a number of protocols, namely, Manufacturing Message Specification (MMS), General Object Oriented Substation Event (GOOSE), Sampled Measured Value (SMV), and so on. As IEC 61850 uses different object models from CIM, efforts are currently underway to harmonize these complementary standards [36]. The IEC 61850 communication protocols are mapped to appropriate CIM data classes. However, not all data fields can be mapped directly as there are differences in data types and data formats defined in IEC 61850 and CIM.

Home Area Network Structure
The HAN structure is shown in Figure 5.10. All the HAN devices, after joining a HAN, must register with an Application Trust Center to obtain specific authorization for specific function sets. Devices in the network can have one of four relationships

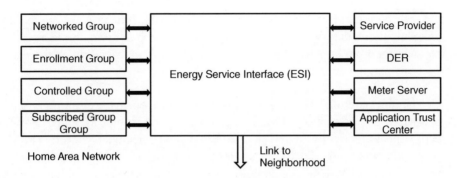

Figure 5.10   The home area network architecture

with the service provider: Networked (or unauthorized), Subscribed, Enrolled, or Controlled. These are analogous to the OpenHAN 2.0 defined states Noncommissioned, Commissioned, Registered, and Enrolled, respectively. Unauthorized devices may be in the network, but have no relationship with the service provider. Subscribed devices have been granted access to customer-specific information by the service provider, but may not be enrolled in any particular service-provider program. Enrolled devices have been granted access to data and are enrolled in a service-provider program, but the customer has not granted control permissions to the service provider. A controlled device is one where the customer has enrolled in a program and granted the service provider the right to manage or control the device. It is noted that controlled devices are assumed to also be enrolled, and enrolled devices are assumed to also be subscribed.

The ESI makes service-provider resources available to the devices within the HAN. Each HAN device enrolled with a utility or service provider shall be provided with the list of enrollment groups registered by the customer and shall act on messaging received that cites those enrollment groups. In general, ESI is responsible for relaying messages between the service provider or trusted control authority and HAN devices. ESI shall be able to receive a message from a service-provider's back office and deliver to devices in the HAN at the beginning of the valid period. Devices that have negotiated to receive messages shall accept those messages. If unable to deliver to a certain device, ESI must retry sending with exponential back-off time until successful or message expiration. Directed messages shall be able to be addressed to a specific premise. These messages shall be delivered to all devices configured to receive messages from the ESI through which they are delivered.

Consumers may grant authorization to the service provider to communicate with or control a HAN device. These authorizations are provided via Out of Band (OOB) communications, for example, phone or Internet, with proper authentication, to confirm the identity of the device.

Function Set
A Function Set defines the set of behaviors necessary to support a given application requirement. Requirements associated with a given Function Set may be either

mandatory or optional. Devices implementing a particular Function Set are required to implement all mandatory requirements for that Function Set. Vendors may choose to implement multiple Function Sets within a single device depending on the market needs. At present, the defined Function Sets are the following:

DRLC: The objective of DRLC applications is to reduce energy consumption during peak load times effectively. DRLC signals are sent from ESI to HAN either directly or indirectly through Premise Energy Management System (PEMS). Robust authentication of DRLC messages and security is required. In some cases, feedback from HAN devices is required to notify the service provider about the failure to respond to the DRLC messages for a given DRLC event. It is also required that consumers should always be able to override a DRLC event.

Messaging: Messaging is sent by service providers to notify consumers about the information such as public service announcements, commodity reliability events, marketing, and conservation advice. An example is shown in Figure 5.11.

Price: The function of pricing is to publish the pricing information of the commodity from the utility or a service provider. The price information should contain information for event ID, Identity Document, provider identifier, currency, start and end date/time, unit of measure, maximum instantaneous demand, price to consume/supply, alternate price, and so on.

Prepayment: Prepayment allows customers to pay for energy usage credit transferred to the meter, where the accounting function is performed. It supports three modes in which the accounting function is performed within service provider systems, within ESI, or within the meter as defined in the IEC 62055 series of standards.

Metering: Metering information provides energy usage information for the customers. It is sent to Metering Servers from End Use Metering Devices (EUMDs), ESIs, or other devices, and then from the Metering Servers to specific display devices. It includes information such as consumption, time of last meter read, demand, time of use, and units of measurement identifier.

Mirroring: Mirroring provides the function of pushing data from a metering device to the ESI for reporting back to the service provider. A set of controls to manage and secure the mirroring process is required.

PEV: PEVs combine many functional areas (e.g., DRLC, pricing, energy storage/DER, and messaging) and logical components with mobility. In particular, bidirectional energy flow to and from PEVs can be treated as metering and DER functionality. PEV requirements defined in SEP 2.0 are derived in part from Society of Automotive Engineers (SAE) documents, that is, SAE J2836 and SAE J2847 series of standards.

DER Management: DER is a small-scale power-generation technology such as a solar panel, a small wind turbine, or PEVs located close to the consumer domain to provide electricity along with the traditional electric power system.

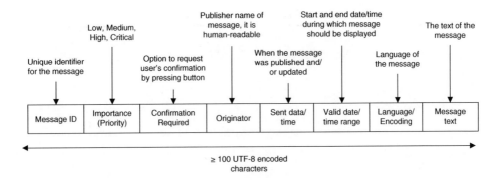

Figure 5.11   A message example

Billing: Billing information gives consumers a financial view of their consumption. The Billing Function Set shall support the communication of the following information: service provider billing characteristics, current bill-to-date, days remaining in billing cycle, forecast bill estimate, billing determinants, usage comparisons, previous sequential and annual billing period data, and so on.

5.4.3   OpenHAN 2.0

The OpenHAN Task Force was established by the UtilityAMI in 2007 to develop guiding principles, use cases, and platform independent requirements for HAN [37].

The OpenHAN Task Force initiated the development of the initial version of the UtilityAMI 2008 HAN System Requirements Specification (SRS). Later, since additional use cases and requirements have been identified, the UCAIug Committee reestablished the OpenHAN Task Force in 2009. The OpenHAN 2.0 Working Group was formed by OpenHAN Task Force to create the next version of HAN SRS, which was named as Unclog HAN SRS v2.0.

By providing common architectures, language, and requirements, the use of the UCAIug HAN SRS v2.0 is able to ensure a competitive market place by driving down costs, increasing interoperability, and maximizing longevity and maintainability. The main purpose is to define the system requirements for an open-standard HAN system and ensure reliable and sustainable HAN platforms. The audience for the HAN SRS includes utilities, vendors, service provides, policy makers, government panel activities, standard working groups, and testing and certification committees.

NIST recognized the OpenHAN as a set of requirements for the data communication interface between the Smart Grid and customer devices [8]. The HAN SRS was developed from the eight-layer interoperability framework provided by the Grid-Wise Architecture Council (Figure 5.12) [37]. The focus of HAN SRS is on

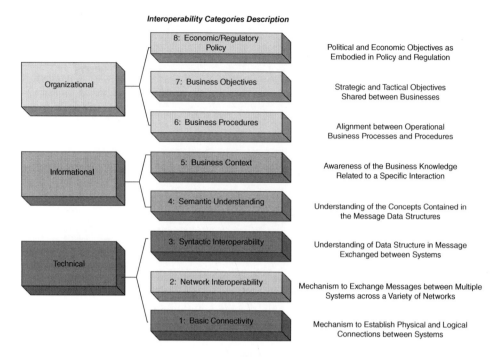

Interoperability Categories Description

Figure 5.12 UCAIug HAN SRS v2.0 interoperability framework. Reproduced from the "GridWise®" Interoperability Context-Setting Framework," March 2008, page 5, with permission from the GridWise Architecture Council. The "GridWise®" Interoperability Context-Setting Framework" is a work of the GridWise Architecture Council.

organizational and informational interoperability category 8 through 4. High-level political and regulatory activities set the political and economic objectives by creation of new policies, laws, and regulations. The remaining interoperability categories (i.e., layer 7 through 4) have been addressed by the multiple organizations who have participated in the standardization activities.

5.4.3.1 Overall Description

The UCAIug HAN SRS provides the foundation on how HAN devices should engage with the service provider. Some of the guiding principles proposed are summarized as follows:

1. Secures Two-way Communication
   HAN devices, including AMI meters, must be capable of communicating with the HAN by utilizing ESI. ESI enables the two-way communication between HAN devices and service providers.

2. Supports Load Control Integration
   Load Control means that load is being managed, for example, deferred, eliminated, cycled or reduced. Therefore, Load Control enables the HAN devices to reduce the peak power consumption under either direct or indirect control.
3. Accesses to Consumer-specific Usage Data
   The HAN should have access to consumer-specific usage data, (e.g., instantaneous usage, interval usage, volts, amps and power factor), through the AMI meter.
4. Supports Three Types of Messaging: Public, Consumer-Specific, and Control
   To support the anticipated growth in the HAN market, the system must provide for various types of messaging, including public, consumer-specific, and control messaging.
5. Supports Open and Interoperable Standards
   The standards should be developed and maintained through a collaborative process that is open to participation by all relevant groups and not dominated by or under the control of a single organization. Two or more HANs are able to directly exchange information securely and seamlessly.

5.4.3.2 Architecture

In HAN SRS, no specific requirement is given regarding the HAN architecture. The HAN architecture allows for more than one ESI in consumer premises, which provides a particular logical function in the HAN.

Utility ESI is important because it provides the real-time energy-usage information from the AMI meter to HAN devices and is protected with cryptographic methods. In some jurisdictions, the Utility ESI may provide for two modes of communications to the premises, either requiring Registration for two-way communication over the UtilityAMI communication network, or not requiring Registration, and only allowing one-way communication (i.e., public broadcast).

Figure 5.13 shows the four states in which a HAN device may exist. To create a HAN, the installer powering on the HAN devices will initiate a process called Commissioning, which allows devices to exchange a limited amount of information (e.g., network keys, device type, device ID and initial path) and to receive public broadcast information. Once the Commissioning process is completed, a HAN device may go through an additional process called Registration.

The Registration process is a further step that creates a trust relationship between the HAN device and the ESI and grants the rights to the HAN device to exchange secure information with other registered devices and with an ESI. In some cases, Commissioning and Registration are combined into one process called Provisioning.

The final process is Enrollment, in which the consumer selects a service provider program and grants the service provider certain rights to communicate with or control their HAN devices. A HAN device must be Commissioned and Registered prior to initiating the Enrollment process.

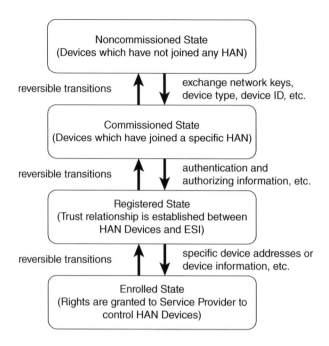

Figure 5.13   The four operation states of a HAN device [37]

If a one-way communication channel requires Commissioning of the HAN device but not Registration, this channel can serve as a public broadcast channel and should be used only for general conservation information.

The HAN SRS supports external interfaces to HAN besides the Utility ESI. Pricing information, control signals, and messaging may be provided from a nonutility service provider. Customers may also have an external interface to communicate with devices on their HAN to enable remote configuration, monitoring, and other applications.

5.4.3.3   HAN Systems Requirements

The HAN system requirements are the main focus of HAN SRS, which enables successful functionality for various HAN stakeholders. Each set of requirements is mapped to functional HAN devices in a table. The table indicates which requirements are needed for the Commissioning Process (CP), Registration Process (RP), Security (S), and Basic Functionality (BF). The requirements can also be Optional (O) or Not Applicable (NA) for the function of the device.

The categories used for the mapping of requirements to the logical device types are defined below:

Commissioning Process (CP) – Minimum requirement needed to support the process of commissioning a HAN device on the HAN.

Registration Process (RP) – Minimum requirement needed to support the process of registering a HAN device on the ESI.

Basic Functionality (BF) – Minimum requirement needed to support the basic functionality of the logical HAN device. This assumes that the device at a minimum has been commissioned.

Security (S) – Minimum requirement needed to protect the HAN against compromises to the confidentiality, integrity, and availability of the HAN.

Optional (O) – An optional requirement that may be included to support a service provider program or allow a vendor to differentiate their product.

Not Applicable (NA) – This requirement is not applicable to this logical HAN device.

The logical devices and their primary functionality are shown in Table 5.3. These devices may include multiple functions but the mapping was decided by the primary function.

The requirements framework includes the following categories:

1. **HAN Applications** – Requirements define what HAN devices should do, including Control, Measurement and Monitoring, Processing, and Human-to-Machine Interfaces (HMIs).
2. **Communications** – Requirements are designed to ensure reliable message transmissions between the authorized parties (e.g., utility, service provider, EMS) and the consumer's HAN devices. Communication requirements include commissioning and Control.
3. **Security** – Requirements are designed to verify users' identities, maintain user privacy, and assure responsible use. Security requirements include access controls and confidentiality, integrity, accountability, registration, and enrolment.
4. **Performances** – Performance requirements are designed to maintain the quality of the HAN communications, which is characterized by reliability, availability, and scalability.
5. **Operation Maintenance Logistics** – Requirements provide chain standards for HAN devices, including manufacturing, configuring, labeling, packaging, installation assistance, user manual, online support, self-testing, and troubleshooting.

   An example is shown in Table 5.4 about how to map the following Communication Commission 1 (Comm. Commission. 1) Requirement to the HAN devices.
6. **Comm. Commission. 1** – HAN device shall accept network configuration data which allows for admission to a new or existing network (e.g., network ID).

5.4.4   Z-Wave

Z-Wave is a proprietary standard for home automation developed by Z-Wave Alliance, which is a group that was established around the proprietary wireless networking protocol developed by a Danish company called Zensys [38]. Z-Wave standard is intended

Table 5.3 Logical HAN device types

Logical device	Primary functionality
Energy Services Interface (ESI)	Network control and coordination
Utility ESI	Network control and coordination
Programmable Communicating Thermostat (PCT)	HVAC control
In-Home Display (IHD)	Display of energy information
Energy management system	Controlling end-device energy
Load control	Resource control
AMI meter	Energy measurement
HAN nonelectric meter	Resource measurement
Smart appliance	Intelligent response
Electric Vehicle Supply Equipment (EVSE)	Charging a PEV
Plug-in Electric Vehicle (PEV)	Electric transportation
End-Use Metering Device (EUMD)	Metering of an end-device load

Table 5.4 Commissioning requirements mapping

ID	HAN system requirements	Utility ESI	ESI	PCT	IHD	EMS	Load control	AMI meter
1	Communication Commission. 1	CP	CP	CP	CP	CP	CP	CP

to provide a simple but reliable method to wirelessly control appliances in houses. It works in the Industrial Scientific and Medical (ISM) band; (868 MHz (Europe) and 908 MHz (United States) bands), using a Frequency Shift Keying (FSK) modulation scheme. The throughput is 40 kbps in both bands. Each network based on Z-Wave standard can include up to 232 nodes. Each node may be configured to retransmit the data to guarantee connectivity in multipath environments of a residential house [39].

The Z-Wave standard protocol stack can be seen in Figure 5.14, where there are PHY/MAC layers that are responsible for control access to RF media, a Network Layer that is responsible for packet routing and a Transport Layer that handles frame-integrity checks, acknowledgments, and retransmissions.

The Z-Wave protocol structure is similar to the ZigBee protocol structure, which consists of four layers, namely, PHY, MAC, Network, and Application Layers. The protocol structure can be seen in Figure 5.15. The PHY layer consists of radio technology that is proprietary to Zensys. It uses ISM bands and is free for public usage. In the next-generation integrated chips, Zensys have mentioned that they will also be supporting 2.4-GHz operation, although the details are not yet published [40].

The MAC layer controls the radio-frequency medium, where the data stream is Manchester coded. It consists of a preamble, Start of Frame (SOF), frame data, and an End of Frame (EOF), as shown in Figure 5.15. The data frame is basically a part of

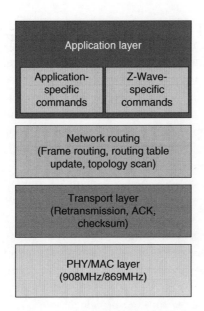

Figure 5.14   Z-Wave standard protocol stack

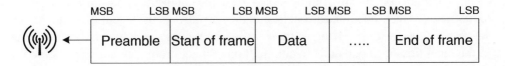

Figure 5.15   Z-Wave protocol structure

the frame that is transferred to the Transport Layer. The MAC layer supports ACKs and retransmissions. There are four types of frames in Z-Wave standard that are used at the MAC layer, namely, unicast, multicast, broadcast, and ACK. All of them use a similar structure, consisting of header, payload, and checksum. They are sent using a simple collision avoidance mechanism in such a way that if traffic is detected, then it will perform a random backoff and try to detect again. For collision avoidance, the MAC layer uses a mechanism that prevents nodes from starting to transmit while other nodes are transmitting. The collision avoidance is achieved by letting nodes stay in receiving mode when they are not transmitting, and then delaying a transmission if the receiver is busy with receiving data. This mechanism is active on all types of nodes when the radio is activated on them.

The transfer of data between two nodes, including retransmission, checksum check, and acknowledgments, is controlled by the Transport Layer. In the Transport Layer, there are four basic frame formats used for transferring commands in the network. The Routing Layer of Z-Wave controls the routing of frames from one node to another.

In the Z-Wave protocol stack, the Application Layer is responsible for decoding and executing commands in a network. It accomplishes this task by using command classes and application frames. These command classes are similar to ZigBee profiles. Actually they are a group of related commands that are designed for a particular application. Typical examples are advanced lighting control and thermostat control. Application frames are embedded inside the Z-Wave frames, which are decoded to send commands to the Application Layer. These frames contain information about the command, the command class, and the command parameters. Comparison of Z-Wave with contemporary standards is drawn in Table 5.5.

There are two types of Z-Wave devices, namely, controllers and slaves. The controllers are master devices in a Z-Wave network. They are responsible for initiating transmissions and have a complete knowledge of the network. They also maintain the network and the control provisioning. There can only be one primary controller in a network, which is responsible for all of the provisioning and maintenance of the routing tables. The primary controller first builds the routing table and then controls it, based on the devices that join the network. It will request the node's neighbor lists that consist of all the nodes within the broadcast range upon joining the network. It uses this information to build the routing table. On the other hand, slave devices are reactive devices. They cannot take the initiative for any type of transmissions on their own and cannot contain any routing tables. Both the controllers and slaves are able to route frames if their positions are static and they are always in "listening mode." They are responsible for transferring a frame with a correct repeater list and for ensuring frame repetition from one node to another node. In addition, they are responsible for scanning the network topology and maintaining the routing table in the controller.

5.4.4.1 Current Status of Z-Wave

The Z-Wave Alliance is basically involved in various home control devices nowadays. For instance, the Z-Wave Alliance member *Z-Wave.Me* demonstrated in March 2012 that the organization operation and set up of home control devices based on Z-Wave through the web is very easy and convenient. The main focus was on the demonstration of the Z-Way controller software, which can be operated by a user from a local control unit, for example, PC (Personal Computer), laptop and iPad [40].

5.4.5 ECHONET

The ECHONET is a Japanese Consortium that was founded in 1997 to promote the development of both software and hardware for home control and monitoring. ECHONET is designed to control home appliances directly and connect to home electronic devices through a gateway [41]. This design enables industry participants to develop a variety of systems having different communication rates and levels of technological sophistication while maintaining optimum cost performance. The goal

Table 5.5 Comparison of Z-Wave with other standards [42]

	Z-Wave	ZigBee	HomePlug	Ethernet	Wi-Fi
Communication type	Wireless	Wireless	Wired (PLC)	Wired	Wireless
Standard type	Proprietary	IEEE 802.15.4	IEEE P1901	IEEE 802.3	IEEE 802.11a/b/g/n
Communication range	30 m (outdoor) <30 m (indoor)	10–100 m	300 m	100 m	100 m (indoor)
Data Rate	40 kbps 868 MHz (EU) 908 MHz (US)	250 kbps (2.4 GHz) and 40 kbps (915 MHz)	200 Mbps 14 Mbps	10–1000 Mbps	11–300 Mbps
Security	128-bit advanced encryption standard (AES) cryptography	128-bit AES encryption	56-bit Data Encryption Standard (DES) cryptography		Wi-Fi (protected access) WPA2
Properties	IP support, free from household interference, and low data rate	Low power usage, longer battery lifetime, and low cost	Communication through existing power line (low cost) lower speed, IP support	Relatively simple, inexpensive, noise immunity, specialized cables	Wide support, available with every other device
Topology	Mesh	Star, tree, mesh	Bus	Star, tree, mesh, bus, ring	Star, tree, mesh
Application in Building and Home Automation	Home and building lighting and automation	Automation, sensing and control for residential, commercial, and industrial sites	Automation using existing AC power line for two-way communication	Use in computer networks (wired)	Wireless communication within home appliances and computing devices

of ECHONET is to develop a standard which requires no special rewiring and can be applied to existing homes and be easily controlled by a wide range of devices. The standard specifies the use of wireless technology and ordinary home electric wiring, which eliminates the need for special rewiring and allows the network to be used in existing homes as well as new housing. In addition, equipment built by different vendors can be interconnected and controlled. Furthermore, the integration of plug-and-play functionality enables new devices to be easily added to the existing network. The ECHONET publishes specifications, prepares middleware, and provides development and supporting tools to encourage the development of highly reliable application software and network-compatible devices. The network provides advanced services by connecting to external systems via communication lines and by working in collaboration with outside associations.

5.4.5.1 ECHONET Development Scope

The overall development scope of ECHONET can be seen in Figure 5.16. It can be seen from the figure that the open disclosure of API and protocol standards will promote application development and result in an open system architecture that facilitates

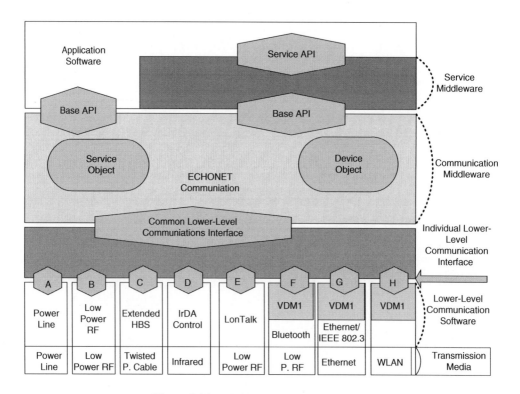

Figure 5.16 ECHONET development scope

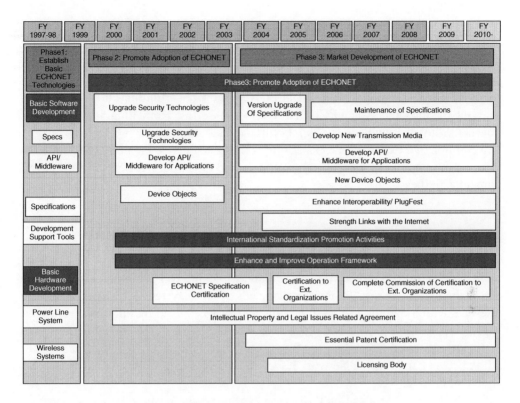

FY 1997-98	FY 1999	FY 2000	FY 2001	FY 2002	FY 2003	FY 2004	FY 2005	FY 2006	FY 2007	FY 2008	FY 2009	FY 2010-

Figure 5.17 Development phases of ECHONET

external expansion and new entries. The PHY is designed to accept various transmission media.

The development schedule for ECHONET has been divided into three phases, which can be seen in Figure 5.17. The first phase is to establish basic ECHONET technologies; the second phase is to promote adoption of ECHONET, while the third phase is to create a market for ECHONET.

5.4.5.2 Current Status of ECHONET

In the fiscal year 2011, the ECHONET Lite standard was recognized as a "commonly known standard interface in Home Energy Management System (HEMS)" by the Japan Smart Community Alliance (JSCA)'s Smart house Standardization Study Group. Grabbing this opportunity, the ECHONET Consortium uses various methods, including applying the results of efforts to develop standards and develop assistance tools to promote ECHONET. At the same time, it is working to design concrete instances of ECHONET application and continue activities to improve the standard in order to meet practical needs. The ECHONET Consortium aims to develop device objects for devices that create, store, and conserve energy along with achieving

IPv6 compatibility and completing standardization of ECHONET *Lite* that is developed to be a user-friendly standard. In the fiscal year 2012, policies were developed for application of the ECHONET Lite standard in smart meter procurement, and projects to promote the development of the Building Energy Management System (BEMS)/HEMS subsidy systems. The ECHONET Consortium has also focused on studying the liaison between ECHONET Lite and ZigBee/HomePlug SEP 2.0, KNX, Internet Protocol Television (IPTV), Forum, and so on.

In addition, the ECHONET Consortium is trying to engage in publicity intended to increase the number of Consortium members and to support related standard development activities.

5.4.6   ZigBee Home Automation (ZHA) Public Application Profile

ZigBee targets a wide variety of applications including home automation, environmental and industrial monitoring, surveillance, tracking of goods and personnel, and automatic meter reading. To address the needs of these systems, ZigBee provides an interface with a common set of functions and services used by different profiles. A ZigBee application profile is defined as a standard set of attributes, functions, and parametric values that can be used by a given class of devices. The interface acts as a link between the ZigBee Network Layer and the applications programs (profile), as shown in Figure 5.18.

The ZHA Public Application Profile is a ZigBee profile for home automation applications. With the introduction of the ZHA profile, Home Automation (HA), can transit from currently limited implementations from the hobbyist and high-end homes to the higher-volume products in the conventional market. The ZHA Profile supports a variety of devices for the home including lighting, heating, and cooling, and even window blind control. It provides interoperability from different vendors, allowing a greater range of control and integration of different devices in the home. It mainly deals with sporadic real-time control of devices.

The ZHA Networks consist of nodes that are based on either the ZigBee Feature Set, or the ZigBee PRO Feature Set or both. Consumers are expected to build an HA system based on a single manufacturer's certified ZHA product suite and then expand the system with certified ZHA products from other vendors. It is also possible that not all products in a HA system will be ZHA devices. In such a situation, ZHA-certified bridge devices are recommended that can link with the non-ZHA network. For instance, a ZHA-certified device can be connected to a computer equipped with a ZHA-certified dongle [13].

It is recommended that any ZigBee device connecting to a ZHA network must be ZigBee certified. ZHA products do not require support of the ZigBee commissioning cluster, although it is highly encouraged. Commissioning is the process through which a HAN device gets access to a particular network. It allows the device to be available on that network. It involves the exchange of information based on security credentials

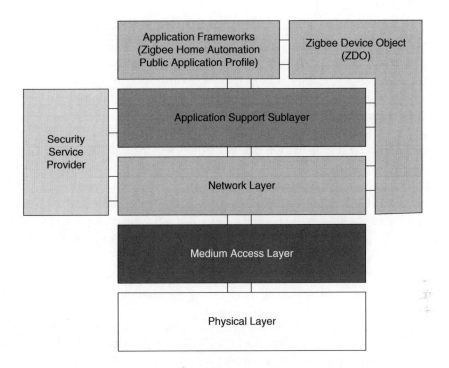

Figure 5.18   ZigBee layer protocol stack with application profiles

required to establish network coordination, assign device addresses, and to route the network packets. In the ZigBee Alliance, three different commissioning modes typically discussed are Automatic (A) mode, Easy (E) mode, and System (S) mode. In ZHA, all the products are required to support the E-mode commissioning. The E-mode commissioning typically involves a button push or two but could also use an Original Equipment Manufacturer (OEM) which is a simple tool like a remote control. It is expected that all ZHA-certified devices will interoperate with other ZHA-certified devices. ZHA devices may interoperate with other ZigBee Public Application Profile devices such as ZigBee/HomePlug Smart Energy and ZigBee Health Care.

There are various mandatory requirements for ZHA public application profile. For example, support for application link keys (interfaces) is mandatory. In addition, source binding and groups shall be implemented on a device-type basis. ZHA end devices in their normal operating state shall poll (a message for getting transmit permission) no more than a specific duration of time, that is, once every 7.5 s. However, there are some exceptions. For example, ZHA end devices may operate with a polling rate higher than 1/7.5 Hz during network maintenance, alarm states, and during commissioning. They may also operate for a lower rate after transmitting a message to allow for acknowledgments and responses to be received quickly. However, they must return to the standard rate specified previously during normal operation.

It is recommended that all devices shall support the mandatory stack profile interoperability. Since fragmentation is not supported in ZHA, it is therefore recommended that a device does not ask for a larger response than what can be fitted into a non-fragmented packet, especially during reading/writing of multiple attributes. When fragmentation is enabled the device shall first inquire the node descriptor of the device it will communicate with, in such a way to determine the maximum incoming transfer size unless manufacturer specific packets are sent. The sending device must use a message size during fragmentation that is smaller than the specified value.

It is recommended that the routing table size of the ZHA device should be increased as much as possible to accommodate the typically dense topology of a ZHA deployment, when a ZHA device is intended to be primarily deployed in a network that does not support many-to-one routing. On the other hand, it is recommended that if possible, the devices primarily installed into many-to-one deployments also increase their own routing tables. It is recommended that a ZHA Coordinator should indicate to the installer whenever a new device joins the network. This indication can be through a PC client, LCD, liquid crystal display, screen, or other simple Light-Emitting Diode (LED) indications.

To ensure interoperability, it is recommended that all ZHA devices should implement compatible Start-up Attribute Sets (SAS). This does not imply that the set must be modifiable through a commissioning cluster, but that the device must internally implement these stack settings in order to ensure consistent user experience and compatibility. To specify a HA start-up set, the SAS parameters described by the commissioning cluster provide a good basis.

In addition to the above recommendations, there are some requirements and best practices for HA specified in the ZHA Profile. When forming a new network, HA devices should do channel scans using the channel mask before scanning the rest of the channels in order to avoid occupying the most commonly used Wi-Fi channels. This is to improve the user experience during installation, which also possibly improves bandwidth.

In ZHA, except for controlling groups or invoking scenes, packet broadcasts are discouraged for HA devices. Devices are limited to a maximum broadcast frequency of nine broadcasts in 9 s. However, it is highly recommended to implement broadcasts much less frequently.

In ZHA, the device description is provided according to the end application area which is addressed by each device. These devices are categorized as generic, closures, lighting, intruder alarm systems, and Heating Ventilation and Air Conditioning (HVAC).

The ZHA Public Profile utilizes the clusters specified in the ZigBee Cluster Library (ZCL). The ZCL gives a mechanism for clusters to report changes to the value of various attributes. It also specifies commands to configure the reporting parameters. In the ZCL specification, the attributes that a particular cluster is capable of reporting for each cluster are listed. Products shall support the reporting mechanism for all attributes mentioned in the ZCL that a product implements itself within a given cluster [44].

5.4.7   BACnet

BACnet is a Building Automation and Control Networking protocol developed by the American Society of Heating, Refrigeration, and Air-Conditioning Engineers (ASHRAE). BACnet has been designed specifically as the data communication protocol of BAC systems for applications such as Heating, Ventilating, Air-conditioning, and Refrigerating (HVAC&R) control, lighting control, access control, and fire detection systems. The purpose is to define data communication service and protocols for computer equipment used for monitoring and controlling of HVAC&R and other building systems. In addition, it defines an abstract, object-oriented representation of information communicated among those devices. BACnet became the ASHRAE/ANSI Standard 135 in 1995, and was acquired by the international ISO 16484-5 standard in 2003 [45].

BACnet is an open multivendor standard that allows building automation systems from different manufacturers to interoperate. Similar to the development in computer networks, open protocols such as TCP/IP enable end users to choose hardware (network cards/adaptors, modems) and software (operating systems, application programs) components from different manufacturers. However, the cost of an Open System Interconnect (OSI) seven-layer model-based protocol is prohibitively high for most building automation applications. Instead of using all seven layers of the OSI model, only the OSI functionality that is actually needed should be included, thereby the seven-layer architecture is collapsed. In the collapsed architecture, only selected layers of the OSI model are employed, while other layers are not used, which reduces message overhead, lowers cost, and increases system performance.

BACnet is based on a four-layer collapsed architecture as shown in Figure 5.19 [46]. For the Data Link and PHYs, BACnet provides six options for the combinations of different communication protocols. Option 1 is the LLC protocol defined by ISO 8802-2 Type 1, combined with the ISO 8802-3 MAC/PHY Layer protocols. Option 2 is the ISO 8802-2 Type 1 protocol combined with ARCNET, Attached Resource Computer NETwork, (ATA 878.1). Option 3 is the Master–Slave/Token-Passing (MS/TP) protocol combined with EIA-485 PHY protocol. Option 4 is the Point-To-Point (P2P) protocol, which provides mechanisms for hardwired or dial-up serial, asynchronous communication, combined with EIA-232. Option 5 is the LONTALK protocol. Option 6 is the BACnet/IP combined with UDP and IP. In conclusion, these options provide various options, that is, a master/slave MAC, deterministic token-passing MAC, high-speed contention MAC, dial-up access, star and bus topologies, and a choice of twisted-pair, coax, or fiber optic media. The details of these options are described in [46].

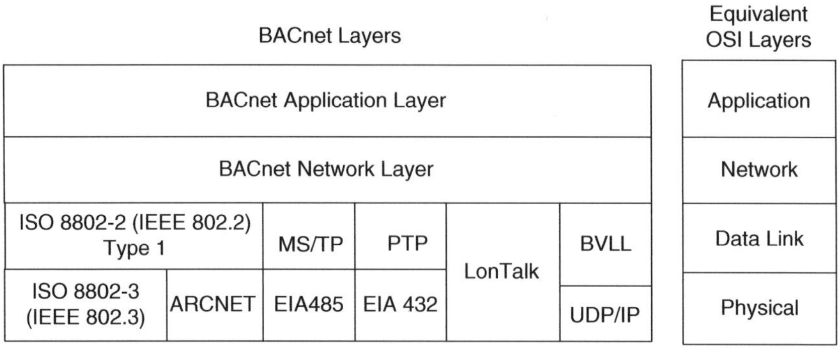

Figure 5.19   BACnet collapsed architecture

A four-layer collapsed architecture was chosen after careful consideration of the particular features and requirements of BAC networks, and to keep the overhead needed as short as possible. The PHY layer provides a means of connecting BAC devices and transmitting signals that convey the data. The Data Link Layer organizes the data into frames or packets, regulates access to the medium, provides addressing, and handles some error recovery and flow control. For Network Layers, in the case of a single network, most Network Layer functions are either unnecessary or duplicate Data Link Layer functions. However, if two or more BACnet internetworks use different MAC layer options, Network Layer functions are needed to differentiate local and global addresses and route messages. Therefore, BACnet provides limited Network Layer capability by defining a Network Layer header that contains the necessary addressing and control information. For the transport layer, since BACnet is based on a connectionless communication model, the scope of the required services is limited enough to justify implementing these at a higher layer. Therefore, the communication overhead of a separate transport layer is saved. The Application Layer of the protocol provides the communication service required by the applications to perform their functions. In this case, monitoring and control of HVAC&R and other building systems can be performed. Since most communications in BAC networks are very brief and do not need to change format or compress data, separate sessions and presentation layers are not needed.

The BACnet standard defines "objects," which are used to communicate with the individual devices and their functions within a BAC network. The object can contain the following information:

- Binary input/output values (e.g., "on/off," "open/close")
- Analog input/output values (e.g., current, voltage)
- Schedule information, alarm, and event information
- Control logic.

Each BAC device is defined as a set of data structures or objects, and each object defines certain properties (e.g., name and current status). Objects enable one device to receive information from a particular device without having to know its internal setup or configuration and are the key to realize interoperability.

5.4.8 LONWORKS

LONWORKS is a distributed control system developed by the American company Echelon, which meets the Peer-to-Peer (P2P) and/or master–slave communication needs of BACnet. LONWORKS is the leading market solution in the United States, whereas KNX has yet to make an impact. KNX will be explained in Section 5.4.10.

LONWORKS uses a flat architecture that supports the address requirements of the entire system but also logical segmentation. Segmentation is achieved through network-level routers that are transparent to the application in the nodes, and that provide direct access by installation, diagnostic, or monitoring tools connected anywhere in the network. Traditional control networks use a closed island of control links with proprietary gateways. These gateways are difficult to install, maintain, and interoperate. Ultimately, the high cost of this design approach has limited the market for control systems. The LONWORKS system is accelerating the trend away from these proprietary control schemes and centralized systems by providing interoperability, robust technology, faster development, and scale economies. Distributing the processing throughout the network and providing open access to every device lowers the overall installation and life-cycle costs. It also increases reliability by avoiding single point of failure, and provides the flexibility to adapt the systems to a wide variety of applications.

LONWORKS has been accepted for the ANSI/CEA-709.1, the ANSI/EIA-852, and the ISO/IEC DIS, Draft International Standard, 14908 series of standards. LONWORKS technology includes five interconnected elements, Neuron Chip, LONTALK protocol, LONWORKS transceiver, LONWORKS tools, and LONMARK.

The core of the LONWORKS system is a microcontroller, named the Neuron Chip, which is manufactured by Toshiba and Cypress. The Neuron Chip comprises three internal processors that each carry out different functions. CPU 1 is responsible for physically accessing the transmission medium, which represents layers 1 (PHY) and 2 (Data Link Layer) of the ISO/OSI model. CPU 2 is responsible for network variables transmission, which represents layers 3–6 of the ISO/OSI model. CPU 3 is responsible for application programs, which exchanges data with the other two CPUs by accessing the shared memory.

The LONTALK protocol is an open standard (also known as ANSI/CEA 709.1 and IEEE 1473-L) which defines how a device can send and receive messages from other devices over the network [47]. The LONTALK protocol is embedded in the Neuron Chip as firmware that ensures that all nodes on the same network are compatible.

LONTALK protocol provides all the services described in the ISO/OSI seven-layer reference model [48].

The LONWORKS transceiver serves as a physical communication interface between LONWORKS devices and a LONWORKS network. Different transmission media including twisted pair, power line, radio frequency, and fiber-optics can be used for communications. Table 5.6 shows the types of transceivers that allow LONWORKS devices to communicate with other devices over a physical network [49]. Other types of allowed transmission media include the RS-485/EIA-485 twisted pair, 900 MHz/2.4 GHz radio frequency, 400–450 MHz radio frequency, 1.25 Mbit/s coaxial cable and infrared. Echelon and other manufacturers have developed various programming and integration tools for LONWORKS technology. For development tools, LONBUILDER and NODEBUILDER are developed by Echelon for programming Neuron Chips. Users can develop their own programs by using the Neuron C programming language on a PC. For network integration tools, the LONMAKER integration tool is a multipurpose LONWORKS network tool that runs on a PC under Windows 2000/NT4.0/98/95 Operating Systems (OSs) and uses Visio 2000 as a graphical interface [50]. The LONMAKER tool can be used as a one-stop solution for all phases of a network's life cycle: from initial design and commissioning to ongoing operation and maintenance, including network design, network installation, network documentation, and network operation.

The LONMARK Interoperability Association, now known as LONMARK International (LMI), is responsible for interoperability of LONWORKS devices. LMI's mission is to enable the easy integration of multivendor systems based on LONWORKS networks. There are millions of LONWORKS technology-based devices installed worldwide. LMI provides an open forum for member companies to work together on marketing and technical programs to promote the availability of open, interoperable control devices [51]. A set of guidelines for defining a device's basic functionalities and minimum requirements has been developed. A device that contains the LONMARK logo and has been certified by the LMI is guaranteed to

Table 5.6 Transceivers for widely used media and network topologies

Medium	Transceiver	Transmission rate	Network topology	Network length (m)	Power supply
Twisted pair	FTT-10A	78 kbit/s	Free topology bus	500, 2700	Separate
Twisted pair	LPT-10	78 kbit/s	Free topology bus	500, 2700	Via the bus
Twisted pair	TPT/XF-78	78 kbit/s	Bus	1400	Separate
Twisted pair	TPT/XF-1250	1.25 Mbit/s	Bus	130	Separate
Power line	PLT-22	5 kbit/s	Free topology	[a]	Via a specialized power adaptor

[a]Depends on attenuation and interference.

be interoperable with other LONMARK-certified devices manufactured by different companies.

5.4.9 INSTEON

INSTEON, a system for connecting lighting switches and loads without extra wiring, is a home automation networking technology designed by SmartLabs, Inc. [52]. INSTEON is an automation protocol enabling appliances to be networked together. This concept can be implemented over power line communication as well as over radio interface. INSTEON messages have a fixed length and are synchronized to the Alternating Current (AC), power line zero crossings. The size of an INSTEON standard message is 10 bytes and can also be extended to 24 bytes. In such a case, 14 bytes of arbitrary user data are used. This creates space for developers of various smart home applications. In addition, if more data have to be transferred, several extended messages can be sent to the destination. User data can be encrypted if necessary (e.g., home security systems). INSTEON devices can send and receive X10 commands, which allows compatibility with the oldest and widely used home automation protocol. This protocol allows more than 65 000 device types and the same number of commands, which creates a vast space for various appliances and functionalities. INSTEON technology defines two protocols, INSTEON Power line protocol and INSTEON RF protocol, which can be used simultaneously. INSTEON technology defines peer-to-peer communication between devices. Each device (peer) can send, receive, and relay messages.

In order to interwork with other home automation technologies, bridging should be implemented. INSTEON can interoperate via bridges with other devices communicating over Wi-Fi, Bluetooth, ZigBee, Z-Wave, HomePlug, HomeRF, INTELLON, KNX, LONWORKS, CEBus, Worldwide Interoperability for Microwave Access (WiMAX), and other technologies [53]. The standard INSTEON message contains "From" and "To" addresses, each occupying 3 bytes, which can identify more than 16 million different devices. INSTEON technology also allows broadcasting to devices of certain device type or to a specified group of devices. In such a case, the "To" address contains the device type or a group number, respectively. The maximum data rate is 13 165 and 2880 bps when instantaneous power line and sustained power line are used over one hop without acknowledgment, respectively.

5.4.10 KNX

KNX, formerly known as the European Installation Bus (EIB), was developed by the KNX Association. As shown in Figure 5.20, KNX is a building control communication system that uses information technology to connect devices such as sensors, actuators, controllers, operating terminals, and monitors [49].

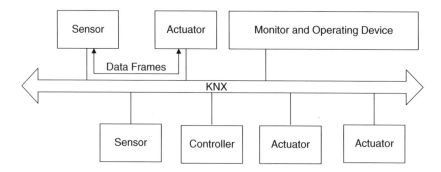

Figure 5.20 Building control devices connected over KNX

The transmission rate of KNX is low compared with LONWORKS, but sufficient for transmitting switching and controlling commands. Unlike LONWORKS, which has a higher transmission rate and advanced processing ability, KNX cannot be used to control operational systems. KNX is usually used in the field level, and is adopted in the standards such as Chinese Standard GB/Z 20965, ISO/IEC 14543-3, 14908, US Standard ANSI/ASHRAE 135, and European Standard EN 500090 and 13321-1.

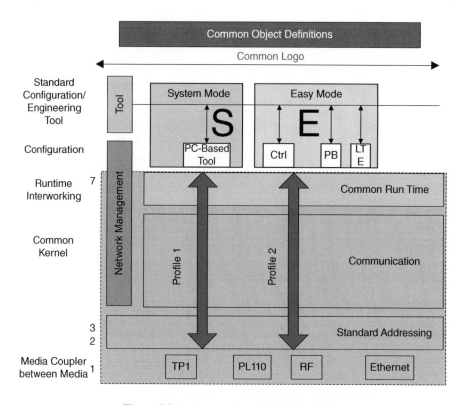

Figure 5.21 An overview of the KNX model

An overview of the KNX model is shown in Figure 5.21 [54]. At the top of the model is the Interworking and (Distributed) Application Models for the Home and Building Automation. KNX models an application as a collection of sending/receiving Datapoints to/from KNX devices. Datapoints may be inputs, outputs, parameters, diagnostic data, and so on [55]. For example, if a local application in a device wants to write a new value to the sending Datapoint, this device sends a "write" message with the corresponding address and the new value. This value will be received by the receiving Datapoint with the same address. The receiving application will act upon this value update, for example, an internal state change or updating or modifying some physical output status, or any combination of these.

Below the Interworking and Distributed Application Models are the Configuration Modes. The different configuration modes are specified by KNX to provide a diverse choice for different markets, local user habits, level of training, and application environment. KNX devices can be divided into the following configuration modes:

1. **System Mode**: The System Mode is used in the creation of backbone-building automation systems and must be programmed and installed by specialist technicians. With the aid of the PC-based Energy Transfer System (ETS) project tools, the System Mode supports the configuration of various features including binding, parameterization, and download of application programs.
2. **Easy Mode**: The Easy Mode, which includes the Controller Mode (Ctrl), Push-Button Mode (PB), and Logical Tag Extended Mode (LTE), is defined to support installation of a limited number of devices on one logical segment of a physical medium. The Easy Mode can be configured according to a structured binding principle or through simple manipulations – without the need for a PC tool. In Ctrl, one special device called controller is needed to be in charge of supporting the configuration process. However, in PB, a specialized device such as controller is not needed for configuration. The exchange of configuration data between devices, for example, sensors or actuators, is enabled through a single Application Layer service. On the other hand, LTE is limited to HVAC applications, which needs a longer set of structured data.

The Common Run Time Interworking is shared by these configuration modes to allow the creation of a comprehensive and multidomain home and building communication system.

The Communication System defines the physical communication media, message protocol, and models for the KNX Network. A Common Kernel model including a seven-layer OSI model is shared by all the devices. The PHY defines the supported different transmission media for KNX, including twisted-pair cable (KNX. TP 1), power line (KNX. PL 110), radio frequency (KNX. RF), and fiber-optic cable for sending data to devices [54]. The Data Link Layer per medium/General provides MAC and LLC. The Network Layer controls the hop count of a frame and provides a segmentwise acknowledged telegram. The Transport Layer defines four types of communication relations: multicast, broadcast, one-to-one connectionless, and one-to-one connection oriented. The Session and Presentation Layers are empty. The Application Layer offers various application services.

KNX can accommodate up to 65 536 devices in a 16-bit Individual Address space. In the KNX specification, a profile is a group of features that enable a device to interwork within a given configuration mode and within the whole network [56]. Profiles provide a top-down view of a consistent set of requirements for any corresponding implementation.

5.4.11 ONE-NET

5.4.11.1 ONE-NET

ONE-NET is an open standard and open source solution for home and building automation based on the proprietary physical interface. It defines physical, network, and message protocols in order to provide a low-power, low-delay, low-cost, and medium-range wireless solution for devices and appliances [57].

At present, it utilizes ISM bands in the range 902–928 MHz in the United States and the 865–868 MHz band in Europe. Other frequencies can also be implemented. At present, it allows raw data rates up to 230 kbps and utilizes Carrier Sense Multiple Access (CSMA) technology on a MAC layer with packet prioritization. Different topologies are defined (peer-to-peer, multihop, and star) and the communication is realized by a Master/Client set up.

The ONE-NET packet ranges from 120 to 472 bits, depending on the packet type. In general, a single-packet payload has 40 bits and consists of four message fields [58], as shown in Figure 5.22.

The Communication protocol offers 14 bits for different message types related to electricity, gas, water, and environment, and also to interoperate with INSTEON and X10 standards.

5.4.11.2   Interoperability between ONE-NET and INSTEON and X10 Technologies

Interoperability between ONE-NET and INSTEON shall be realized by a ONE-NET/INSTEON bridge, which understands a special message sent from a ONE-NET device to an INSTEON device. This special message consists of two

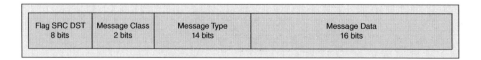

Figure 5.22   Single-packet payload

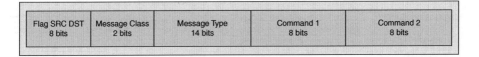

Figure 5.23   INSTEON COMMAND data message format

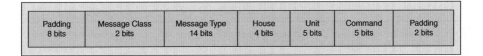

Figure 5.24   X10 SIMPLE data message format

ONE-NET messages, where the first one defines the INSTEON address message (16 bits) and the second one defines INSTEON COMMAND data format. The INSTEON COMMAND data message is shown in Figure 5.23 and contains a Flag field and two Command fields of an INSTEON standard message.

X10 technology is supported by an X10 SIMPLE STATUS message and an X10 SIMPLE COMMAND message. The former is used to report the X10 command from the X10 device to ONE-NET that was heard by the ONE-NET/X10 bridge. The latter message is created by the ONE-NET device in order to send a command to the X10-compatible device over the bridge (to relay an X10 command). The X10 SIMPLE data message format is shown in Figure 5.24.

5.4.12   A Comparison of Smart Home and Building Automation Standards

The comparison of the smart home and building automation standards is shown in Table 5.7. These standards are compared in terms of communication type, radio/communication range, data rate, frequency band, security, and network topology. We have also analyzed these standards on the basis of the unique properties of each standard.

It is expected that the future communication network for the Smart Grid should be IP based, as stated in the NIST roadmap. This is because the migration toward IP-based standards brings the following benefits:

- Simplified system architecture and control
- End-to-end visibility
- Interoperability with different networks
- Support for existing IP-based networks.

Table 5.7 A comparison of the smart home and building automation standards [42]

	Z-Wave	ZHA	INSTEON	LonWorks	ECHONET	OneNet	BACnet
Comm. type	Wireless	Wireless	Wireless/wired	Wireless/wired	Wireless/wired	Wireless	Wired
Standard type	Proprietary	IEEE 802.15.4	Proprietary	Proprietary	Proprietary	Open standard	ASHRAE, ANSI, ISO
Radio/communication range	30 m (outdoor) <30 m (indoor)	10–100 m	45 m	30–120 m	<100 m	500 m	1200 m (for MSTP, Master–Slave–Token Passing)
Data rate/frequency	40 kbps/868 MHz (EU) 908 MHz (US)	250 kbps (2.4 GHz) and 40 kbps (915 MHz)	13.165 kbps 902–924 MHz	5 kbps to 1.25 Mbps/ 400 MHz, 450 MHz, 900 MHz, 2.4 GHz	4 kbps–36 kbps/230 kbps/ 426 MHz, 429 MHz, 2.4 GHz	902–928 MHz, 865–868 MHz	9.6–76.8 kbps (MSTP) 78.8 kbps (LonTalk) 100 + Mbps (BACnet IP)
Topology	Mesh	Star, tree, mesh	Peer-to-peer,	Peer-to-peer, free, bus	Peer-to-peer,	Peer-to-peer, star	Peer-to-peer, star, tree
Security	128-bit Advanced Encryption Standard (AES) cryptography	128-bit AES encryption	Unique 24-bit address, all transmissions are encoded onto the network	256-bit AES encryption and NIST certified FIPS 140-2 level-2	Private key encryption function	Extended tiny encryption algorithm (XTEA2)	BACnet network security for BACnet interoperability building block (NS-SD, Network Security-Secure Device,-BIBB)

	Properties	Application in home and building automation
	IP support, free from household interference, and low data rate	Home and building lighting and automation
	Low power usage, longer battery lifetime, and low cost	Automation, sensing and control, for residential, commercial, and industrial sites
	Not limited to a single physical network technology, and supports both RF and PLC	Home management network technology, that provides a secure, highly available, affordable, robust home management network
	Two physical-layer signaling technologies, twisted pair and PLC, routers, network management software, and so on, from Echelon Corporation	Peer-to-peer and/or master–slave communication for home and building automation and control networks
	Supports both RF and PLC	Developed key software and hardware for home automation. A universal transmission standard for various services in home network system that combine home appliances of different vendors
	Low power, delay, and cost, medium radio range, interoperate with INSTEON and X10	Open source solution, based on the proprietary physical interface
	Designed for heating, ventilating, air conditioning control, lighting control, access control, fire detection, and their associated equipment	
	Data communication protocol for building automation and control networks	

5.5 Conclusion

In this chapter, we introduced the various DR, AMI, and smart home and building automation technologies, specifications, and standards developed by different organizations. In order to successfully enable DR technologies, one of the most important factors is the development and adoption of consistent DR signals. Consistent DR signals are necessary to improve the responsiveness of the entire power generation and delivery systems to take advantage of renewable and other intermittent resources. Several SDOs and industry alliances have been working on the development of interoperable DR standards. Besides technological barriers, there are also social barriers such as the lack of knowledge among users about how to response to complex time-varying electricity prices. For AMI, we introduced the two dominating AMI standards, that is, the IEC 62056 and ANSI C12 standards. The IEC 62056 standards are based on DLMS for standardizing the communication profile, the data objects, and the object identification codes; and COSEM for specifying the procedures for the transfer of information for application association control, authentication, and for data exchange between COSEM clients and servers. On the other hand, the ANSI C12 standards are used for metering protocols in North America instead of the IEC 62056 standards used in Europe. AMI can potentially benefit both utilities and consumers by providing a better service such as real-time billing, energy usage monitoring, and control for customers, and supporting better outage detection, more rapid grid deficiencies treatment, and better management and maintenance of utility assets. However, AMI standards still face interoperability issues such as how to describe a common set of requirements to facilitate exchange of confidential and authentic information across standards developed by different SDOs. For smart home and building automation, we introduced a wide range of standards and technologies developed by different SDOs and industry alliances. The main focus is on the smart home and building automation standards that have been identified by NIST in the *NIST Framework and Roadmap for Smart Grid Interoperability Standards, Release 2.0*. However, those standards that are not identified by NIST but have been already widely deployed by industry and consumers were also covered and explained in detail. All these standards have been developed in parallel by different organizations, and each of them has its own advantages and disadvantages. Therefore, it is important for readers to fully understand the latest achievements and ongoing technical works of smart home and building automation standards for decision making. Continued R&D and standardization efforts are needed to increase customer awareness, and to promote the integration of DR, AMI, and smart home and building automation systems with other Smart Grid systems such as DER and Electric Storage (ES).

References

[1] Hamed, M.R.A. and Alberto, L.G. (2010) Optimal residential load control with price prediction in real-time electricity pricing environments. *IEEE Transactions on Smart Grid*, **1** (2), 120–133.

[2] Quantum Consulting Inc., Summit Blue Consulting, LLC Working Group 2 Measurement and Evaluation Committee, *et al.* (2005) *Demand Response Program Evaluation Final Report*.

[3] Piette, M.A., Ghatikar, G., Kiliccote, S. *et al.* (2009) Design and operation of an open, interoperable automated demand response infrastructure for commercial buildings. *Journal of Computing Science and Information Engineering*, **9** (2), 1–9.

[4] CEN/CENELEC/ETSI Joint Working Group (2011) *Final Report of the CEN/CENELEC/ETSI Joint Working Group on Standards for Smart Grids*, ftp://ftp.cen.eu/CEN/Sectors/List/Energy/SmartGrids/SmartGridFinalReport.pdf (accessed 27 December 2012).

[5] IEC SMB Smart Grid Strategic group (2010) *IEC Smart Grid Standardization Roadmap Edition 1.0*, www.iec.ch/smartgrid/downloads/sg3_roadmap.pdf (accessed 27 December 2012).

[6] Electric Power Research Institute (2010) *A Perspective on Radio-Frequency Exposure Associated with Residential Automatic Meter Reading Technology*, www.ferc.gov/eventcalendar/Files/20070423091846-EPRI%20-%20Advanced%20Metering.pdf (accessed 15 September 2012).

[7] National Electrical Manufactures Association (NEMA) (2009) *NEMA Smart Grid Standards Publication SG-AMI 1-2009- Requirements for Smart Meter Upgradeability*, www.nema.org/standards/Pages/Requirements-for-Smart-Meter-Upgradeability.aspx#download (accessed 20 September 2012).

[8] National Institute of Standards and Technology (2012) *NIST Framework and Roadmap for Smart Grid Interoperability Standards, Release 2.0*, www.nist.gov/smartgrid/upload/NIST_Framework_Release_2-0_corr.pdf (accessed 2 March 2012).

[9] International Organization for Standardization/International Electrotechnical Commission (2002) *Smart Grid Standards for Residential Customers*. ISO/IEC JTC 1/SC, Subcommittee, 25/WG 1 N 1516.

[10] Schoechle, T. (2009) *Energy Management Home Gateway and Interoperability Standards*, The Grid Wise Architecture Council (GWAC), www.smartgridnews.com/artman/uploads/1/Schoechle.pdf (accessed 5 March 2012).

[11] International Organization for Standardization/International Electrotechnical Commission (2004) ISO/IEC 18012-1. *Information Technology-Home Electronic System (HES)-Guidelines for Product Interoperability-Part 1: Introduction*, International Organization for Standardization.

[12] International Organization for Standardization/International Electrotechnical Commission (2004) ISO/IEC 15045-1. *Information Technology-Home Electronic System (HES)-Part 1: A Residential Gateway Model for HES*, International Organization for Standardization.

[13] International Organization for Standardization/International Electrotechnical Commission (2012) ISO/IEC 15045-2. *Information Technology-Home Electronic System (HES)-Part 2: Modularity and Protocol*, International Organization for Standardization.

[14] International Organization for Standardization/International Electrotechnical Commission (2000) ISO/IEC TR 15067-3 *I*nformation Technology-Home Electronic System (HES) Application Mode-Part 3: Model of an Energy Management System for HES, International Organization for Standardization.

[15] Itron, Inc. (2011) *ZigBee/HomePlug Smart Energy Profile 2.0 Technical Requirements Document*. SEP 2.0 TRD.

[16] Open Smart Grid User Group (2008) *Open SG*, http://osgug.ucaiug.org/default.aspx (accessed 20 March 2012).

[17] Consortium for Smart Energy Profile Interoperability (2012) *First Interoperability Plugfest*, www.csep.org/media/uploads/documents/csep_incorporation_pr_120731.pdf (accessed 3 March 2013).

[18] ZigBee/HomePlug Joint Working Group (2009) *Smart Energy Profile Marketing Requirements Document (MRD)*.

[19] IEEE (2006) IEEE 802.15.4 Standard. *PHY/MAC Layer Control for Low Rate Wireless Personal Area Networks (LR-WPANs)*, IEEE.

[20] Montenegro, G., Kushalnagar, N., Hui, J., and Culler, D. (2007) *Transmission of IPv6 Packets Over IEEE 802.15.4 Networks*. RFC 4944.

[21] IEEE (1998) IEEE 802.2. *IEEE Standard for Information Technology-Telecommunications and Information Exchange Between Systems-Local and Metropolitan Area Networks-Specific Requirements-Part 2: Logical Link Control*.

[22] Crawford, M. (1998) *Transmission of IPv6 Packets over Ethernet Networks*. RFC 2464.

[23] Kushalnagar, N., Montenegro, G., and Schumacher, C. (2007) *IPv6 over Low-Power Wireless Personal Area Networks (6LoWPANs): Overview, Assumptions, Problem Statement, and Goals*. RFC 4919.

[24] Deering, S., Hinden, R. (1998) *Internet Protocol, Version 6 (IPv6) Specification*. RFC 2460.

[25] Kent, S. (2005) *IP Encapsulating Security Payload (ESP)*. RFC 4303.

[26] Thomson, S., Narten, T., and Jinmei, T. (2007) *IPv6 Stateless Address Autoconfiguration*. RFC 4862.

[27] Droms, R., Bound, J., Volz, B. *et al.* (2003) *Dynamic Host Configuration Protocol for IPv6 (DHCPv6)*. RFC 3315.

[28] Hinden, R. and Deering, S. (2006) *IP Version 6 Addressing Architecture*. RFC 4291.

[29] Narten, T., Nordmark, E., Simpson, W., and Soliman, H. (2007) *Neighbor Discovery for IP version 6 (IPv6)*. RFC 4681.

[30] Conta, A., Deering, S., and Gupta, M. (2006) *Internet Control Message Protocol (ICMPv6) for the Internet Protocol Version 6 (IPv6) Specification*. RFC 4443.

[31] Dohler, M., Watteyne, T., Winter, T., and Barthel, D. (2009) *Routing Requirements for Urban Low-Power and Lossy Networks*. RPL 5548.

[32] Kent, S. (2005) *IP Authentication Header*. RFC 4302.

[33] Postel, J. (1980) *User Datagram Protocol*. RFC 0768.

[34] Postel, J. (1981) *Transmission Control Protocol*. RFC 0793.

[35] Dierks, T. and Rescorla, E. (2008) *The Transport Layer Security (TLS) Protocol Version 1.2*. RFC 5246.

[36] Naumann, A., Komarnicki, P., Buchholz, B.M., and Brunner, C. (2011) *Seamless Data Communication and Management over All Levels of the Power System*. Proceedings of 21st International Conference on Electricity Distribution (CIRED), paper 0988, 1–5. Frankfurt, June 6–9, 2011..

[37] UCA, Utility Communications and Architecture, International Users Group (2010) *UCAIug Home Area Network System Requirements Specification*. OpenHAN 2.0.

[38] Jorgensen, T. (2006) *Z-Wave as Home Control RF Platform*, www.zen-sys.com (accessed 8 May 2012).

[39] Zensys, A.S. (2006) *Z-Wave Protocol Overview*, Zensys, Copenhagen.

[40] Z-Wave Alliance (2012) www.Z-Wave.com/modules/iaCM-ZW-PR/readMore.php?id=577765376 (accessed 10 May 2012).

[41] ECHONET Consortium (2012) www.echonet.gr.jp/english/index.htm (accessed 15 April 2012).

[42] Tariq, M., Zhou, Z., Wu, J., Macuha, M. and Sato, T. (2012) *Smart grid standards for home and building automation,* Proceedings of IEEE Powercon, 1-6. Auckland, New Zealand.

[43] ZigBee Alliance (2010) *ZigBee Home Automation Public Application Profile*. http://ZigBee.org/Markets/Overview/tabid/223/Default.aspx (accessed 7 December 2012).

[44] Jamieson, P. (2008) *ZigBee Cluster Library Specification*.

[45] International Organization for Standardization/International Electrotechnical Commission (2007) ISO/IEC 16484-5 *Building Automation and Control Systems – Part 5: Data communication Protocol*, International Organization for Standardization.

[46] American National Standards Institute/American Society of Heating, Refrigerating and Air-conditioning Engineering. (2008) *A Data Communication Protocol for Building Automation and Control Networks*. BACnet ANSI/ASHRAE 135-2008.

[47] Palo, A. and Echelon Corporation (1994) *LONTALK Protocol Specification Version 3.0*.

[48] Echelon Corporation (2003) *Building Automation Technology Review*.

[49] Merz, H., Hansemann, T., and Hubner, C. (2009) *Building Automation-Communication Systems with EIB/KNX, LON, and BACnet*, Springer, New York.

[50] Palo, A. and Echelon Corporation. *LONMAKER User's Guide Release 3*.

[51] LONMARK International (2012) *LONMARK International Overview*, www.lonmark.org/about/ (accessed 20 March 2012).

[52] INSTEON (2012) *INSTEON – The Details, Smart Home Technology*, www.INSTEON.net/pdf/ INSTEONdetails.pdf (accessed 24 April 2012).

[53] INSTEON (2006) *INSTEON – Compared, Smart Labs Technology*, www.INSTEON.net/pdf/ INSTEONcompared.pdf (accessed 24 April 2012).

[54] KNX Association (2009) *KNX System Specifications-Architecture Version 3.0-1*.

[55] KNX Association (2010) *KNX System Specifications-Interworking-Datapoint Types Version 1.5.00*.

[56] KNX Association (2010) *KNX Profiles*.

[57] Threshold Corporation (2011) *ONE-NET Specification, Version 1.6.2*, www.ONE-NET.info/ spec/ONE-NET_Specification_v1.6.2.pdf (accessed 2 May 2012).

[58] Threshold Corporation (2011) *ONE-NET Device Payload Format, Version 1.6.2*, www.ONE-NET .info/spec/ONE-NET%20Device%20Payload%20Format%20v1.6.2.pdf (accessed 2 May 2012).

6

Communications in the Smart Grid

6.1 Introduction

If the entities of the Smart Grid system create one consistent body, operation and maintenance can be compared to a brain, responsible for the overall control of the whole system. The transmission and distribution lines are like blood vessels carrying "blood," that is, energy, to the places where it is needed. Then, the communication architecture of the Smart Grid can be considered as the nervous system connecting all the organs and limbs to the brain and enabling interaction with it.

In recent years, utilities have worked on the modernization of the electrical grid with an integrated communication infrastructure. There are numerous communication technologies covered by various standards developed by Standards Developing Organizations (SDOs) and technical specifications created by alliances, consortia, and forums consisting of manufacturers, vendors, and service providers in order to ensure interoperability, cost efficiency, and reliability. The Smart Grid is a very complex system and it is crucial for the whole system that its communication architecture will meet the specific requirements of various applications. The key features of communication technologies are the following:

- **Reliability** – It is crucial to ensure operation of a Smart Grid system 24/7; therefore, its communication architecture must also meet at least the same reliability requirements to prevent overload, blackouts, and so on.
- **Security** – Communication architecture is expected to transport customer privacy information including address and billings and needs to be highly secure to prevent potential attacks and frauds.
- **Scalability** – The Smart Grid is expected to rapidly grow with millions of devices and appliances. At the same time, communication infrastructure is expected to last for decades.

Smart Grid Standards: Specifications, Requirements, and Technologies, First Edition. Takuro Sato,
Daniel M. Kammen, Bin Duan, Martin Macuha, Zhenyu Zhou, Jun Wu, Muhammad Tariq and Solomon Abebe Asfaw.
© 2015 John Wiley & Sons, Ltd. Published 2015 by John Wiley & Sons, Ltd.

- **Low Latency** – The requirements for latency varies from application to application and the lowest latency requirements (i.e., less than 10 ms for teleprotection) are far below the requirements of common data or voice services as well as lower than the practically achievable latencies by most of the current communication technologies.
- **Quality of Service** – Minimum bit rate, low error rate as well as latency limits are important for various applications in the Smart Grid.
- **Interoperability** – Standardized solutions are crucial for global interoperability between different communication technologies.
- **Low Cost** – Communication infrastructure and maintenance should not incur high capital and operation expenditures.

In this chapter, we focus on communication technologies as a means of transportation of data necessary for Smart Grid operation and its applications rather than on the data itself, as the standards related to the communicated content, interoperability, and security are provided in other chapters related to different areas of the Smart Grid architecture. On the other hand, many current communication technologies are application specific and their protocol stack contains specific application layers. This is typical for many home automation and control technologies, which are described in *Section 5.4 Smart Home and Building Automation Standards*. For simplicity and consistency reasons, we omit these application-specific standards for Home Area Networks (HANs) and Building/Business Area Networks (BANs) in this chapter.

We first overview the communication requirements of the Smart Grid applications, then introduce the communication architecture and importance of end-to-end connectivity. In the second part, we provide information on the standards and technologies of wired communication systems, including power line communication (PLC), optical fiber communication, and other technologies. In the third part of this chapter, we survey wireless communication standards and technologies that are being deployed for Smart Grid communication or might play an important role for future deployments.

6.1.1 Communication Requirements for the Smart Grid

The Smart Grid is a very broad concept requiring the support of many different applications with specific requirements for communication links and network topologies. Distributed Energy Resources (DERs) and storage, Electric Vehicles (EVs), Demand Response (DR) and real-time payments, Advanced Metering Infrastructure (AMI) and distributed automation, load management and substation automation, fault management and blackout prevention are key applications for which proper communication links play a crucial role. The communication requirements have been analyzed in detail by OpenSG in [1] as well as by the US Department of Energy in [2], totaling more than 1400 different data flows with specific requirements such as payload size, payload type, frequency of data transmission, required reliability, security, latency, and importance.

AMI is located on the line between a distribution network and consumers' premises. Therefore, the information communicated over this entity is mostly related to metering and measurements, payment information, and outage notification. Typical communication on the premises involves automation of the smart appliances monitoring the energy consumption, historical analysis of the consumed energy, communication with EVs, and energy microgenerators such as on-site solar systems. The information amount per device should likely be 10–100 kbps [1], which does not put a significant constraint on the capacity of the links. However, larger premises such as office buildings and industrial parks should be scaled properly as the number of communicating devices might be very large. Metering does not require very high reliability of the communication. On the other hand, the on-site microgeneration and distribution network overload prevention requires highly reliable links. In addition latency requirements for metering are not strict and are usually of the order of seconds. DR has similar latency and bandwidth requirements and it is not a "mission critical" application. DER might require fairly low latencies due to fault prevention mechanisms (20 ms), however, for nonemergent operations, a delay of 300 ms should be sufficient. DER can be considered as a "mission critical" application operating with high reliability. Bandwidth requirements of DER are similar that for AMI or DR. Wide Area Situational Awareness (WASA) is used for monitoring of the power system across large geographic areas in order to improve the whole system reliability and prevent potential power-supply disruption. Therefore, WASA is a typical "mission critical" application, where low communication latency and high reliability of the communication technology are crucial. Substation automation and distribution automation should have relatively fast responsiveness with latencies less than 100 ms due to life-critical situations involving high-voltage lines and isolating potential faults [2]. This puts limitations on the selection of the proper communication technologies for both applications. The bidirectional communications between EVs and the grid will likely not impose high demands on the bandwidth as the expected communication with the grid will be mostly load balancing and billing. Similarly, EV is not a critical application and does not require very low latency. An overview of the communication requirements for various Smart Grid applications is shown in Table 6.1. The list of communication standards is shown in Table 6.2.

Table 6.1 Communication requirements of various Smart Grid applications

Application	Bandwidth	Latency	Reliability[a]
Advanced metering infrastructure	500 kbps for backhaul (10–100 kbps per device on the premises)	2 s or more	Medium
Demand and response	14–100 kbps per device/node	500 ms – several minutes	Medium
Distributed energy resources	9.6–56 kbps	300 ms to 2 s	High
		Fault protection 20 ms	
Wide-area situational awareness	600–1500 kbps	20–200 ms	High
Substation automation	9.6–56 kbps	15–200 ms	High
Distribution automation	9.6–56 kbps	20–200 ms	High
Electric vehicles	9.6–56 kbps	2 s to 5 min	Medium

[a]Reliability: low ($<$99%), medium ($>$99%), and high ($>$99.99%).

6.1.2 List of Standards

Table 6.2 List of standards

Standard	Body	Details
Power line		
HomePlug AV	HomePlug alliance	HomePlug AV offers a peak data rate of 80 Mbps at the MAC layer. HomePlug AV devices are required to coexist, and optionally to interoperate, with HomePlug 1.0 devices
HomePlug AV2	HomePlug alliance	It is interoperable with HomePlug AV and HomePlug Green PHY devices and is IEEE 1901 standard compliant. It enables gigabit-class data rates, supports MIMO, and utilizes power-saving modes
HomePlug Green PHY	HomePlug alliance	It is interoperable with HomePlug AV and HomePlug AV2 devices and is IEEE 1901 standard compliant. The HomePlug Green PHY specification is a subset of HomePlug AV that is intended for use in the Smart Grid. It has peak rates of 10 Mbit/s and is designed to go into smart meters and smaller appliances such as HVAC thermostats, home appliances, and plug-in electric vehicles
IEEE 1901	ITU	Standard for high-speed ($>$ 100 Mbps at the physical layer) communication over power lines. The key technology for broadband PLC

Table 6.2 (*continued*)

Standard	Body	Details
HD-PLC	HD-PLC alliance	HD-PLC uses a high-frequency efficient wavelet-OFDM modulation method. The theoretical maximum data transmission rate is up to 210 Mbps
ITU-T G.9955	ITU	It defines the physical layer specification for narrowband OFDM power line communications transceivers for communications via alternating current and direct current electric power lines over frequencies below 500 kHz
PRIME	PRIME alliance	It is a narrowband PLC data transmission system over the electricity grid. The whole architecture has been designed to be low cost but high performance. It uses Orthogonal Frequency Division Multiplexing (OFDM) in narrowband frequency ranges
G3-PLC	G3-PLC alliance	Provides high-speed, highly reliable, long-range communication over the existing power line grid. It has the ability to cross transformers and supports IPv6
IEEE P1901.2	IEEE	Standard for Low Frequency (less than 500 kHz) narrow band power line communications for Smart Grid applications
Netricity PLC	HomePlug alliance	Netricity PLC is compliant and interoperable with IEEE 1901.2. It also supports interoperability with PRIME and G3-PLC technologies
IEC 61334	IEC	Standard for low-speed reliable power line communications by electricity meters, water meters, and SCADA
ITU-T G.9960 (G.hn)	ITU	Standard defining networking over power lines, phone lines, and coaxial cables with data rates up to 1 Gbps. It utilizes OFDM and modulation techniques up to 4096 QAM
Optical ITU-T G.651.1	ITU	Characteristics of a 50/125 μm multimode graded-index optical fiber cable for the optical access network
ITU-T G.652	ITU	Characteristics of a single-mode optical fiber and cable
ITU-T G.959	ITU	Optical transport network physical layer interfaces
ITU-T G.693	ITU	Optical interfaces for intraoffice systems
ITU-T G.692	ITU	Optical interfaces for multichannel systems with optical amplifiers
T1.105.07	ANSI	Synchronous Optical Network (SONET) - Sub-STS-1 Interface Rates and Formats Specification

(*continued overleaf*)

Table 6.2   (*continued*)

Standard	Body	Details
ITU-T G.707	ITU	Recommendation G.707/Y.1322, Network node interface for the Synchronous Digital Hierarchy (SDH)
ITU-T G.783	ITU	Characteristics of Synchronous Digital Hierarchy (SDH) equipment functional blocks
ITU-T G.784	ITU	Management aspects of the Synchronous Digital Hierarchy (SDH) transport network element
ITU-T G.803	ITU	Architecture of transport networks based on the Synchronous Digital Hierarchy (SDH)
ITU-T G.983.x	ITU	Broadband optical access systems based on Passive Optical Networks (PON), a series of recommendations for broadband passive optical networks
IEEE 802.3ah	IEEE	Also known as "Ethernet in the First Mile;" defines the physical-layer specifications for Ethernet links providing 1000 Mbps over PONs up to at least 10 km (1000BASE-PX10) and up to at least 20 km (1000BASE-PX20)
ITU-T G.984.x	ITU	Gigabit-capable passive optical networks (GPON); a series of recommendations for GPON access networks
IEEE 802.3av	IEEE	Physical layer specifications and management parameters for 10 Gbps Ethernet Passive Optical Networks (10 GE-PON)
ITU-T G.987	ITU	10-Gigabit-capable passive optical network (XG-PON) systems: definitions, abbreviations, and acronyms; a series of recommendations for gigabit passive optical access networks
NG-PON2	FSAN	NG-PON2: next-generation passive optical network 2
Very short range/contactless		
ISO 10536	ISO	ISO RFID standard for close coupled cards
ISO 11784	ISO	ISO RFID standard that defines the way in which data is structured on an RFID tag
ISO 11785	ISO	ISO RFID standard that defines the air interface protocol
ISO 14443	ISO	ISO RFID standard that provides the definitions for the air interface protocol for RFID tags used in proximity systems – aimed for use with payment systems
ISO 15693	ISO	ISO RFID standard for use with what are termed vicinity cards
ISO 15961	ISO	ISO RFID standard for item management (includes application interface (part 1), registration of RFID data constructs (part 2), and RFID data constructs (part 3)

Table 6.2 (*continued*)

Standard	Body	Details
ISO 18000	ISO	ISO RFID standard for the air interface for RFID frequencies around the globe
ISO 24753	ISO	Air interface commands for battery assist and sensor functionality
UHF Class 1 Gen 2	EPCglobal	Physical and logical requirements for a passive-backscatter, Interrogator-Talks-First (ITF), Radio-Frequency Identification (RFID) system operating in the 860–960 MHz frequency range
EPC Class-1 HF	EPCglobal	Physical and logical requirements fora passive-backscatter, Interrogator-Talks-First (ITF), Radio-Frequency Identification (RFID) system operating in 13.65 MHz frequency
ASTM D7434	ASTM	Standard test method for determining the performance of passive Radio-Frequency Identification (RFID) transponders on palletized or unitized loads
ASTM D7435	ASTM	Standard test method for determining the performance of passive Radio-Frequency Identification (RFID) transponders on loaded containers
ASTM D7580	ASTM	Standard test method for rotary stretch wrapper method for determining the readability of passive RFID transponders on homogeneous palletized or unitized loads
JIS X 6319-4	JIS	Also known as "Felica," the Japanese Industrial Standard (or JIS) specifying the physical characteristics, air interface, transmission protocols, file structure, and commands of high-speed contactless proximity integrated circuit cards
ISO/IEC 21481	ISO	This standard specifies the communication mode-selection mechanism, which is designed to not interfere with any ongoing communication at 13.56 MHz, for devices implementing ECMA-340, ISO/IEC 14443, or ISO/IEC 15693
ISO/IEC 18092	ISO	Also known as "ECMA-340." The standard defines communication modes for Near Field Communication Interface and Protocol (NFCIP-1) using inductive coupled devices operating at the center frequency of 13.56 MHz for interconnection of computer peripherals

(*continued overleaf*)

Table 6.2   (*continued*)

Standard	Body	Details
WLAN and WPAN		
IEEE 802.11	IEEE	Wireless Local Area Network (WLAN) technologies developed by IEEE containing many amendments for improving security, quality of service, data rates, interworking, and so on. It is a base standard for the widely spread Wi-Fi technology, which is specified by the Wi-Fi Alliance
IEEE 802.15.1	IEEE	Wireless Personal Area Network (WPAN) technology for connecting peripherals is developed by IEEE and is a base for an initial version for the widely used Bluetooth technology, which has been further enhanced by the Bluetooth Special Interest Group
IEEE 802.15.4	IEEE	Wireless Persona Area Network (WPAN) technology for low-rate transmission is developed by IEEE and a base for many widely used technologies for sensor networks and M2M communication including ZigBee
IEC 62591	IEC	Industrial communication networks – Wireless communication network and communication profiles – WirelessHART™
Cellular/WAN/MAN		
GSM	ETSI	Standard developed by ETSI and maintained under 3GPP as TS 45.001 (PHY) and TS. 23.002 (network architecture). It belongs to second-generation cellular systems
EDGE	ETSI	Enhanced data rates for GSM evolution. It belongs to the GSM family and is backward compatible. The physical layer is maintained under the same standard as GSM (TS 45.001) by 3GPP
CDMAone	TIA	Developed by Qualcomm Inc. as IS-95. It defines the compatibility standard for 800-MHz cellular mobile telecommunications systems and 1.8–2.0 GHz Code Division Multiple Access (CDMA) Personal Communications Service (PCS) systems
CDMA2000	TIA	Approved standard for ITU's IMT-2000 (also called 3G). Defined by TIA and backward compatible with CDMAone (IS-95)
UMTS	3GPP	Universal Mobile Telecommunications System. Defined by 3GPP in Release 99 and approved by ITU's IMT-2000 (also called 3G). Radio network technology is W-CDMA
HSPA	3GPP	High-Speed Packet Data Access is an enhancement of W-CDMA technology. The downlink (HSDPA) enhancement is defined in Release 5 and uplink enhancement (HSUPA) is defined in Release 6. Further enhancements of HSPA technology are in Release 7 and later releases, known as HSPA+

Table 6.2   (*continued*)

Standard	Body	Details
WiMAX	IEEE/WiMAX Forum	Worldwide Interoperability for Microwave Access (WiMAX) is a technology based on the IEEE 802.16 family of standards and is maintained and promoted by the WiMAX forum. Mobile WiMAX is a candidate for ITU's IMT-advanced (also called 4G) technology together with LTE
LTE	3GPP	Long-Term Evolution (LTE) is a technology developed by 3GPP specified in Release 8 and further enhanced in later releases. It is a candidate for ITU's IMT-advanced (also called 4G) as its further enhancements (LTE advanced) satisfy the requirements defined in ITU's IMT-advanced technology
Satellite		
GMR	ETSI	Geostationary earth orbit Mobile Radio Interface (GMR) is developed by ETSI and maintained by 3GPP, to support access to GSM/UMTS core networks
EN 302 977	ETSI	Satellite Earth Stations and Systems (SES); Harmonized EN for Vehicle-Mounted Earth Stations (VMESs) operating in the 14/12-GHz frequency bands covering the essential requirements of article 3.2 of the R&TTE directive
TS 102 856-1	ETSI	Satellite Earth Stations and Systems (SES); Broadband Satellite Multimedia (BSM); Multiprotocol Label Switching (MPLS) interworking over satellite Part 1: MPLS-based Functional Architecture
TS 102 856-2	ETSI	Satellite Earth Stations and Systems (SES); Broadband Satellite Multimedia (BSM); Multiprotocol Label Switching (MPLS) interworking over satellite; Part 2: Negotiation and management of MPLS labels and MPLS signaling with attached networks
EN 302 550-1	ETSI	Satellite Earth Stations and Systems (SES); Satellite Digital Radio (SDR) Systems; Part 1: Physical Layer of the Radio Interface
TS 102 550	ETSI	Satellite Earth Stations and Systems (SES); Satellite Digital Radio (SDR) Systems; Outer Physical Layer of the Radio Interface
EN 302 574	ETSI	Satellite Earth Stations and Systems (SES); Harmonized Standard for satellite earth stations for MSS operating in the 1980–2010 MHz (earth-to-space) and 2170–2200 MHz (space-to-earth) frequency bands

(*continued overleaf*)

Table 6.2   (*continued*)

Standard	Body	Details
TR 101 865	ETSI	Satellite component of UMTS/IMT-2000; general aspects and principles; parts 1–6
TS 101 851	ETSI	Satellite component of UMTS/IMT2000; G-family
TS 102 442	ETSI	Satellite component of UMTS/IMT-2000; Multimedia Broadcast/Multicast Services; parts 1–6
ITU-R M.1854	ITU	Information on the range of radio frequencies for mobile-satellite service (MSS) systems
ITU-R S.1001-2	ITU	Information on the range of radio frequencies that can be used by fixed-satellite service (FSS) systems for emergency and disaster-relief operations

6.2   Architecture of the Communication System in the Smart Grid

Communication systems in the Smart Grid are hierarchical and are also slightly different from the typical Information and Communication Technology (ICT) network architecture. A simplified architecture with selected candidate technologies suitable for different types of area networks is shown in Figure 6.1. HAN, recognized as the smallest network type, is the area behind the customer's premises, connecting various devices such as Personal Computers (PCs), entertainment equipment, security devices, smart home appliances, and smart meter. Similarly, BAN includes building management systems, Heating Ventilation and Air Conditioning (HVAC) systems,

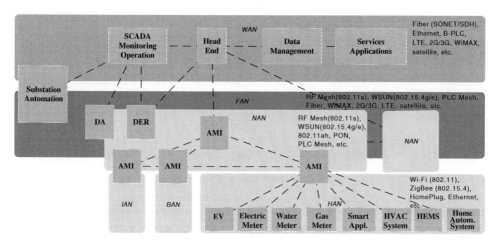

Figure 6.1   Simplified architecture with selected technologies as candidates for different types of area networks

local power generators and storages, and Industrial Area Network (IAN) includes machinery and industrial automation. On the edge of the premises resides the smart meter, an essential part of AMI. In the neighborhood area, several AMI can be connected together to create the Neighborhood Area Network (NAN), effectively aggregating traffic from HANs. Similarly, several NANs can be aggregated in the Field Area Network (FAN) that also connects DER and Distribution Automation and the substation network. On top of this hierarchy is a Wide Area Network (WAN), which connects all separated parts together and provides connectivity to the centralized entities such as the AMI head end, operation centers, and application servers.

6.2.1 IP in the Smart Grid

The Internet Protocol (IP) plays a crucial role in the Smart Grid design and is one of the key protocols to enable the end-to-end connectivity and interoperability of the Smart Grid networks. There are several reasons for Smart Grid functions to operate over IP. For example, controllability, network visibility, and addressability of various sensors in the distribution system, automation, and control of DER; even control of energy generators, smart meters, and thermostats within AMI can be achieved by end-to-end connection over IP. IP makes the Smart Grid applications independent of the physical media and data link communication technologies, provided these technologies meet the requirements for the given application. This greatly reduces the complexity for developing upper-layer applications and enables interoperability. IP offers good scalability, which is another important requirement for a Smart Grid network integrating millions of devices.

The well-known problem of IP is the currently insufficient addressing range of IPv4, which imposes the need for support of IPv6 routing protocol. This is in the design a contradiction to the current technologies for sensor networks as well as technologies for home automation, which try to minimize overheads by using their own proprietary addressing schemes. In order to support IPv6 on many wired and especially wireless communication technologies, various adaptation layers have been designed by the Internet Engineering Task Force (IETF) with IPv6 header compression, neighbor discovery optimization, and various other functionalities.

The overview of layered architecture with unified network layer for end-to-end connectivity and interoperability together with various communication technologies and adaptation layers is shown in Table 6.3.

6.2.1.1 IPv6 over Low-Power Wireless Networks and Routing Protocol for Low-Power and Lossy Networks

IPv6 over Low-Power Wireless Personal Area Networks (6LoWPANs) is an open standard that is defined by IETF. The key role of this document is to ensure

Table 6.3 Layered architecture of the Smart Grid communication

Application	IEC 61850, IEC 60870, CIM, SNMP, and so on							
Network	TCP/UDP							
	RPL							
	IPv4/IPv6							
	EAP-TLS							
Adaptation	6LoWPAN				RFC 2464		RFC 5072	RFC 5121
MAC	IEEE 802.15.4	IEEE 802.15.4e		Wave2M	IEEE 802.11 b/g/n (2.4 GHz), a/n/ac (5 GHz) ah (subgigahertz)	IEEE 802.3 Ethernet	2G/3G/LTE	IEEE 802.16/WiMAX
		IEEE 802.15.4	IEEE P1901.2					
PHY		IEEE 802.15.4g						
Media	Radio	Radio	Powerline	Radio	Radio	Coax./twist. pair/fiber	Radio	Radio

interoperability of implementations of 6LoWPAN networks among different applications [3]. 6LoWPAN was originally aimed to provide an adaptation layer for the IEEE 802.15.4 PHY (Physical Layer)/MAC (Medium Access Control) layer technologies and defines necessary optimization schemes in order to transport communication over IPv6 [4]. Besides IEEE 802.15.4, 6LoWPAN adaptation layer has been later adopted also by Wave2M, IEEE P1901.2, Bluetooth Low Energy, and some other technologies. 6LoWPAN is considered as a necessary standard for deploying IP-based wireless sensor networks and extending the global connectivity to various devices in order to provide end-to-end connectivity [5]. Special attention is paid to a reduction of the need for multicast compression of IPv6 addresses [6], neighbor discovery optimization in IPv6 [7], and routing protocols for low-power links [8]. The default routing protocol for a low-power link is RPL (Routing Protocol for Low-Power and Lossy Networks) and is defined by ROLL (Routing Over Low-power and Lossy networks) Working Group of IETF [9]. It is a distance vector routing protocol over IPv6 designed for Low-Power and Lossy Networks (LLNs). It combines multiple metrics to select the best path. The distance vector protocol creates logical topology over a physical network and can include Quality of Service (QoS) of the traffic, and various constraints for the specific graph creation. The RPL plays an important role in the mesh networks, where the proper path selection over several hops has great impact on the end-to-end throughput and latency.

6.3 Wired Communication

6.3.1 Power Line Communication

6.3.1.1 Overview

Compared to wireless and other wired communication technologies such as Digital Subscriber Line (DSL) and fiber optic, PLC offers a cost-effective solution by transmitting data over the existing electrical infrastructure. PLC substantially reduces system cost by eliminating the need to install additional wires such as twisted pair or coaxial cables to interconnect devices. PLC has the natural advantage in applications of electricity monitoring, demand response, and load control, and AMI as nearly every home appliance is connected to the power line.

PLC can be classified into two groups: Narrowband Power line Communication (NB-PLC) and Broadband Power line Communication (BB-PLC). NB-PLC, which is usually operated in a low-frequency band such as 3–500 kHz, is suited for wide-area access applications or Smart Grid applications, where low cost and high reliability are essential. On the other hand, BB-PLC, which is usually operated in a high-frequency band such as 2–100 MHz, is suited for in-home broadband applications such as HPTV, VoIP, Video Game, and Internet where high-speed broadband network connections are essential. The various PLC standards and technologies have been classified into these two groups and are shown in Figure 6.2. The latest BB-PLC technologies, for example, HomePlug AV2, offers a peak rate of more than 1 Gbps by employing advanced signal processing technologies such as Multiple Input Multiple Output (MIMO) and precoding which are achieved in the field of wireless communications. However, these technologies also increase system complexity, power consumption, and capital cost. In contrast, NB-PLC technologies are developed with the main goal to meet the requirements of long-range, outside-the-home applications, while maintaining low complexity, low power consumption, and high reliability. Therefore, BB-PLC and NB-PLC technologies should be combined together to enable both the in-home broadband applications and wide-area access applications.

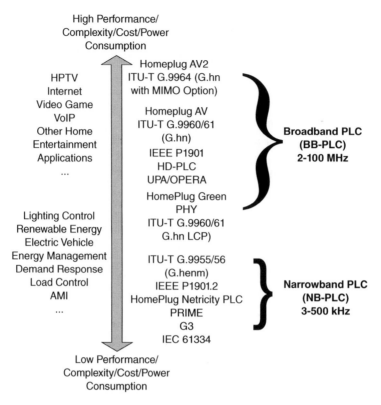

Figure 6.2 A classification of various PLC technologies and standards into two groups: NB-PLC and BB-PLC

Despite the significant benefits of PLC technologies, PLC faces many challenges that are particularly caused by the physical characteristics of the transmission medium, that is, power line cables. Power line cables were not designed to carry communication signals. While the PLC channel characteristics depend heavily on the application scenario and use case, in general, multipath fading, frequency selective fading, time-varying channel, and interference are major technical challenges for deploying PLC. Multipath fading arises when PLC signals reach a receiver over more than one path, especially when many branches exist between the transmitters and the receiver. Owing to the different length of each path, the transmitted signal in each path experiences different delays and attenuations and are recombined at the receiver. Frequency-selective fading is caused by partial cancellation of a signal by itself, that is, the signals arrive at the receiver by different paths and cancel each other when recombined. Time variation of the PLC channel is caused by loads being connected or disconnected into power lines. Interference is caused by amateur radio or TV broadcasting signals that are operated in the same frequency band as the PLC signal. Electromagnetic Compatibility (EMC) regulations defined for NB-PLC and BB-PLC are summarized in Table 6.4.

Table 6.4 EMS regulations defined for NB-PLC and BB-PLC

EMC regulations related to NB-PLC

Countries/regions	Frequency bands	Responsible institutions/SDOs
United States	10–490 kHz	Federal Communications Commission (FCC)
Japan	10–450 kHz	Association of Radio Industries and Business (ARIB)
European Union	3–148.5 kHz	European Committee for Electrotechnical Standardization (CENELEC)
China	3–90 kHz	China Electric Power Research Institute (CEPRI)

EMC regulations related to BB-PLC

EU	1.6–30 MHz	CENELEC (CENELEC EN EN 50561-1)
Worldwide	Limits specified for 9 kHz to 400 GHz [10]	International Electrotechnical Commission (CISPR 22 ed6.0)

Various popular PLC technologies have been proposed by different regional SDOs or alliances to meet the unique needs of that region. Several national or international SDOs have been trying to merge those different PLC technologies into a single international standard to guarantee interoperability between multiple vendors. In this section, the interrelationships between those PLC technologies will be introduced, and each PLC technology will be compared with others in technical details.

6.3.1.2 PLC Technologies, Specifications, and Standards

NB-PLC technologies are widely used for smart home, building automation, and smart metering applications. For smart home and building automation applications, NB-PLC technologies have been included in various regional/international standards, such as the ISO/IEC 14543 (KNX) standard series, the BACnet protocol (included in the ASHRAE/ANSI 135 and ISO 6484-5), the LONWORKS system (included in ANSI/CEA 709, ANSI/EIA 852, and ISO/IEC DIS 14908). A more detailed introduction and explanation about the above standards can be found in *Section 5.4 Smart Home and Building Automation Standards*.

For smart metering applications, a broad set of technologies and standards such as PoweR line Intelligent Metering Evolution (PRIME), G3, HomePlug Netricity PLC, ITU-T G.henm (ITU-T G.9955/9956), and IEEE P1901.2, have been developed by different alliances or SDOs. A detailed comparison of NB-PLC standards used for smart meters is provided in the technical details in Table 6.5.

PRIME was developed by the PRIME alliance to define PHY and MAC layers of an NB-PLC system over the electric grid [11]. Orthogonal Frequency Division

Table 6.5 A comparison of NB-PLC standards used for smart metering in technical details

Technology	PRIME	G3	ITU-T G.henm	IEEE P1901.2	IEC61334
Modulation	OFDM	OFDM	OFDM	OFDM	SFSK
Coding	Conv.	RS + Conv.	RS + Conv.	Rs + Conv.	No
Peak data rate	130 kbps	34 kbps	1 Mbps	500 kbps	2.4 kbps
Bands (kHz)	3–95	3–95/ 150–490	32–490	10–490	20–95
Access mechanism	CSMA/CA + TDM	CSMA/CA	CSMA/CA + TDM	CDMA/CA	Repeater call (IEC 61334-5-1)
Security	AES 128	AES 128	AES 128	AES 128	No
ROBO mode	No	Yes	Yes	Yes	No
Convergence sublayer	IEC 61334-4-32/IPv4	6LoWPAN/ IPv6	IPv4/IPv6/ ethernet/L3 protocol flexible	6LoWPAN/ IPv6	IEC 61334-4-32

Multiplexing (OFDM) was selected as the basic modulation scheme for PRIME due to its high data rate and robustness against multipath fading and frequency-selective fading. A service-specific Convergence Sublayer (CS) was defined to map different types of traffic to be properly defined in MAC Service Data Units SDUs. Another OFDM-based NB-PLC specification is G3-PLC, which was developed by the G3-PLC Alliance. G3-PLC was designed to support true IPv6 addressing and also the ROBO mode [12]. In a ROBO mode, the same information is transmitted redundantly on multiple subcarriers to provide reliability under severe channel conditions. Both PRIME and G3-PLC form the basis of major international NB-PLC standards such as IEEE P1901.2 [13] and ITU-T G.hnem (also referred to as ITU-T G.9955/9956). ITU-T G.hnem is developed by ITU-T in cooperation with ISO (International Organization for Standardization)/IEC (International Electrotechnical Commission) and SAE (Society of Automotive Engineers) to be the next generation NB-PLC standard for the Smart Grid [14]. ITU-T G.hnem is compliant with IEEE P1901.2 via legacy Annexes. IEEE P1901.2 was developed in parallel with ITU-T G.hnem with mainly the same goal as ITU-T G.hnem to support Smart Grid applications such as AMI, EV, and solar panel. In order to promote adoption and provide compliance and interoperability testing of products based on IEEE P1901.2, the HomePlug Alliance has announced the Netricity PLC certification brand and marketing program [15]. Unlike the above-mentioned NB-PLC standards that employ OFDM as the modulation scheme, IEC 61334 uses Spread-Frequency Shift Keying (S-FSK) as the modulation scheme to provide robust low data rate communication over power lines [16]. IEC has also defined the application layer standard (IEC 62056) based on IEC 61334 standards to allow development of interoperable solutions. Despite the widespread popularity of S-FSK, the increasing data demand requires more robust and multicarrier modulation schemes such as OFDM.

On the other hand, BB-PLC technologies are widely used for broadband networking applications such as Internet, VoIP, and HPTV. A broad set of BB-PLC technologies and standards have been developed in parallel by different SDOs or alliances. The HomePlug AV specification was developed by the HomePlug Alliance to be completely interoperable with the IEEE P1901 standard. HomePlug AV has a signal bandwidth of 28 MHz and provides a peak PHY data rate of 10 Mbps for the ROBO mode and 200 Mbps for the adaptive bit loading mode, respectively [17]. Adaptive bit loading enables each subcarrier to use the most suitable modulation scheme for the given channel conditions. HomePlug AV2 was developed to support HD (High Definition)/3D video and other bandwidth-hungry applications, while maintaining full interoperability with HomePlug AV and HomePlug Green PHY (GP) [18]. By employing MIMO with beamforming and the additional 30–86 MHz bandwidth, HomePlug AV2 achieves a peak PHY data rate of more than 1 Gpbs. HomePlug GP was developed as a simplified version of HomePlug AV to reduce both cost and power consumption [19]. A comparison of the PHY parameters for HomePlug BB-PLC specifications is given in Table 6.6. Besides HomePlug BB-PLC technologies, other BB-PLC technologies include High Definition Power Line Communication (HD-PLC) using the wavelet OFDM and M-PAM modulation schemes [20], and the Universal Power line Association (UPA)/Open PLC European Research Alliance (OPERA) specification [21].

In order to enable interoperability between multiple PLC vendors, ITU-T and IEEE (Institute of Electrical and Electronics Engineers) have been working to develop internationally adopted BB-PLC standards, that is, ITU-T Gigabit Home Networking (G.hn) (G.9960/9961) and IEEE P1901. ITU-T G.hn is the first standard to define a single standard for all major wired communication media including power lines, phone lines, and coaxial cables [22]. The support of MIMO is included in the G. 9964 extension specification. IEEE P1901 is based on two different PHY/MAC specifications – HomePlug AV and HD-PLC [23]. In order to harmonize the coexistence between ITU-T G.hn and IEEE P1901, the National Institute of

Table 6.6   A comparison of the PHY parameters for HomePlug BB-PLC specifications

Technology	HomePlug AV	HomePlug Green PHY	HomePlug AV2
Bands (MHz)	2–30	2–30	2–86
Modulation	OFDM	OFDM	OFDM
Subcarrier modulation	BPSK, QPSK, 16/64/256/ 1024QAM	QPSK only	Up to 4096QAM
Coding	Rate 1/2 or rate 16/21 turbo code	Rate 1/2 turbo code only	Up to rate 8/9 turbo code
Peak data rate	ROBO: 10 Mbps Adaptive bit loading: 200 Mbps	ROBO mode only: 10 Mbps	1.8 Gbps (2 streams)

Standards and Technology (NIST) Smart Grid Interoperability Panel (SGIP) initiated the PAP 15 *Harmonize Power Line Carrier Standards for Appliance Communications in the Home*. A coexistence mechanism named Inter System Protocol (ISP) has been developed to allow for frequency bands being shared among up to four noninteroperable systems in the time domain and frequency domain, or both. Devices that implement noninteroperable protocols such as IEEE 1901 wavelet, IEEE 1901 FFT, Low-Rate Wide-Band Services (LRWBSs), and ITU-T G.hn are able to coexist by employing ISP.

6.3.2   Optical Communication

Fiber-optic communication has been used for high-demand, high-reliability, and long-distance applications since the beginning of the 1990s. However, the real boom of deployment of optical networks started 10 years later, driven by the decrease in prices of optical fiber with other components also becoming much more cost efficient. Currently, optical-fiber networks are largely used for core networks. As for the Smart Grid, one particular reason for favoring fiber optics against some other broadband technologies like LTE (Long-Term Evolution), WiMAX (Worldwide Interoperability for Microwave Access), or distributed radio frequency (RF) Mesh, is the aggregate data bandwidth for two-way communication for Smart Grid-related data in large cities and dense populations. When serving hundreds of thousands of households or even more, the aggregate bandwidth rises to several gigabytes per second [24]. Other use cases, when fiber-optic network deployment is of particular interest, are smart city deployments and multiservice solutions (e.g., community-owned utilities, combining various services including voice, video Internet, and Smart Grid services by service or utility providers)[25]. Another aspect and advantage of fiber optics is its very high bandwidth that can accommodate various bandwidth-demanding applications in the future without the need for replacement or upgradation of the infrastructure.

In general, current fiber-optic networks consist of an optical transmitter, optical amplifiers, and an optical receiver. An optical transmitter is usually a device that emits light by using a Light-Emitting Diode (LED) or laser diode, and also converts an electrical signal into an optical signal to be sent over an optical fiber. Optical amplifiers are placed in between the transmitter and receiver in order to cope with attenuation and distortion of light over the long distances of transmission. An optical amplifier amplifies directly the signal in the form of light when a length of fiber is doped with erbium and pumped with light, with its wavelength shorter than the communication signal. An optical receiver is a device that contains a photodiode to detect the light and convert it back into the electrical signal by the photoelectric effect.

There are two types of optical fibers – multimode optical fibers and single-mode optical fibers. The difference between multimode and single-mode optical fibers is the size of the fiber core. Multimode optical fibers have larger core diameters, allowing several modes of various wavelengths to be transmitted. The fiber core is typically

50–100 μm. In general, larger core diameter requires less expensive and less precise equipment and multimode fiber is also highly affected by multimode distortion from multiple spatial modes, which limits the length of the optical link and the bandwidth. For multimode transmitters, an LED can be used for light emission, while a single mode one requires a precise laser-based transmitter. Characteristics of the most common multimode optical fiber with diameter 50/125 μm are defined in ITU-T Recommendation G.651.1 [26]. Single-mode optical fibers and their characteristics are defined by G.652 ITU-T recommendation [27]. Single-mode fibers typically have a core diameter that is very close to the wavelength of the light used for carrying information, between 8 and 10.5 μm. Multimode optical fibers are used for relatively short distance transmissions such as backbone application for buildings, while single-mode optical fibers are used for very long distance high bandwidth transmissions over tens or hundreds of kilometers.

Wavelength Division Multiplexing (WDM) is a commonly used technology that allows use of multiplex parallel channels, where a wavelength of light is dedicated to each channel. Before transmission, channels are multiplexed in the transmitter and then demultiplexed in the receiver. WDM is very common on single-mode fiber optical cables and equipment that can work as multiplexer and demultiplexer simultaneously is named an add-drop multiplexer. The International Telecommunication Union (ITU) recognizes six wavelength bands [28]:

O (Original) band, 1260–1360 nm	C (Conventional), 1530–1565 nm
E (Extended) band, 1360–1460 nm	L (Long-wavelength) band, 1565–1625 nm
S (Short-wavelength) band, 1460–1530 nm	U (Ultralong-wavelength) band, 1625–1675 nm

In the past, the ITU also recommended the first phase of optical systems to operate near 850 nm. However, this band (800–900 nm) is out of the scope of the current standard and it is being used by private networks for short-distance communication because of very high losses at this wavelength.

6.3.2.1 SONET/SDH

Synchronous Optical NETworking (SONET) and Synchronous Digital Hierarchy (SDH) were originally designed to support circuit-switched communications, especially for transmitting voice signals in Pulse-Code Modulation (PCM) format [29]. The SONET is standardized by the American National Standards Institute (ANSI) as T1.105 [30] and SDH is considered as its counterpart, which was first defined by the European Telecommunication Standards Institute (ETSI), and standardized by ITU [31–34].

SONET and SDH are almost identical standards using different terminologies for the same features. SONET/SDH are multiplexing structures, defining a set of transport containers for various technologies such as Asynchronous Transfer Mode (ATM), Ethernet, TCP (Transmission Control Protocol)/IP, or traditional telephony. SONET/SDH supersedes an earlier standard named Plesiochronous Digital Hierarchy (PDH). As might be obvious from the name, while PDH networks are nearly synchronized, tight synchronization by atomic clocks in the entire network is required in SONET/SDH networks.

While SDH has a basic unit named STM-1 (Synchronous Transport Module, level 1), which operates at 155.52 Mbps, SONET has a basic unit of transmission named STS-1 (Synchronous Transport Signal 1), which operates at 51.84 Mbps. SDH's STM-1 is equal to STS-3c/OC-3c, which is realized by multiplexing three STS-1 signals (interleaving the byte of the STS-1 frames) to form the next level of the SONET hierarchy.

6.3.2.2   FTTx and PON

FTTx refers to Fiber-to-the-x, where x can be replaced by any letter, in dependence on the deployment and end point of the optical fiber network installation. It is mostly used to refer to the optical networks for last-mile communication and out of many "FTT"s (e.g., FTTP (Fiber-to-the-Premises), FTTC (Fiber-to-the-Curb), FTTD (Fiber-to-the-Desk)) the most widely known term is Fiber-to-the-Home (FTTH) and Fiber-to-the-Building (FTTB). FTTx architecture can be realized by various ways of distribution of the optical fiber links at the end points. In Active Optical Networks (AONs), the optical signal is usually distributed between several endpoints by active (powered) equipment realizing optical-electrical transformation, switching, or routing to the proper interface and again converting back to the optical light in order to be sent to the Optical Network Terminal (ONT). Passive Optical Network (PON) realizes a communication distribution by unpowered optical splitters. Many PON types have been standardized by ITU [35–37], and before that were specified by the forum of major telecommunication service providers named Full Service Access Network (FSAN) [38]. The comparison of different types of PON networks is shown in Table 6.7.

6.3.3   Digital Subscriber Line (DSL) and Ethernet

DSL has a long history and enables data or Internet access over a local telephone line. Over the years many DSL systems have been developed such as HDSL (High bit rate Digital Subscriber Line) or SDSL (Symmetric Digital Subscriber Line) and the more recent ADSL (Asymmetric Digital Subscriber Line) [39], VDSL (Very high bit rate Digital Subscriber Line) [40] systems and their evolutions ADSL2 [41], ADSL2+ [42], VDSL2 [43]. Data rates have evolved from a few Mbps for an ADSL link to 24 Mbps peak rate on downlink for ADSL2+ and 100 Mbps for VDSL2 at 500 m from the source. DSL is commonly deployed for the last-mile access and connects the DSL modem at the customers' premises, with Digital Subscriber Line Access Multiplexer (DSLAM) unit that is usually from a few hundred meters to

Table 6.7 Comparison of various types of Passive Optical Networks

	A-PON	B-PON	E-PON	G-PON	10 GE-PON	XG-PON1	NG-PON2	WDM-PON
Standard	ITU-T G.983.1	ITU-T G.983.x	IEEE 802.3ah	ITU-T G.984.x	IEEE 802.3av	ITU-T G.987	FSAN NG-PON2	No std.
Multiplex	TDM	TDM	TDM	TDM	TDM	TDM	TDM	WDM
Framing	ATM	ATM	Ethernet	GEM	Ethernet	GEM	GEM	variable
Bandwidth	155 Mbps	622 Mbps	1.25 Gbps	2.5/1.25 Gbps	10 Gbps	10/2.5 Gbps	10 Gbps	1–10 Gbps
Bandwidth per user	10–20 Mbps	20–40 Mbps	30–60 Mbps	40–80 Mbps	> 100 Mbps	> 100 Mbps	> 100 Mbps	1–10 Gbps
Users	16–32	16–32	16–32	32–64	≥ 64	≥ 64	≥ 64	16–32
Cost	Low	Low	Low	Medium	High	High	High	High

several kilometers far. DSLAM is used to aggregate links from a large number of subscribers. Aggregated bandwidth requirements from DSLAM are usually very high and optical fiber access technology is used toward the central office of the telecommunication provider. Although DSL is an aging technology, it still takes a major portion of the fixed broadband market worldwide by providing its services to 400 million subscribers [44]. Owing to its high penetration in many countries, DSL can provide efficient backhaul for the Smart Grid data from homes to the utilities.

Ethernet was originally a computer networking technology for Local Area Networks (LANs) and was standardized in the 1980s by IEEE as the 802.3 family of standards. While the first standard assumed coaxial cables, later standards included other physical media such as twisted pair, optical fiber, and even wireless radio. Ethernet is very popular for its simplicity and this technology operates at 10 Mbps, 100 Mbps, 1 Gbps, 10 Gbps, and recently also at 40 and 100 Gbps [45]. While speeds of 100 Mbps are usually deployed within LANs over twisted pair cable, gigabit rates are used to connect LANs together carried over the backbone networks. Ethernet allowed many specific services with LAN networks and in order to provide these services over the WAN networks, a carrier class Ethernet has been defined, named Carrier Ethernet. It has been defined as a ubiquitous, standardized, carrier-class service and network defined by five attributes that distinguish Carrier Ethernet from the familiar LAN-based Ethernet [46]: standardized services, scalability, reliability, quality of service, and service management. Carrier Ethernet fits in a utility network where the capacity exceeds 10 Gbps, where SONET is too expensive or not flexible enough to be deployed [47, 48].

6.4 Wireless Communication

6.4.1 Introduction

Wireless communication technologies and wireless standards have been here for many decades. However, broad utilization of wireless technologies for the consumer market started only a few decades ago and is still gradually growing. Wireless technologies have experienced tremendous development and capacity increase in recent years together with minimizing hardware size, cost, and increasing energy efficiency. Wireless technologies also have principal advantages against wired technologies in the cost for building infrastructure, simplicity and speed of deployment, flexibility, mobility, and accessibility to remote sites.

An application of wireless technologies to the electrical grid and power system is also not a new topic and it has been used for home and building automation, monitoring, data collection, and metering. However, the Smart Grid applications differ widely in the requirements as explained at the beginning of this chapter. Therefore, there is no solution that "fits all" and technical and economical feasibility shall be analyzed before the selection of any technology. For this purpose, the national and international organizations and SDOs are elaborating the characteristics of the wireless communication technologies that could be potentially used in the Smart Grid system. The typical example is the NIST Priority Action Plan 2: Guidelines for Assessing Wireless Standards for Smart Grid, which provides business functional and application quantitative requirements, wireless technologies' characteristics, mathematical and simulation models, and test beds [49].

In this section, we present major wireless communication technologies and standards that can find potential utilization in the Smart Grid applications. First, we introduce the Machine-to-Machine (M2M) concept and its utilization for the Smart Grid. Then, we provide an overview of the wireless technologies, their characteristics, and potential areas of application in the Smart Grid. We split wireless technologies into further categories based on the communication range and describe the features of each technology or standard.

6.4.1.1 M2M Communication

M2M communication is a term referring to devices, which can communicate between each other without the need for human interactions. The key characteristics of M2M are autonomous operation, power efficiency, reliability, self-organization, and scalability.

In the world of wireless communication, a scientific area focused on Wireless Sensor and Actuator Networks (WSANs) provides various different communication technologies that satisfy the key characteristics of M2M communication and M2M is considered the main driving application for WSAN. The idea is that various devices whose main function is not to communicate, but to perform other activities ranging from home appliances, sports, and fitness goods to industrial and heavy machinery, and structures and buildings, have communication capability and share their information about the status of activity or some other information. These devices usually provide measurements by employing various sensors and the communication of the sensed data takes place either at some periodical interval or is triggered when the status of the sensed information has changed. The occurrence of communication might vary significantly from relatively frequent updates of near-real-time utility (water, gas, electricity) measurements to very infrequent updates, for example on the bridge stability when sensors that are embedded in the bridge structure. From this point, there are many situations where a communicating device is powered by battery and energy efficiency is a critical factor in such situations. It might be very ineffective, costly, or even impossible to exchange the battery during the lifespan of the device, where a communicating sensor is embedded.

Besides WSAN, which can provide standardized communication platform for various M2M systems, there have been significant activities and progress in realizing M2M communication over cellular systems in Europe [50], the United States [51] as well as in Japan [52]. This is mostly driven by cellular network operators and service providers. One of the key advantages of cellular M2M systems is the long range and very good reachability also in sparse and rural areas which are already covered by the cellular network infrastructure. In addition, a cellular network is principally designed for connecting millions of devices and communicating at relatively high data rates. Moreover, by QoS provisioning, which is an essential part of any cellular network, it can offer also relatively low latencies for real-time control and monitoring.

Although there are still many challenges in M2M communication, such autonomous systems can bring intelligence into the network, thus opening the space for optimization of many processes and large-scale systems, and solutions. M2M communication is considered a basic building block for Smart Grid systems, especially for Smart Metering.

6.4.1.2 Taxonomy of the Current Wireless Communication Technologies and Standards

There are various characteristics of wireless communication technologies that greatly affect the Smart Grid operations, functionality, efficiency, and reliability. Improper communication technology selection when designing the Smart Grid can have deleterious effects on the whole system. In Table 6.8, we provide an overview of various wireless technologies and their characteristics that are considered to play important roles in the communication between various elements and nodes in the Smart Grid. In Table 6.9, we provide an overview of those technologies in the context of Smart Grid applications.

6.4.2 *Wireless Very Short Distance Communication*

Radio-Frequency Identification (RFID) technology is widely used for identifying and tracking assets, because of its simplicity and low cost. In particular, the most common applications of RFID are store/warehouse product identification, production control, livestock identification, and vehicle tracking. The RFID concept recognizes two entities: reader/writer and a tag. The RFID reader/writer communicates with RFID tags and reads/writes the information from/to the tag. There can be three different types of RFID tags: passive, semipassive, and active. The passive RFID tags take the major market share due to the extremely low cost without the need for any battery and with almost limitless lifespan. The passive RFID tag is powered by the antenna of a reader/writer device. Semipassive tags are battery powered for internal operation. However, the read or write process utilizes the same method as a passive tag (powering from the antenna of the reader/writer). Finally, an active RFID tag is powered by battery and enables significantly larger distances for communication.

There are several standards defining RFID technologies and the way in which two entities interact (an antenna-coupling mechanism). The most common methods are RFID inductive coupling, RFID capacitive coupling, and RFID backscattering. Every coupling method has different features and varies in range, frequencies, and

Table 6.8 Overview of various wireless technologies and their characteristics

Technology	SDO/consortium	Wireless technology					Communication characteristics						
		Spectrum	Frequency	Data rates	Range/coverage	Security	Latency	QoS	Scalability	Cost	Market penet.	Network type	Mobility
RFID	ISO/IEC, ASTM, EPCGlobal	Unlicensed	125 kHz to 5.8 GHz	100 kbps	10 cm to 200 m	Medium–high	Low	No	No	Low	High	PAN, BAN	No
NFC	NFC forum	Unlicensed	13.56 MHz	~424 kbps	1–10 cm	High	Low	No	No	Low	Medium	PAN	No
IEEE 802.15.4	IEEE	Unlicensed	868 MHz, 915 MHz, 2.4 GHz, and so on	20–250 kbps	10–100 m	High	Medium	No	Yes	Low	Medium	PAN	Low
ZigBee	ZigBee alliance	Unlicensed	868 MHz, 915 MHz, 2.4 GHz, and so on	20–250 kbps	10–100 m	High	Medium	No	Yes	Low	Medium	PAN	Low
Wave2M	Wave2M	Unlicensed	433, 868, 915 MHz	5–20 kbps	~100 m	High	Medium		Yes	Low	Low	PAN, LAN	Low
Wireless HART	HART	Unlicensed	2.4 GHz	See 802.15.4	~100 m	High	Medium	No	Yes	Low	Low	LAN	Low
IEEE 802.11	IEEE	Unlicensed	subgigahertz, 2.4, 3.6, 5.8, 60 GHz	~Mbps, ~Gbps	~10 m to 1 km	Medium.–high	Low	Yes	Yes	Low	High	LAN	Low
Wi-Fi	Wi-Fi alliance	Unlicensed	See 802.11	See 802.11	See 802.11	Medium High	Low	Yes	Yes	Low	High	LAN	Low
Bluetooth (BT)	Bluetooth SIG	Unlicensed	2.4 GHz	~1 Mbps, ~10 Mbps	~1–100 m	Medium	Low	Yes	No	Low	High	PAN	Low

(continued overleaf)

Table 6.8 (continued)

| Technology | Wireless technology | | | | | | Communication characteristics | | | | | | |
	SDO/consortium	Spectrum	Frequency	Data rates	Range/coverage	Security	Latency	QoS	Scalability	Cost	Market penet.	Network type	Mobility
2G (GSM, GPRS)	ETSI	Licensed	900 MHz, 1800 MHz, and so on	~ kbps, ~ 10 kbps	~ 10(s) km	High	Medium	No	Yes	High	High	WAN	High
3G (UMTS, HSPA)	3GPP	Licensed	2100 MHz, and so on	~ 100 kbps, ~ Mbps	~ km	High	Low – medium	Yes	Yes	High	High	WAN	High
LTE	3GPP	Licensed	700 MHz to 2.7 GHz	~ 10–100 Mbps	~ km, ~ 10 km	High	Low	Yes	Yes	High	Medium	WAN	High
IEEE 802.16	IEEE	Licensed/ unlicensed	700 MHz to 66 GHz	~ 10–100 Mbps	~ km, ~ 10 km	High	Low	Yes	Yes	High	Low	MAN	High
WiMAX	WiMAX forum	Licensed/ unlicensed	700 MHz to 66 GHz	~ 10–100 Mbps	~ km, ~ 10 km	High	Low	Yes	Yes	High	Low	MAN/ WAN	High
GMR-1 3G	ETSI/ITU/TIA	Licensed	L band, S band	~ 100 kbps	N. Amer.	Medium	Medium	Yes	Yes	High	Low	GEO	High
DVB-S2	ETSI/TIA	Licensed	C, Ku, Ka	~ 10 Mbps	Global	High	Medium	Yes	Yes	High	Low	GAN	No
RSM-A	ETSI/TIA	Licensed	Ka	~ 10 Mbps	N. Amer.	High	High	Yes	Yes	High	Low	NGSO	No
Inmarsat BGAN	ITU/ETSI/TIA	Licensed	L band	~ 100 kbps	Global	High	High	Yes	Yes	High	Low	GAN	High

Table 6.9 An overview of various wireless technologies and their applications in the Smart Grid

Wireless technology Standard family	Smart Grid characteristics Communication type	Sample application in the Smart Grid
RFID	IAN, BAN, HAN	Smart metering
NFC	IAN, BAN, HAN	Smart metering
IEEE 802.15.4	HAN, BAN, IAN, NAN	Control and automation of home appliances, direct load control
ZigBee	HAN, IAN, BAN, NAN	AMI, home and building automation and control, sensor and mesh networks, direct load control
Wave2M	HAN, IAN, BAN, NAN	Metering, sensor networks
IEEE 802.15.4	HAN, IAN, BAN, NAN	AMI, home and building automation and control, sensor and mesh networks, direct load control
WirelessHART	HAN, IAN, BAN, NAN	Control and automation of industrial appliances, direct load control, mesh sensor networks
IEEE 802.11	HAN, IAN, BAN, NAN	AMI, home and building automation and control, sensor networks, AMI-to-AMI communication
Wi-Fi	HAN, IAN, BAN, NAN	AMI, home and building automation and control, sensor networks
Bluetooth (BT)	HAN, BAN, IAN	AMI, home and building automation and control, sensor networks
RF Mesh	NAN, FAN	AMI, distribution automation, work force automation, AMI-to-AMI communication
2G	WAN, NAN, FAN	Monitoring and metering of remote DERs, SCADA interface for remote distribution substation
3G (UMTS, HSPA)	WAN, NAN, FAN	Monitoring and metering of remote DERs, SCADA interface for remote distribution substation
LTE	WAN, NAN, FAN	Monitoring and metering of remote DERs, SCADA interface for remote distribution substation, site remote video surveillance
IEEE 802.16	WAN, NAN, FAN	Monitoring and metering of remote DERs, SCADA interface for remote distribution substation, site remote video surveillance
WiMAX	WAN, NAN, FAN	Monitoring and metering of remote DERs, SCADA interface for remote distribution substation, site remote video surveillance
GMR-1 3G	WAN, FAN	Remote sensing, SCADA interface for remote distribution substation
DVB-S2	WAN, FAN	Remote sensing, SCADA interface for remote distribution substation
RSM-A	WAN, FAN	Remote sensing, SCADA interface for remote distribution substation
Inmarsat BGAN	WAN, FAN	Remote sensing, SCADA interface for remote distribution substation

transmission data rates. There are more than 20 international standards defining RFID technologies. The ISO 18000 set of standards defines air interfaces for RFID, with allowed frequencies in different regions. Besides various ISO standards, there is also the EPCglobal standardization body defining Electronic Product Code (EPC) Class-1 HF RFID standard [53] and the newer UHF Class-1 "Gen 2" standard [54].

The EPC Class-1 HF air radio interface is another important ISO standard that is compatible with ISO 15693 and utilizes RFID inductive coupling. Inductive coupling utilizes the near-field effect. In order to power the tag circuitry, the distance (range) between reader/writer and the tag must be within 0.15 of the wavelength of the used frequency. There are two frequencies defined in the standard, 135 kHz and 13.56 MHz.

A Near-Field Communication (NFC) offers a very short range (or contactless) communication up to 4 or 5 cm. NFC technology operates at 13.56 MHz, which is an unlicensed band and utilizes inductive coupling. The NFC Forum groups manufacturers, vendors and service providers, and certifies NFC-enabled devices in order to maintain interoperability. The current NFC defines three modes of operation: NFC card emulation mode, peer-to-peer mode, and reader/writer mode. In the card emulation mode, NFC behaves like RFID and the NFC standard can be seen as an extension of RFID technology. NFC was originally developed by Sony and NXP Semiconductors. Currently, there are several NFC technologies and standards. NFC has been standardized under ISO/IEC 18092 international standard (mirrored in ECMA-340) and is compatible with several proprietary as well as open RFID technology standards for communication. FeliCa is Japanese Industrial Standard (JIS) X 6319-4, (also called NFC-F), which is developed by Sony and widely used in Japan for payments via feature phones and smart cards at convenience stores, shopping centers, buses, subway stations, and so on. MIFARE (MIkron FARE collection system) was developed by NXP Semiconductors and it is mirrored in ISO/IEC 14443 standards for Type A. ISO/IEC 14443 also defines the Type B standard that mainly differs from Type A in modulation and coding schemes. ISO/IEC 21481 standard (mirrored in EMCA-352) specifies the communication mode-selection mechanism, designed to not interfere with any ongoing NFC communication in 13.56 MHz, for devices implementing ECMA-340, ISO/IEC 14443, or ISO/IEC 15693.

NFC has been utilized for a wide variety of applications demanding high security such as contactless payments [55], different wireless technology authentication [56], and location access. One great advantage of NFC technology is the rapid growth of smartphones and other devices with embedded NFC capabilities, which opens new opportunities for various services offered directly to the end users.

RFID and NFC technologies are considered a key solution for access of reading data from smart meters by consumers and utility companies, and many other applications such as tracking battery-charging information, Plug-in Hybrid Electric Vehicle (PHEV)-charging information, and secure and convenient prepayment of utilities.

6.4.3 Wireless Personal and Local Area Networks and Related Technologies in the Unlicensed Spectrum

Most of the currently used wireless communication technologies operating in unlicensed spectrum are often categorized as Wireless Local Area Networks (WLANs) and Wireless Personal Area Networks (WPANs). WLANs provide connection between two or more devices in the vicinity, usually within a home, office, warehouse, or building. Such networks usually range from tens of meters to about a hundred meters. In contrast, WPANs provide connection between devices within personal area, covering distances from a few centimeters to tens of meters.

Recently, there was a huge development in both areas and the separation between WLANs and WPANs is becoming less clear. Moreover, WLAN and WPAN technologies are continuously improving not only in their offered data rates, reliability, security, and quality of service, but also significantly increasing coverage. Therefore, many originally WLAN and WPAN technologies are currently offering long-range communication and can be categorized as Wireless Metropolitan Area Networks (WMANs) or even Wireless Regional Area Networks (WRANs). Typical examples are IEEE 802.11 and IEEE 802.15.4 families of standards with new amendments, and other technologies, which offer PHY layers based on the sub-one gigahertz wireless radio with transmission radius often up to several kilometers or technologies utilizing unoccupied bands of former analog TV channels.

As for Smart Grid systems, WLAN and WPAN technologies and standards are excellent candidates for various applications in the customer's premises including smart home and building automation, home energy management systems (EMSs), and so on. Similarly, WMAN and WRAN technologies can play a key role in communication within a neighborhood or communities, AMI-to-AMI communication, remote metering and monitoring, communication between various logical blocks within NANs and FANs.

This section covers major wireless communication standards and specifications utilizing unlicensed frequency bands.

6.4.3.1 WPAN – IEEE 802.15.4 Family of Standards and ZigBee Technology

The IEEE 802.15.4 standard defines PHY and MAC requirements and is designed for Low-Rate Wireless Personal Area Networks (LoWPANs). It is aimed for low-power, low data rate, short-range communication. The communication range is usually from 10 to 75 m with low data rate of tens of kbps up to 250 kbps [57].

Except for Ultrawide Band (UWB) PHYs that use ALOHA medium access, IEEE 802.15.4 standard utilizes Carrier Sense Multiple Access with Collision Avoidance (CSMA/CA) at the MAC layer. Communication within Personal Area Network (PAN) is controlled by a PAN coordinator, which issues periodically short frames named beacons for synchronization and medium access. In addition, the communication can be structured by superframes, which are large time slots consisting of smaller time slots and bounded by beacons. The PAN coordinator can allocate time slots in every superframe for communication of the network nodes in order to guarantee short packet delays as well as the required data rate. This structure allows a certain level of QoS when needed. Another important factor of the superframe structure is that in a dependence on the traffic demand, the superframe is divided into an active period and an inactive period, allowing the PAN coordinator to enter a power-safe mode.

IEEE 802.15.4 PHY and MAC Amendments

Besides the originally defined PHY layer specifications for 2.4 GHz, 915 MHz, and 868 MHz bands, various alternate PHYs have been added to cover different applications and use cases for 802.15.4 technology as well as to extend the utilization of this technology to the available bands in China and Japan. In the following, we will briefly introduce these radio technologies as well as specifications that are being standardized at the time of writing of this book.

IEEE 802.15.4a specifies another two different high-rate radio technologies, which can be optionally used instead of PHYs defined in IEEE 802.15.4-2006. The first one is based on chirp modulation, which has been used for radar systems since the middle of the twentieth century. Chirp Spread Spectrum (CSS) PHY operates at 2450 MHz and utilizes Differential Quadrature Phase-Shift Keying (DQPSK). It offers 1 Mbps data rate and an optional data rate of 250 kbps.

The other alternate PHY defined in IEEE 802.15.4a amendment is UWB, which is based on the impulse radio. UWB PHY supports operations in three separate bands. The first band is the subgigahertz band that has one 500-MHz wide channel and operates between 249.6 and 749.6 MHz. The second band is the spectrum from 3.1 to 4.8 GHz that can be divided into four channels. The third band consists of 11 channels covering the spectrum from 6.0 to 10.6 GHz.

The IEEE 802.15.4c-2009 amendment specifies the operation of two alternate PHYs, each having eight channels in the 779–787 MHz band. Both have a data rate of 250 kbps and one uses Offset Quadrature Phase Shift Keying (OQPSK), while the other one uses the M-ary Phase Shift Keying (MPSK) modulation technique, respectively [58]. This standard is limited to China.

IEEE 802.15.4d-2009 defines alternate PHY layers in the 950-MHz band (950–956 MHz) in Japan. Binary Phase Shift Keying (BPSK) and Gaussian Frequency Shift Keying (GFSK) modulation techniques are employed. BPSK offers a 20-kbps data rate while GFSK offers a 100-kbps data rate [59].

IEEE 802.15.4g – Wireless Smart Utility Network

In 2012, a 920-MHz radio band was standardized as an amendment IEEE 802.15.4g-2012. IEEE 802.15.4g is a global standard harmonizing the PHY layer, power levels, data rates, modulations, and other technical properties, as well as frequency bands for large-scale Smart Utility Networks (SUN). The IEEE802.15.4g PHY radio enables interoperable communication between smart meters and also various other Smart Grid devices and smart home appliances. It is also known as a Wireless Smart Utility Network (Wi-SUN) standard and the enabled devices are being certified by the recently established Wi-SUN Alliance, which is a consortium of major corporations and companies involved in the Smart Utility markets. The Alliance is conducting extensive testing to demonstrate interoperability in subgigahertz frequency bands appropriate for Japan, North America, Australia/New Zealand, Latin America, and other regional markets. In line with these efforts, the Japanese Ministry of Internal Affairs and Communications allocated the 920-MHz band for smart meters to bring the benefit of the international cooperation and competitiveness [60]. In addition, the IEEE 802.15.4 MAC layer has been enhanced by a 802.15.4e MAC sublayer and the combination of 802.15.4g PHY and 802.15.4e MAC sublayer creates energy-efficient and reliable solutions for SUN.

IEEE 802.15.4ak – Low Energy Critical Infrastructure Monitoring (LECIM)

The aim of this technical specification is to cover the large amount of wireless critical infrastructure applications, whose requirements cannot be satisfied by the currently existing technologies [61]. Low Energy Critical Infrastructure Monitorings (LECIMs) focus is on the outdoor environment with thousands of communicating nodes with low data traffic requirements. Also, the nodes are assumed to be deployed for the very long term and the application is tolerant to the communication latency. Examples of critical infrastructure applications are electricity and gas production, transport and distribution, water supply, and so on. The key technological advancement of the IEEE 802.15.4k is a simplified MAC layer to handle a large number of devices and prioritize emergency communication if needed.

ZigBee

ZigBee is a widespread wireless technology aimed for the low-power devices and sensor communication. It is a low-cost technology with large potential to be embedded in many devices and appliances for control and monitoring applications. As it is a low-power technology, the communication chip can be powered by small batteries like button-cell batteries for several years. Thus, it is also suitable for devices where replacement of the battery during the device's lifetime is difficult (e.g., radiation leakage), costly (e.g., ocean underwater sensors), or impossible (e.g., bridge monitoring).

ZigBee can form a mesh network topology or can create a tree or a star topology. ZigBee wireless technology utilizes Industrial Scientific and Medical (ISM) bands. While the subgigahertz ISM radio band in Europe is 868 MHz, the 915-MHz band is used in the United States and Australia. Also, the 2.4-GHz band is used worldwide similarly to several other WPAN and WLAN technologies. There are 16 channels in the 2.4-GHz band and each channel has a bandwidth of 5 MHz. While the European 868-MHz band offers a maximum data rate of 20 kbps, the US and Australian 915-MHz band is limited by a data rate up to 40 kbps. The public 2.4 GHz allows data rates up to 250 kbps.

ZigBee is mainly a suite of high-level communication protocols rather than a pure wireless technology standard. It is based on the application profile usage and application-specific information transferring and processing.

There are public profiles and manufacturer-specific profiles. Public profiles are specified by the ZigBee Alliance and ensure the interoperability of the products from various manufacturers. Widely known public application profiles are Home Automation, Commercial Building Automation, Advanced Metering Initiative, Health Care, Telecommunication Services, and Smart Energy Profile (SEP) 1.0 and 2.0. More details about the application profiles aimed for the Smart Grid can be found in the related book chapters (e.g., SEP 2.0 is discussed in *Section 5.4 Smart Home and Building Automation*).

The IEEE 802.15.4 standard is considered as a basic radio technology for wireless sensor networks and especially M2M communications. Therefore, besides the ZigBee Alliance and other consortia with open initiatives and standards, there are also several proprietary solutions utilizing IEEE 802.15.4 radio (e.g., SynkroRF or IEEE 802.15.4 PHY with Simple Medium Access Control) [62].

6.4.3.2 WirelessHART

WirelessHART is a wireless complementary enhancement to the Highway Addressable Remote Transducer (HART) protocol [63] and has been standardized as the IEC 62591 standard [64]. It supports the 2.4 GHz band operation with IEEE 802.15.4 PHY/MAC wireless technology. WirelessHART employs the channel-hopping scheme to avoid interference and utilizes transmission power adaptation. It also monitors a path for degradation and has self-healing features. Another feature is an adjustment of the communication path for optimal network performance and utilization of a mesh network. WirelessHART is designed for industrial applications with high requirements on reliability and security [65]. More information on the security aspects of WirelessHART technology are analyzed in Chapter 7.

6.4.3.3 ISA100.11a

ISA100.11a is a wireless networking technology standard developed by the International Society of Automation (ISA) [66]. It provides reliable and secure communication designed for noncritical monitoring, supervisory control, and alerting, where higher latencies (several hundreds of ms) can be tolerated. ISA100.11a is robust against interference found in harsh industrial environments and can coexist with other technologies such as IEEE 802.11 and IEEE 802.15 network technologies. The standard defines radio link operation as well as wired operation over Ethernet and field buses. It can form various topologies such as star or mesh and defines security architecture, which is described in more detail in Chapter 7.

6.4.3.4 WLAN – IEEE 802.11 Family of Standards and Wi-Fi Technology

The most commonly used wireless standard for WLANs is IEEE 802.11 and its amendments. Well-known and widely deployed standards from this family are

802.11a, 802.11b, 802.11g, and 802.11n, as well as the newly released standard 802.11ac. There are two basic network architectures defined in IEEE 802.11 standards: the infrastructure network architecture and the ad hoc network architecture. In the IEEE 802.11 network, the device that is capable of communication is called a station (STA) and the device, which can create infrastructure network, is called an Access Point (AP).

Infrastructure network requires that an STA in a network should associate with an AP in order to communicate. Therefore, generally all the communication between two stations in the network is relayed over the AP. This network is aimed to provide connectivity to resources that are behind the AP, such as Internet connectivity or connection to an enterprise network. All stations (including the AP) in the network create a Basic Service Set (BSS). This service set is identified by a Basic Service Set Identifier (BSSID), which is a MAC address of the AP in the infrastructure network. In addition, every WLAN network is represented by a human-readable text-based Service Set Identifier (SSID), which is a string value of maximum 32 characters. In scenarios that require larger coverage than conventional 802.11 networks, several physical WLAN networks can be part of one logical wireless network to create an Extended Service Set (ESS). An ESS that consists of one or more BSSs, is identified by one SSID. In contrast, the ad hoc network, named Independent Basic Service Set (IBSS), allows stations to communication in a peer-to-peer manner without involvement of the AP.

IEEE802.11 networks were aimed to operate in unlicensed bands and are operating in ISM bands with carrier frequencies of 2.4 and 5 GHz. Also, the 60 GHz as well as sub-one gigahertz bands are currently being considered for the PHY layer standardization process. The millimeter wave (60 GHz) standard, IEEE802.11ad, is aimed for ultrahigh-throughput short-range communication [67] and the sub-one gigahertz standard, IEEE 802.11ah, is aimed at low-power and long-range communication [68].

IEEE802.11 networks use MAC technology known as CSMA/CA, which is a modification of Carrier Sense Multiple Access (CSMA) technology used in Ethernet. This channel access is based on carrier sensing and transmitting when there is no transmission on the channel detected (channel is idle). In addition, when there is transmission detected on the channel (channel is busy), the station waits until the ongoing transmission ends and for an additional interval, named the contention period. This brings more fairness into the system when there are several stations willing to transmit at the same time and helps to avoid collisions caused by simultaneous transmission of several stations right after the channel becomes idle. After the contention period elapses, the above process is repeated until the channel is in an idle state.

Another feature that is optionally used in IEEE 802.11 networks is exchange of Ready-To-Send (RTS) and Clear-To-Send (CTS) messages prior to actual data transmission. This is used to prevent the hidden terminal problem and a sample scenario is shown in Figure 6.3. This problem occurs when there are two or more stations (e.g., Stations A and C), and another station or AP in the middle (Station B). Then, if those stations are far from each other, they cannot communicate between each other, but all can communicate with the station in the middle (Station B). Therefore, when Station A and Station C transmit data to Station B at the same time, collision occurs.

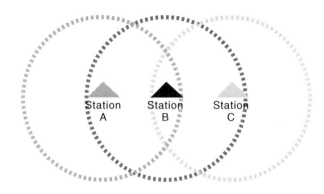

Figure 6.3   Hidden terminal problem

This collision can be avoided by RTS and CTS messages. If Station A is willing to transmit, it sends an RTS message and then Station B confirms receiving the RTS message by sending a CTS message, and the CTS message is received by all stations within the radius of Station B. Following this, all other stations (in this case Station C) will refrain from transmission.

IEEE 802.11e-2005 is an amendment of the IEEE 802.11 standard, which defines QoS on the MAC layer. It is an important set of enhancements critical for delay constrained traffic. Originally, the MAC layer had a Distributed Coordination Function (DCF) and a Point Coordination Function (PCF), and this amendment introduces a new Hybrid Coordinated Function (HCF). This function has two different methods of channel access, HCF Controlled Channel Access (HCCA) and Enhanced Distributed Channel Access (EDCA). HCCA is similar to the original PCF in that it splits the period between two beacons into a contention period and a contention-free period, which can be allocated to the station in order to guarantee resources. The HCCA method is not used in most of the current implementations of IEEE 802.11 networks. Both HCCA and EDCA use the feature called Transmit Opportunity (TXOP), which allows a station to send several frames in a row, bounded by the maximum value of the TXOP period. Received frames are usually acknowledged by the receiver at the end of a TXOP period through a Block ACK message. While the key feature of HCCA is resource allocation and scheduling, EDCA is based on traffic prioritization. There are four Access Categories (ACs) defined in EDCA: background, best effort, video, and voice. The difference between these four categories is a Contention Window (CW) minimum and maximum value, which gives the highest priority (short CW) to voice traffic and the lowest (long CW) to background traffic.

IEEE 802.11s is an amendment based on the main PHY/MAC standards within the 802.11 family (i.e., IEEE 802.11a/b/g/n). It extends the MAC layer functionality to support multicast and broadcast base on the radio propagation properties and defines a multihop routing protocol for message dissemination over the mesh network. 802.11s supports Hybrid Wireless Mesh Protocol (HWMP), which is based on an Ad hoc On-demand Distance Vector (AODV). AODV is a mandatory feature of 802.11s and other mesh and ad hoc routing protocols are optionally supported. IEEE 802.11s networks are usually used for outdoor scenarios, where a mesh network can provide good coverage, and is deployed in the NANs for smart meter-to-smart meter communications [69].

IEEE 802.11ah is an ongoing standardization effort for the amendment of IEEE 802.11 networks with low-power, long-range communications at subgigahertz frequency bands. This emerging standard is likely to play an important role in M2M, and the Internet of Things (IoT). The key use case of IEEE 802.11ah standard is in the Smart Grid, that is, by connecting metering devices with Data Aggregation Points (DAPs) and backhaul for IEEE 802.15.4g mesh sensor networks. As subgigahertz frequency bands differ between regions and countries, it defines channelization for each region separately. The currently specified channel widths and number of channels (in brackets) for the main regions [70], which may be extended before the standard release, are listed below:

- Europe (863–868 MHz): 1 MHz (5c.) or 2 MHz (2c.).
- China (755–787 MHz): 1 MHz (24c.) or 2 MHz (4c.) or 4 MHz (2c.) or 8 MHz (1c.).
- Japan (917–927 MHz): 1 MHz (11c.).
- South Korea (917.5–923.5 MHz): 1 MHz (6c.) or 2 MHz (3c.) or 4 MHz (1c.).
- USA (902–928 MHz): 1 MHz (26c.) or 2 MHz (13c.) or 4 MHz (6) or 8 MHz (3c.) or 16 MHz (1c.).

The PHY layer utilizes OFDM with 64 subcarriers and supports modulations from BPSK up to 256 QAM. Both multiuser MIMO and single-user beam forming are supported. In the MAC layer, improvements are mostly about power-saving modes, where some devices might be battery powered, and also about the scalability of the channel access method to handle a large number of connected devices. IEEE 802.11ah is aimed for star topology of infrastructure mode. However, the IEEE 802.11s mesh network standard might be updated in the future to also support this sub-one gigahertz standard in order to create large mesh networks with very wide coverage.

IEEE 802.11af is another amendment and ongoing standardization effort that completely redefines PHY and MAC layers. A key target is utilization of the TV White Space (TVWS) spectrum fragments originally allocated to analog TV channels and no longer used because of digitalization of the TV transmission. It utilizes OFDM technology and offers long-range communication. The standard is expected to be completed by 2014 and it is considered as one of the most promising technologies operating in the TVWS spectrum [71].

Wi-Fi

The term Wi-Fi has been coined by the Wi-Fi Alliance for the purpose of giving a less technical and more appealing name to IEEE 802.11 technology [72].

The Wi-Fi Alliance is a consortium of a large number of manufacturers, vendors, service providers, carriers, and other companies. Devices fulfill the requirements of the Wi-Fi Alliance defined in the technical specification by successful testing process which is followed by certification. Besides the promotion of Wi-Fi technology, the Wi-Fi Alliance is also responsible for ensuring interoperability between products from different manufacturers.

Although Wi-Fi network and IEEE 802.11 network are terms that are very often used interchangeably, there are slight differences between those two. While IEEE 802.11 is a set of standards that defines PHY/MAC layers, protocols, association, authentication, QoS, and many other relative basic functionalities, the Wi-Fi Alliance issues technical specifications, which mostly refer to IEEE 802.11 standards and focus on bridging the gaps between standards and functionalities (i.e., functionalities out of scope of IEEE 802.11 set of standards) and developing simple and user-friendly features. The Wi-Fi Alliance has simplified the way in which two Wi-Fi-certified devices directly communicate with each other by introducing the Wi-Fi Direct technology, which is currently embedded in almost every smartphone or tablet [73]. The Wi-Fi Alliance is also very active in improving the way in which two devices can establish connection simply and securely with each other [74], how the QoS is provisioned in the network, how to increase energy efficiency [75], and how to enable high-definition video streaming over a wireless channel [76].

The Wi-Fi Alliance has various Technical and Marketing task groups focusing on different aspects, scenarios, and use cases, where the Wi-Fi technology will play an important role. It also maintains liaison with various consortia and alliances such as the ZigBee Alliance, in order to provide better interoperability between complementary technologies.

From the aspect of the Smart Grid, a very important activity is on Wi-Fi Smart Energy Profile (WSEP) 2.0 specification and interoperability with SEP2.0 compliant ZigBee and Homeplug implementations. In order to achieve that, the Wi-Fi Alliance has to define the technical specification for Wi-Fi-certified devices, which will satisfy SEP2.0 requirements on the transport technology. SEP2.0 is described in more detail in *Section 5.4 Smart Home and Building Automation*.

One of the key advantages for this WLAN technology is the high penetration rate, with more than nine billion devices [77] shipped since 2009. It is noted that one quarter of worldwide households already has Wi-Fi connectivity [78]. Wi-Fi technology will likely play an important role in smart home automation and control solutions, smart metering and smart energy applications, and services.

6.4.3.5 WAVE2M

The WAVE2M open standard, originally named Wavenis Wireless Technology, is a multiradio standard dedicated to M2M communication [79]. The PHY layer is comprised of three main ISM bands used for wireless sensor networks and M2M communication, that is, 433, 868, and 915 MHz. WAVE2M offers a wide range of data rates, based on the application need. In general, a lower data rate, which is typically of several kilobytes per second allows a long transmission range by utilizing narrowband receivers with high sensitivity.

WAVE2M technology supports a mesh network topology configuration with very high scalability, although the end-to-end delay grows with the number of clusters in the network. It is based on the Fast FHSS (Frequency Hopping Spread Spectrum) with data interleaving and forward error correction. The ultralow power consumption is realized by an MAC layer, and a network layer supports IPv6 as well as RPL protocol for routing. The main target market segments for WAVE2M open standard is AMI-to-AMI communication, home and building automation, and smart cities.

6.4.3.6 Weightless

Weightless in another long-range technology designed for M2M communication. This technology standard is maintained by the Weightless Special Interest Group (SIG) and although the standard is available, the patents are licensed only to qualified devices, thus this technology can be considered as a proprietary solution. It uses low-frequency unlicensed spectrum and the key target is the TVWS bands. A PHY layer currently operates between around 470 and 790 MHz frequencies in the TV channels band in the United Kingdom and United States. Other interesting features of the Weightless standard are very low power consumption and scalability to a large number of connected terminals. Weightless has a very similar design to cellular technologies and uses Time Division Duplex (TDD) with frequency hopping. For the downlink (communication from base station to terminal) it uses Time Division Multiple Access (TDMA) and the channel bandwidths are 6 and 8 MHz. For the uplink, Frequency Division Multiple Access (FDMA) technology is used.

The key applications of Weightless communication standards are typical M2M applications such as smart metering, industrial machine monitoring, vehicle tracking, smart appliances, health and fitness and traffic sensors, asset tracking.

6.4.3.7 WRAN – IEEE 802.22

This standard belongs to the category of the technologies operating in the TVWS spectrum. It is WRAN standard and utilizes cognitive radio technology in order to avoid the interference with TV signals. The transmission coverage of IEEE 802.22 can reach up to 100 km and the PHY layer offers very high flexibility in modulation and coding techniques. The MAC layer enables a large number of connected devices. On the other hand, very large transmission coverage incurs propagation delays and limits the use of effective access schemes for dense environments. The IEEE 802.22 technology can be used for Smart Grid applications such as remote smart metering, as well as for livestock and animal monitoring, infrastructure monitoring, broadband access within communities, and so on.

6.4.3.8   Bluetooth

The Bluetooth technology was developed to replace cables between devices within PANs. It is a master–slave-based technology and includes two types of connections: point-to-point and point-to-multipoint. In a point-to-point connection, there are only two devices connected for communications. In comparison, in a point-to-multipoint connection, one master can connect up to seven slaves. This network type is named piconet and offers various different network topologies. Every piconet can be connected to other piconets to create a scatternet, which is achieved when a slave in one piconet network serves as the master for another piconet. In practice, most of the Bluetooth networks have point-to-point connection and consist of only two devices (e.g., PC and mouse/keyboard, mobile phone and headset).

Similar to the PHY of 802.11 networks, Bluetooth occupies the unlicensed spectrum at 2.4 GHz (2.402–2.480 GHz). It is based on the IEEE 802.15.1 standard and the PHY layer utilizes FHSS technology to mitigate the cochannel interference in this frequency band. This is realized by separation of the 79 MHz bandwidth into 79 channels with 1 MHz for each channel, and frequently hopping from one channel to another. The sequence of the channels for hopping is pseudorandom and is determined by the address of the master Bluetooth device [80]. The GFSK modulation scheme was used originally for the Bluetooth. However, in later versions (since Bluetooth 2.0) the $\pi/4$-DPSK and 8DPSK modulation schemes are used in order to achieve higher data rates. In addition, as the IEEE 802.11 network technology has been adopted as the High Speed (HS) alternate PHY/MAC layer in Bluetooth 3.0, this increases data rates from previous 3 to 24 Mbps [81].

Bluetooth 4.0 is a newer standard without backward compatibility with previous versions of Bluetooth [82]. Therefore, the dual mode, which supports Bluetooth 4.0 and previous standards, is required to be supported in order to communicate seamlessly. Bluetooth 4.0 is also known as Bluetooth Low Energy or Bluetooth Smart Energy. Supporting devices are certified as Bluetooth Smart or Bluetooth Smart Ready devices by the Bluetooth SIG. Bluetooth 4.0 supports a communication range up to 50 m and data rates of about 200 kbps (the maximum raw data rate over the air is 1 Mbps) and operates in the same ISM band as previous Bluetooth versions (i.e., the 2.4-GHz band). The main feature of Bluetooth 4.0 is the very low power consumption and ability to operate for months or even years with just a button-cell battery. This feature is realized by utilizing intelligent power saving and sleep management system.

Bluetooth 4.0 targets various markets, such as healthcare (i.e., heart rate monitor), sport (i.e., vital monitoring, training, fitness), and consumer (i.e., electronic leash). Besides these markets, it is expected to play an important role in the Smart Grid, to enable smart meters with this low-cost and low-power features.

6.4.3.9   Interference Issues of WLAN and WPAN Technologies

The ISM frequency bands have been used by various technologies and the most crowded frequencies are in the 2.4-GHz band. This is due to the extensive use of this frequency band by various widely deployed devices such as microwave ovens, cordless phones, car alarms, baby monitors, and security cameras. Also, peripherals powered by the Bluetooth technology have spread and this technology is embedded in most of the smartphones on the market. The same frequency range is occupied by sensing devices based on the ZigBee technology and the most widely used portable and mobile devices provide connectivity to the Internet as well as peer-to-peer communication (i.e., Wi-Fi Direct) based on the Wi-Fi technology.

All these technologies are already causing severe problems in the dense deployment of Wi-Fi APs with a limited number of channels. Moreover, most systems often use the same default channel for communication. In the future, the number of Wi-Fi- and Bluetooth-powered devices is expected to grow rapidly and it is realistic that many households and offices will have several Wi-Fi and Bluetooth networks (e.g., screens connected over Wi-Fi Direct based Miracast, at the same time smartphones and tablets with synchronized content connected over Wi-Fi, and connected smart appliances).

Current standards of Wi-Fi, Bluetooth, and ZigBee offer various ways to avoid or minimize the mutual interference when using the same 2.4-GHz band. Wi-Fi devices can also use the 5-GHz band, which is much less crowded than the 2.4-GHz band. Classical Bluetooth utilizes frequency hopping to rapidly hop among a total of 79 channels (up to 1600 hops per second). Channel bandwidth is 1 MHz, which does not cause significant interference to Wi-Fi's 22-MHz wide channels. Moreover, the latest Bluetooth specifications utilize adaptive frequency hopping to avoid channels with interference from Wi-Fi or other technologies in the 2.4-GHz frequency band and to negotiate channel map with other communicating Bluetooth devices.

6.4.4   Cellular Networks in the Licensed Spectrum and WiMAX Technology

Cellular networks are recently considered as the key technologies for smart metering, distribution automation, smart homes, and other applications in Smart Grid solutions [83]. Some cellular operators and service providers are already providing Smart Grid solutions and services as there are several factors that make these technologies very suitable for the Smart Grid.

- A cellular network offers advanced and widely deployed infrastructure, which can significantly reduce costs for utility companies.
- Cellular network technologies successfully meet the strict requirements on the network reliability, latency, security, and overall performance of most of the Smart Grid applications.
- The latest technologies provide broadband access, and offer even greater performance suitable for real-time video surveillance and other bandwidth-demanding services.
- A licensed band ensures network reliability.
- Cellular technologies are scaled for very large number of connected devices and varying densities therefore eliminating potential scalability issues for large-scale Smart Grid deployments.

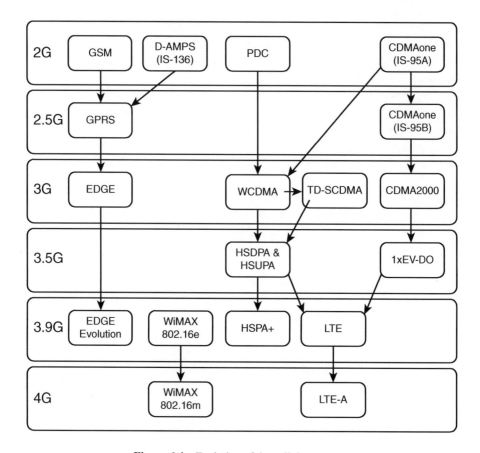

Figure 6.4 Evolution of the cellular systems

A simplified evolution of the cellular systems over time is shown in Figure 6.4. In the following parts we will introduce the modern cellular technologies with their characteristics.

The second-generation cellular system is considered as a breakthrough in Cellular system utilization by creating the first digital system and at the same time terminals of small size and fairly low cost for mass market production.

However, the beginnings of 2G cellular networks were not simple and the 2G network evolved from analog systems in different paths in the United States, Japan, and Europe. In the United States, a 2G cellular network was standardized by digitalizing Advanced Mobile Phone System (AMPS) and developing D-AMPS, which separated 30-kHz channels into three time slots and tripled the cell capacity compared to previous analog cellular systems by compressing voice. Similar to D-AMPS, Japan has developed its own 2G system based on TDMA, called Personal Digital Cellular (PDC). PDC used Frequency Division Duplex – Time Division Multiple Access (FDD-TDMA), 25 kHz carrier spacing and three full rate (or six half-rate) channels per carrier. PDC operated in the 800-MHz and 1.5-GHz bands, offering voice and data services, supplementary services such as call waiting, call forwarding, and voice mail. The maximum data rate for packet-switched data was 28.8 kbps and for circuit-switched data service 9.6 kbps.

At the same time, the Global System for Mobile communication (GSM) has been deployed in Europe and later adopted also in the United States. While D-AMPS and PDC digital cellular systems have already been shut down, GSM survived and it is still actively used by service providers worldwide.

6.4.4.1 GSM Family of Standards

GSM is currently the most widely deployed and globally accepted system. It is serving the vast majority of mobile subscribers in the world today [84]. The GSM standard set was developed in Europe by ETSI [85]. It was originally proposed to operate in the 900-MHz band and was later extended to 1800-MHz bands in Europe, Asia, and Australia, and the 850- and 1900-MHz bands in the Unites States and Canada. GSM uses narrowband TDMA for voice and Short Messaging Service (SMS). From the point of the architecture of GSM systems, there are several key entities consisting of various devices. The Base Station Subsystem covers the base stations and radio network controllers. The Network Switching Subsystem is also called a core network and it is the brain of the GSM network for switching networks, storing user profiles, signaling, and so on. Maintenance of the network is realized by an Operation Support Subsystem. Originally, the GSM network could handle only voice traffic and SMS service by utilizing signaling channels. Later, it was extended to support data traffic by a **General Packet Radio Service** (GPRS) core network, known also as a packet core.

GPRS is a packet data service on the cellular networks compatible with GSM systems. It was originally standardized by ETSI and currently is maintained by the 3rd Generation Partnership Project (3GPP) [86]. GPRS provides data rates of 56–114 kbps and is considered as 2.5G cellular technology. In contrast to GSM circuit-oriented systems, which are charged per unit of time (minutes or seconds), GPRS is charged per unit of data transferred (bytes). Every data service offered by the GPRS system is defined by a specific Access Point Name (APN). Typical and well-known GPRS services are Multimedia Messaging Service (MMS) and Wireless Application Protocol (WAP).

Enhanced Data Rates for GSM Evolution (EDGE) is a further enhancement of the GRPS technology. It is backward compatible and also known as the **Enhanced General Packet Radio Service** (EGPRS) system. EDGE allows higher data rates by utilizing a higher modulation scheme – 8 Phase Shift Keying (8PSK), with data rates up to 384 kbps. EDGE fulfills the requirements of the ITU for IMT-2000 and it is one of the 3G standards, although sometimes referred to as 2.75G technology as the typical EDGE implementations do not obtain data rates of 3G networks. Similar to GPRS,

3GPP maintains EDGE standards and also its more recent enhancement in Release 7, named Evolved EDGE [87], which further increases data rates and decreases latency.

6.4.4.2   CDMA Family of Standards

Code Division Multiple Access (CDMA) is another major access technology, which, in contrast to TDMA, allows terminals to be active all the time and send information simultaneously over the same communication channel. CDMA utilizes the spread-spectrum technology, which spreads the overall bandwidth uniformly for the same transmitted power. In addition, in order to share the same frequency, transmitted data has to be "coded" in such a way that these data do not interfere with the signals of other users. Codes that are used for transmitted signal can be either orthogonal or pseudorandom. Walsh codes, which have orthogonal properties, are being used in synchronous CDMA systems. For asynchronous CDMA, Pseudonoise (PN) sequences are adopted. PN sequences are quasirandom and therefore appear to be random (just like a noise) to the terminals without knowledge of the PN sequence.

A **CDMAone**, which is also know by communication community as IS-95, has been proposed by Qualcomm and standardized by the Telecommunications Industry Association (TIA). The CDMAone belongs to the 2G standards and utilizes QPSK modulation in downlink. Walsh codes of length 64 and PN sequences of length 2^{15} are used to create 64 possible communication channels. In the uplink OQPSK modulation is used. Apart from circuit-switched voice traffic, CDMAone offers data rates up to 14.4 kbps. The standard IS-95B for high-speed data allows data rates up to 115 kbps by combination of seven data channels.

CDMA2000 is the evolution of CDMAone which meets ITU's IMT-2000 requirements. It has been standardized by TIA. There are several versions that were evolved from CDMAone. CDMA2000 1xEV-DO is often referred to as Evolution-Data Optimized (EV-DO) and is widely used in many countries and is part of the 3GPP2 family of CDMA 3G standards. EV-DO utilizes multiplexing and combines Time Division Multiplex (TDM) and CDMA. A channel bandwidth is identical as in CDMAone, thus 1.25 MHz. EV-DO offers various data rates from 38.4 kbps with QPSK modulation up to 3.1 Mbps with 16-QAM.

6.4.4.3   WCDMA/HSPA/HSPA+

Wideband Code Division Multiple Access (WCDMA) is another radio technology belonging to the family of 3G networks. WCDMA was developed by NTT Docomo in Japan and it is known in Japan under the name Foma. WCDMA has been standardized by 3GPP as part of the Release 99 and is the most widely deployed radio interface of Universal Mobile Telecommunication Systems (UMTSs). UMTS Terrestrial Radio Access (UTRA) recognizes three different radio technologies and two of these use TDD and the third one, WCDMA, is based on the Frequency Division Duplex (FDD), sometimes referred to as UTRA-FDD. WCDMA is based on the Direct Sequence Spread Spectrum (DSSS) with CDMA and has a 5-MHz channel bandwidth. It offers a maximum data rate of 384 kbps for users. The WCDMA standard evolved in the later releases of 3GPP (since Release 5) into High-Speed Packet Access (HSPA). HSPA introduces several changes to the PHY of the original WCDMA including a higher modulation scheme, shorter Transmission Time Interval (TTI), MIMO antenna technology, and several other features, resulting in a significant increase of data rates. In the downlink (HSDPA), the maximum data rate is 14 Mbps and on the uplink (HSUPA) it is 5.76 Mbps. HSPA and the evolved version named HSPA+ (3GPP standard since Release 7) are considered as transition cellular technologies toward a fulfillment of ITU's requirements of IMT-Advanced, referred to sometimes also as the fourth generation of cellular systems. However, the term "4G" is not officially recognized by ITU and is only used for technologies that can significantly improve performance compared to the original family of 3G technologies [88]. HSPA+ cellular technology further evolves with every new release of 3GPP standard. The latest release (Release 11) offers peak data rates up to 672 Mbps in the downlink which is theoretically achievable by using 64QAM modulation, and using multicarrier aggregation to combine eight carriers (40 MHz) and MIMO technology (4×4) at the same time [89]. Although the deployment of these new extensions offers very high data rates, it requires usually not only software upgrade in the operator's network but also might be difficult to widely deploy and maintain owing to the lack of regulations and frequency spectrum in many countries.

6.4.4.4 Long-Term Evolution (LTE)

LTE was defined in Release 8 of 3GPP [90] and has further evolved in later releases. LTE inherits many functionalities and features from GSM and HSPA cellular technologies. However, it is a new standard aimed to further evolve into LTE-Advanced and to meet the requirements of ITU for IMT-Advanced. LTE is also referred to as 3.9G, considered to be the final step before realizing "true 4G" cellular systems. LTE provides several major improvements compared to previous 3G technologies and technologies evolved from 3G. The core network of LTE is known as the Evolved Packet Core, which offers very low latencies, functionalities for seamless handover of voice as well as data services to other cellular systems, and interoperability between LTE and non-3GPP technology such as CDMA or WiMAX. It also allows creation of a policy in order to offload data traffic from the LTE network to non-3GPP networks such as Wi-Fi. Another important functionality is the QoS provisioning for services requiring very low latency or a guarantee of a certain data rate. LTE's air interface is known as Evolved UMTS Terrestrial Radio Access (E-UTRA) and makes significant changes compared to the 3G technologies. First, while CDMA2000 and also WCDMA utilized FDD based CDMA, in China and some other countries, LTE has evolved from the TD-SCDMA 3G standard, which utilizes TDD. Therefore, the LTE standard defines both modes, FDD and TDD (which is referred to as TD-LTE). LTE uses Orthogonal Frequency Division Multiple Access (OFDMA) on the downlink and Single-Carrier Frequency Division Multiple Access (SC-FDMA) on its uplink. Other important features are MIMO technology taking advantage of multiple antennas on the transmitter and receiver sides, and dynamic channel allocation.

OFDMA separates the spectrum into several narrowband channels, that is, subcarriers, which are orthogonal to each other in order to avoid crosschannel interference. Each subcarrier is then modulated by selected modulation scheme (e.g., QPSK) at lower symbol rate with guard interval in the time domain, which makes the overall system robust against selective-frequency fading, intersymbol interference, and other effects. A subset of carriers is assigned to each user. While OFDMA uses a multicarrier transmission scheme, SC-FDMA on the uplink uses a single-carrier transmit signal. This is an advantage as the single-carrier transmit signal has lower Peak-to-Average Power Ration (PAPR) and low sensitivity to carrier-frequency offset.

6.4.4.5 WiMAX

WiMAX is an Metropolitan Area Network (MAN) technology developed by IEEE within the IEEE 802.16 family of standards. The WiMAX Forum certifies the products with WiMAX technology in order to ensure interoperability and compatibility and also promotes the technology [91]. IEEE 802.16 standards have several major releases. The first one, the IEEE 802.16-2004 or 802.16d standard, is aimed for fixed wireless communication as a replacement of DSL lines. It can provide data rates up to 70 Mbps with a range of tens of kilometers and therefore is also suitable as a wireless backhaul technology. A nomadic version of this standard is defined in release IEEE 802.16-2005, which is also known as 802.16e or WiMAX Release 1. It can have a peak rate of 15 Mbps and supports handover. Release IEEE 802.16m-2011 defines the advanced air interface, which is referred to as Mobile WiMAX Release 2. It is considered as the main competing technology to LTE technology as it has been approved by ITU-R as one of the IMT-Advanced technologies [92].

WiMAX PHY utilizes OFDM technology similarly to LTE's radio. WiMAX has a very wide spectrum range of operation from 2 to 66 GHz. There are 128 subcarriers and the bandwidth may vary from 1.25 to 20 MHz. In real deployments, the actual licensed bands differ by countries in the center frequency as well as in the bandwidth. The most commonly used frequencies are 3.5 and 5.8 GHz for fixed wireless (802.16d) and 2.3, 2.5, and 3.5 GHz for mobile WiMAX (802.16e).

One of the key roles of the MAC layer of WiMAX is an adaptation of different PHYs to the network layer. WiMAX supports point-to-multipoint communication, which is realized by the CSMA/CA channel access method, similarly to IEEE 802.11 WLAN technology. Another important feature of the MAC layer is QoS support. There are QoS classes defined as follows: Unsolicited Grant Service, real-time Polling Service, Extended real-time Polling Service (only in mobile WiMAX), nonreal-time Polling Service, and Best Effort for non-QoS traffic.

6.4.5   Satellite Communication

Satellite communication has been utilized to connect remote substations and offer Supervisory Control and Data Acquisition (SCADA) and other Smart Grid-related applications for years. However, the major limitation is very high cost, which forced satellite communication to be used only in the edge situations, where there is no better cost-efficient option.

Current satellite technologies have evolved significantly by reducing costs and latency, and increasing reliability and data rates. These improvements bring this very long-range communication technology back into the Smart Grid market. The major advantage of satellite communication remains area coverage, and being able to provide communication services during disasters such as earthquakes and tsunamis. The properties of satellite communication link depend on the satellite's orbit.

Geostationary Earth Orbit (GEO) satellite systems appear stationary as they move with the same angular velocity as the earth. The orbital period is therefore 24 h. There are three satellites in constellation necessary for global coverage and the orbit is at the altitude of 35 786 km. This global visibility makes GEO satellite systems very popular. On the other hand, the very high cost of installation at the high altitudes and very long communication latency make GEO orbit less suitable for two-way communications. The most known satellites at GEO orbit are broadcasting systems for TV and radio, such as Astra, Inmarsat, and Hispasat.

Medium Earth Orbit (MEO) ranges from 2000 to 35 000 km and the orbital period is up to 10 h, thus requiring fewer satellites in constellation for the coverage of the whole Earth. The MEO satellite system is more expensive on the terminal side for two-way communication. The widely used Global Positioning System (GPS) system is served by a constellation of GPS satellites at nearly 27 000 km altitude.

Low Earth Orbit (LEO) is between 160 and 2000 km of altitude and the LEO satellites do not require terminals with antenna pointing, which makes these systems very suitable for handheld and portable terminals. As the orbital period is very short (less than 2 h), LEO satellites create satellite constellations, which consists of tens of satellites to deliver required coverage. Examples of such systems are Globalstar, Iridium, and Iridium Next, which is planned to be launched in 2015 and replace its ancestor [93]. Another LEO satellite system, Orbcomm, operates in the 137–150 MHz VHF band and provides a global M2M platform for asset tracking, management, and remote control.

Highly elliptical orbit (HEO) is an elliptic orbit with a perigee of 1000 km and an apogee of about 40 000 km. The main advantage of an elliptical orbit is long dwell time and visibility, which can exceed 12 h at apogee. The origins of this orbit utilization date back to the Soviet era and the Molniya satellite system. Currently, the HEO is used by satellites for radio broadcasting by the US Sirius Satellite radio.

Satellites operate in various frequency bands ranging from 1 to 40 GHz, defined by IEEE. In general, low-frequency bands require less-expensive RF equipment and lower pointing accuracy of the terminal antenna than high frequencies. Also, for the same reason, it is more suitable for Mobile Satellite Service (MSS) systems.

Higher frequencies are commonly used for satellite TVs. These frequencies are susceptible to rain fade and require higher pointing accuracy of the antenna, which increases the cost of the terminal equipment. IEEE recognizes L-band (1–2 GHz), S-band (2–4 GHz), C-band (4–8 GHz), X-band (8–12 GHz), Ku-band (12–18 GHz), K-band (18–26.5 GHz), and Ka-band (26.5–40 GHz).

A **Very Small Aperture Terminal** (VSAT) is the common satellite ground antenna with a dish up to 3 m, usually around 1 m in diagonal. VSATs are usually configured in star topology. This configuration utilizes dedicated uplink nodes to relay communication from other satellites. As the communication is relayed via an earth ground station, it can incur large delays for communication involving several satellites. On the other hand, mesh topology allows for directly connecting terminals via the satellites. This is commonly realized by connecting satellites in constellation by intersatellite links, which increases the cost of the satellite system, but at the same time allows relatively low latency for communication of remote terminals. It is especially important for LEO satellite systems as their coverage is rather small and the communication is likely to go over several intersatellite hops.

Satellite systems have been used for utilities to provide communication between remote places (e.g., SCADA and remote monitoring) for many years. Current changes in cost of deployment and advancements in satellite communication technology are moving satellite systems into the Smart Grid market as a potential candidate to provide connectivity with high reliability, security, and global coverage. This is in line with requirements for communication in many Smart Grid applications such as AMI backhaul from meter aggregation nodes in remote areas, monitoring and control of remote renewable generation sites, and video surveillance of remote substations. These are just some examples of many potential applications enabled by satellite systems for the Smart Grid [94]. Many current satellite systems have recently started to provide their M2M services to utilities. For example, Inmarsat has launched BGAN M2M service [95], Iridium's Short Burst Data service is being used for M2M and remote sensing [96], and Orbcomm launched Satellite M2M service [97] recently. Besides a role of a primary connectivity provider, satellite systems provide an excellent solution for robust communication as a backup infrastructure for critical communications, emergency services, and services in disaster-prone areas and areas without sufficient communication infrastructures.

6.5 Conclusion

Requirements for different applications of the Smart Grid vary significantly from low bandwidth and delay-tolerant traffic for periodical metering to "mission critical" low-latency traffic for blackout prevention and emergency situations. The decision-making process of how to select the most suitable communication technology for each application requires an in-depth analysis, precise modeling, and validation against current and future traffic expectations for each segment of Smart Grid systems. As is clear from the requirements for Smart Grid applications, a communication network may not consist of one communication technology, but more likely will be a combination of several technologies together to meet the given application requirements. At the same time it is essential to take the overall environment, security and safety, system complexity, economic feasibility, and future trends into account. Regardless of the used technologies, a tight integration with energy transportation and distribution systems should be taken into consideration. Moreover, the design of a communication system must provide a very high level of interoperability and will likely require optimization. With these constraints in mind, many international and national councils and standards development organizations such as NIST, IEEE, CEN (European Committee for Standardization)/CENELEC (European Committee for Electrotechnical Standardization), and NEDO (New Energy and Industrial Technology Development Organization) are trying to precisely identify priority action plans, requirements, complexity, and strategies for communication systems of the Smart Grid in order to develop new standards and specifications, or to propose recommendations for existing standards.

One principal requirement is to enable end-to-end bidirectional communication between different entities and nodes in the system. This becomes very urgent for new functionalities such as distributed smart energy storage, renewables at the premises plugged into the distribution grid and demand response systems. In order to achieve end-to-end connectivity, the natural choice is the IP, which is the core part of the Internet and many ICT systems. IP enables interoperability and makes the actual communication technology and transport media completely transparent for actual Smart Grid-related data flow that is carried over these communication links.

Wired communication technologies such as Fiber, DSL, or Ethernet might require high installation expenses, however, depending on the transport media, many wired technologies can offer very reliable, high-capacity, and low-latency communication links. Moreover, utilization of already deployed infrastructures for data communication as well as for electric energy transport can lead to highly efficient and economically feasible solutions.

Without any doubt, we are currently living in the wireless technology society, where portability, mobility, and out-of-the box wireless solutions and deployments are affecting every aspect of our life and many industries and social areas, such as medicine, agriculture, transportation. It is also very likely that wireless communication will play a key role in many applications in the Smart Grid. Scalable cellular networks and wireless networks behind the premises have already many years of technology advancement and commercial deployment. Mesh networks offer low-cost solutions with relatively high system capacity out of the-box. Satellite networks are becoming more economically feasible with recent technology improvements.

A large number and variety of communication technologies might give the false impression that the complete architecture of the Smart Grid is only a matter of precise design and technology selection. As described at the beginning of this chapter, the system complexity, security, economic feasibility, and application-specific requirements are very important factors, which are leading to the development of new communication technologies specifically designed and optimized for the Smart Grid and its application requirements.

References

[1] OpenSG (2010) *SG Network System Requirements Specification*, OpenSG, SG-Network Task Force Core Development Team.

[2] DOE (2010) Communication Requirements of Smart Grid Technologies, Department of Energy, United States of America, Washington, DC.

[3] Kushalnagar, N., Montenegro, G., Schumacher, C. (2007) *IPv6 over Low-Power Wireless Personal Area Networks (6LoWPANs)*. RFC4919, IETF.

[4] Montenegro, G., Kushalnagar, N., Hui , and D. Culler, (2007) *Transmission of IPv6 Packets Over IEEE 802.15.4 Networks*, RFC 4944. IETF.

[5] Sarwar, U., Sinniah, G.R., Suryady, Z., and Khosdilniat, R. (2010) Architecture for 6LoWPAN mobile communicator system. *International Multi Conference of Engineers and Computer Scientists, Hong Kong*.

[6] Kim, E., Kaspar, D., Vasseur, J.P. (2012) *Design and Application Spaces for IPv6 over Low-Power Wireless Personal Area Networks (6LoWPANs)*. RFC6568, IETF.

[7] Kim E. (2012) *Neighbor Discovery Optimization for IPv6 over Low-Power Wireless Personal Area Networks (6LoWPANs)*, RFC6775, IETF.

[8] Winter, T. (2012) *RPL: IPv6 Routing Protocol for Low-Power and Lossy Networks*. RFC6550, IETF.

[9] IETF (2012) *Routing Over Low Power and Lossy Networks (Roll): Description of Working Group*, http://datatracker.ietf.org/wg/roll/charter/ (accessed 09 January 2013).

[10] IEC (2008) IEC CISPR22. *Edition 6. Information Technology Equipment-Radio Disturbance Characteristics-Limits and Methods of Measurement*. International Electrotechnical Commission.

[11] PRIME (2008) *Technology Whitepaper: PHY, MAC and Convergence Layers*, PRIME.

[12] G3-PLC *G3-PLC Overview*, www.g3-plc.com/content/g3-plc-overview (accessed 10 January 2013).

[13] IEEE (2010). P1901.2. *Standard for Low Frequency (less than 500 kHz) Narrow Band Power Line Communications for Smart Grid Applications*, Institute of Electrical and Electronics Engineers.

[14] ITU (2011) ITU-T G.9955. *Narrow-band OFDM Power Line Communication Transceivers - Physical Layer Specification*, International Telecommunication Union.

[15] HomePlug Alliance (2011) *HomePlug® Alliance Announces Netricity™ Power line Communications Targeting Smart Meter to Grid Applications*, www.homeplug.org/tech/Netricity/ (accessed 10 January 2013).

[16] IEC (2001) *Distribution Automation Using Distribution Line Carrier Systems – Part 5–1: Lower Layer Profiles – The Spread Frequency Shift Keying (S-FSK) Profile*, International Electrotechnical Commission.

[17] HomePlug (2005) *HomePlug AV White Paper*, HomePlug Alliance.

[18] HomePlug (2012) *HomePlug® AV2 – The Specification for Next-Generation Broadband Speeds Over Power line Wires*, HomePlug Alliance.

[19] HomePlug (2012) *HomePlug Green PHY™ Specification*, HomePlug Alliance.

[20] HD-PLC *Originalities of HD-PLC*, HD-PLC Alliance, www.hd-plc.org/modules/about/original.html (accessed 10 January 2013).

[21] EKOPLC (2007) *Universal Power line Association (UPA) and OPERA Announce Joint Agreement on Power line Access Specification*, EKOPLC.

[22] ITU-T (2011) *Unified High-Speed Wireline-Based Home Networking Transceivers – System Architecture and Physical Layer Specification*. ITU-T Recommendation G.9960.

[23] IEEE (2010) IEEE Std 1901. *IEEE Standard for Broadband over Power Line Networks: Medium Access Control and Physical Layer Specifications*, Institute of Electrical and Electronics Engineers.

[24] Berger, L.T., and Iniewski, K. (2012) Smart Grid: Applications, Communications, and Security, John Wiley & Sons, Inc., Hoboken, NJ.

[25] Baker, L. (2012) *EPB Deploys America's Fastest Fiber-optic Smart Grid*, www.electricenergy online.com/?page=show_article&mag=68&article=550 (accessed 5 January 2013).

[26] ITU (2007) *Characteristics of a 50/125 μm Multimode Graded Index Optical Fibre Cable for the Optical Access Network*. ITU-T G.651.1.

[27] ITU (2009) *Characteristics of a Single-Mode Optical Fibre and Cable*, G.652 ITU-T.

[28] ITU (2009) *Optical Fibre, Cables and Systems*, ITU-T Manual.

[29] Horak, R. (2007) Telecommunications and Data Communications Handbook, John Wiley & Sons, Inc., Hoboken, NJ.

[30] ANSI (1996) T1.105.07-1996. *Synchronous Optical Network (SONET) – Sub-STS-1 Interface Rates and Formats Specification*. American National Standards Institute, New York.

[31] ITU (2007) *Network Node Interface for the Synchronous Digital Hierarchy (SDH)*. Recommendation G.707/Y.1322, ITU, Geneva.

[32] ITU (2006) *Characteristics of Synchronous Digital Hierarchy (SDH) Equipment Functional Blocks*. Recommendation G.783, ITU-T, Geneva.

[33] ITU (2008) *Management Aspects of the Synchronous Digital Hierarchy (SDH) Transport Network Element*. Recommendation G.784, ITU-T, Geneva.

[34] ITU (2000) *Architecture of Transport Networks based on the Synchronous Digital Hierarchy (SDH)*, Recommendation G.803, ITU-T, Geneva.

[35] ITU (2005) *Broadband Optical Access Systems Based on Passive Optical Networks (PON)*. Recommendation G.983.1, ITU-T, Geneva.

[36] ITU (2006) *A Broadband Optical Access System*. Recommendation G.983.x, ITU-T, Geneva.

[37] ITU (2008) *Gigabit-capable Passive Optical Networks (GPON)*. Recommendation G.984.x, ITU-T, Geneva.

[38] FSAN (2012) *Full Service Access Network*, www.fsan.org/ (accessed 5 January 2013).

[39] ITU (1999) *Asymmetric Digital Subscriber Line (ADSL) Transceivers*. ITU-T G.992.1.

[40] ITU (2004) *Very High Speed Digital Subscriber Line Transceivers (VDSL)*. ITU-T Recommendation G.993.1.

[41] ITU (2009) *Asymmetric Digital Subscriber Line Transceivers 2 (ADSL2)*. ITU-T Recommendation G.992.3.

[42] ITU (2009) *Asymmetric Digital Subscriber Line 2 Transceivers (ADSL2)– Extended Bandwidth ADSL2 (ADSL2plus)*. ITU-T Recommendation G.992.5.

[43] ITU (2012) *Very High Speed Digital Subscriber Line Transceivers 2 (VDSL2)*. ITU-T Recommendation G.992.5.

[44] Snyder, B. (2012) *Readwrite: The Future of Broadband is … DSL*. SAY Media Inc., http://readwrite.com/2012/06/28/the-future-of-broadband-is-dsl (accessed 10 January 2013).

[45] IEEE (2010) 802.3ba-2010. *Media Access Control Parameters, Physical Layers, and Management Parameters for 40 Gb/s and 100 Gb/s Operation*, IEEE Standard, New York.

[46] MEF *What is Carrier Ethernet?* Metro Ethernet Forum, http://metroethernetforum.org/page_loader .php?p_id=140 (accessed 10 January 2013).

[47] LightRiver *Carrier Ethernet for Smart Grid Communications Modernization*. LightRiver Technologies, www.lightriver.com/index.php?p=Carrier_m (accessed 10 January 2013).

[48] CEN (2010) *Why the Smart Grid needs Carrier Ethernet*, Carrier Ethernet News, 23 December 2010, www.carrierethernetnews.com/articles/158846/why-the-smart-grid-needs-carrier-ethernet/ (accessed 10 January 2013).

[49] NIST (2010) *NIST Priority Action Plan 2: Guidelines for Assessing Wireless Standards for Smart Grid Applications*, NIST.

[50] Orange *Innovate and be competitive with M2M*, Orange Business Services, www.orange-business .com/en/machine-to-machine (accessed 2 February 2013).

[51] Wallen, J. (2013) *M2M is One of Verizon's Key Business Tech Trends*, Techrepublic (Jan. 7, 2013), www.techrepublic.com/blog/smartphones/m2m-is-one-of-verizons-key-business-tech-trends/6068 (accessed 13 January 2013).

[52] NTT Docomo (2012) *DOCOMO to Launch Global M2M Platform*, 5 December 2012, www .nttdocomo.com/pr/2012/001622.html (accessed 13 January 2013).

[53] EPCglobal (2011) *EPC Class-1 HF RFID Air Interface Protocol for Communications at 13.56 MHz*, GS1 EPCglobal.

[54] EPCglobal (2008) *Class-1 Generation-2 UHF RFID Protocol for Communication at 860 MHz – 960 MHz*, GS1 EPCglobal.

[55] Google (2012) *Google Wallet*, www.google.com/wallet/how-it-works/in-store.html (accessed 2 January 2013).

[56] NFC (2011) *Bluetooth Secure Simple Pairing Using NFC*. NFCForum and Bluetooth SIG.

[57] IEEE (2011) IEEE 802.15.4. *Low-Rate Wireless Personal Area Networks (WPANs); Standard*, Institute of Electrical and Electronics Engineers, New York.

[58] IEEE (2009) IEEE 802.15.4c. *Amendment 2: Alternative Physical Layer Extension to Support One or More of the Chinese 314–316 MHz, 430–434 MHz, and 779–787 MHz bands; Standard*, Institute of Electrical and Electronics Engineers, New York.

[59] IEEE (2009) IEEE 802.15.4d. *Amendment 3: Alternative Physical Layer Extension to support the Japanese 950 MHz bands; Standard*, Institute of Electrical and Electronics Engineers, New York.

[60] NICT (2012) *World's First Small-sized, Low-power "Smart Meter Radio Device" Compliant with IEEE 802.15.4g/4e Standards for Japan's new 920 MHz Band Allocation*, 5 April 2012, www.nict.go.jp/en/press/2012/04/05en-1.html (accessed 5 January 2013).

[61] IEEE (2013) IEEE 802.15.4. *Amendment: Physical Layer (PHY) Specifications for Low Energy, Critical Infrastructure Monitoring Networks (LECIM); Draft Standard*, Institute of Electrical and Electronics Engineers, NewYork.

[62] Freescale (2010) IEEE® 802.15.4. *Technology from Freescale*, Freescale Semiconductor, Inc.

[63] WirelessHART *The First Simple, Reliable and Secure Wireless Standard for Process Monitoring and Control* HART Communication Foundation, Austin, TX, 2009.

[64] IEC (2010) IEC 62591. *Industrial Communication Networks – Wireless Communication Network and Communication Profiles – WirelessHART™ Standard*. International Electrotechnical Commission, Geneva.

[65] Song, J., Han, S., Mok, A.K., *et al.* (2008) WirelessHART: Applying Wireless Technology. *IEEE Real-Time and Embedded Technology and Applications Symposium, St. Louis, Mo.*

[66] ISA (2011) ANSI/ISA-100.11a-2011. *Wireless Systems for Industrial Automation: Process Control and Related Applications*, ANSI/ISA, Durham, NC.

[67] IEEE (2012) *IEEE 802.11ad, draft*, www.ieee802.org/11/Reports/tgad_update.htm (accessed 2 January 2013).

[68] IEEE (2013) *Status of Project IEEE 802.11ah*, 16 January 2013, www.ieee802.org/11/Reports/ tgah_update.htm (accessed 22 January 2013).

[69] RedpineSignals (2010) *WinergyNet™: Wireless Communications Architecture for the Smart Grid*, http://redpinesignals.com/Solutions/Reference_Designs/Smart_Grid_Comm/index.html (accessed 20 January 2013).

[70] IEEE (2013) IEEE 802.11ah. *Proposed Specification Framework for TGah*, IEEE, New York.

[71] NICT (2013) *World's First TV White Space Prototype Based on IEEE 802.22 for Wireless Regional Area Network*, 30 January 2013, www.nict.go.jp/en/press/2013/01/30-1.html (accessed 12 February 2013).

[72] Graychase, N. (2007) *ITBusinessEdge: "Wireless Fidelity" Debunked*, 27 April 2007, www.wi-fiplanet.com/columns/article.php/3674591 (accessed 2 January 2013).

[73] In-Stat (2011) *Wi-Fi Direct: It's All About the Software*, In-Stat/MDR.

[74] WFA (2011) *Wi-Fi Simple Configuration Technical Specification*, Wi-Fi Alliance.

[75] WFA (2012) *Wi-Fi Multimedia Technical Specification*, Wi-Fi Alliance.

[76] WFA (2012) *Wi-Fi Display Technical Specification*, Wi-Fi Alliance.

[77] ABIresearch (2012) *Wi-Fi Enabled Device Shipments will Exceed 1.5 Billion in 2012, Almost Double that Seen in 2010*, 11 October 2012, www.abiresearch.com/press/wi-fi-enabled-device-shipments-will-exceed-15-bill (accessed 5 January 2013).

[78] Wu, J. (2012) *Strategy Analytics: A Quarter of Households Worldwide Now Have Wireless Home Networks* 4 April 2012, www.strategyanalytics.com/default.aspx?mod=pressreleaseviewer&a0= 5193 (accessed 2 January 2013).

[79] Wave2M *Wave2M Specification: Technology Overview*, www.wave2m.com/the-specification (accessed 2 January 2013).

[80] Bluetooth (2004) *Specification of the Bluetooth System; Covered Core Package version: 2.0 + EDR*, BluetoothSIG.

[81] Bluetooth (2009) *Covered Core Package Version: 3.0 + HS*, BluetoothSIG.

[82] Bluetooth (2011) *Covered Core Package Version: 4.0*, BluetoothSIG.

[83] Torchia, M. and Sindhu, U. (2011) *Cellular and the Smart Grid: A Brand-New Day*, IDC Energy Insights, Framingham, MA.

[84] ZTE (2010) *Overview of Global GSM Market*, 19 April 2010, http://wwwen.zte.com.cn/endata/ magazine/ztetechnologies/2010/no4/articles/201004/t20100419_182951.html (accessed 2 January 2013).

[85] 3GPP (2000) *Digital Cellular Telecommunications System (Phase 2+); Physical Layer on the Radio Path*. 3GPP TS 45.001.

[86] ETSI/3GPP (2011) *General Packet Radio Service (GPRS) Enhancements for Evolved Universal Terrestrial Radio Access Network (E-UTRAN) Access*. ETSI TS 123 401/3GPP TS 23.401 Version 8.14.0 Release 8, ETSI, Sophia Antipolis Cedex.

[87] 3GPP (2007) *Feasibility Study for Evolved GSM/EDGE Radio Access Network (GERAN)*. 3GPP TR 45.912; Release 7, 3GPP.

[88] ITU (2011) *IMT-Advanced*, www.itu.int/net/newsroom/wrc/2012/reports/imt_advanced.aspx. (accessed 7 January 2013).

[89] 3GPP (2011) *High Speed Downlink Packet Access (HSDPA); Overall description*. 3GPP TS 25.308; Release 11, 3GPP.

[90] 3GPP (2010) *Evolved Universal Terrestrial Radio Access (E-UTRA) and Evolved Universal Terrestrial Radio Access Network (E-UTRAN); Release 8*, 3GPP Specification, TS 36.300.

[91] WiMAX (2012) *WiMAX Forum Certification Program*, www.wimaxforum.org/certification (accessed 2 January 2013).

[92] IEEE (2011) *IEEE approves IEEE 802.16m™ – Advanced Mobile Broadband Wireless Standard*, 31 March 2011, http://standards.ieee.org/news/2011/80216m.html (accessed 5 January 2013).

[93] Iridium *Iridium NEXT*, Iridium Communications Inc., www.iridium.com/About/IridiumNEXT .aspx (accessed 6 January 2013).

[94] Gohn, B. and Wheelock, C. (2010) Smart Grid Network Technologies and the Role of Satellite Communications, Pike Research LLC, Boulder, CO.

[95] Inmarsat (2012) *BGAN M2M*, Inmarsat plc, www.inmarsat.com/services/bgan-m2m (accessed 5 January 2013).

[96] Iridium (2010) *Iridium Short Burst Data Service*, Iridium Communications Inc.

[97] Orbcomm (2013) *Satellite M2M*, www.orbcomm.com/services-satellite.htm, (accessed 6 January 2013).

7

Security and Safety for Standardized Smart Grid Networks

7.1 Introduction

Control and monitoring networks of Smart Grids form the infrastructure of a country that can meet the requirements of the production and life of humans. Smart Grid control networks are used in application domains such as discrete, batch and process manufacturing, electric power generation and distribution, gas and water supply, and transportation. Depending on the type and purpose of the Smart Grid control networks, its components are distributed on a local, wide-area, or even global scale. Communication links and facilities are an important element of such Supervisory Control and Data Acquisition (SCADA) systems. In the past, automation and power systems were not linked to each other and were not connected to public networks such as the Internet. Today, the market puts pressure on companies to make fast and cost-effective decisions. For this purpose, accurate and up-to-date information about the plant and the process status have to be available not only on the plant floor but also at the management level in the enterprise and even for supply-chain partners. This results in increasing interconnection between different automation systems as well as between automation and office systems. Initially, such interconnections were based on specialized proprietary communication mechanisms, and protocols. Today, open and standardized wireless and wired network technologies are increasingly used for that purpose.

Security/safety is one of the key concerns for both wireless and wired Smart Grid control networks. In this report, we overview and analyze security technologies in the existing security and communication standards of Smart Grids. Related standards that include security/safety are shown in Table 7.1.

The rest of this chapter is organized as follows. Section 7.2 presents the threats and vulnerabilities of Smart Grids. In Section 7.3, existing wireless and wired network

Smart Grid Standards: Specifications, Requirements, and Technologies, First Edition. Takuro Sato,
Daniel M. Kammen, Bin Duan, Martin Macuha, Zhenyu Zhou, Jun Wu, Muhammad Tariq and Solomon Abebe Asfaw.
© 2015 John Wiley & Sons, Ltd. Published 2015 by John Wiley & Sons, Ltd.

Table 7.1 Standard list of security/safety in Smart Grids

Function field	Standard name	Short introduction
Wireless control data exchange	WirelessHART	A wireless sensor networking technology based on the Highway Addressable Remote Transducer Protocol (HART)
Wireless control data exchange	ISA100.11a	Wireless systems for industrial automation: process control and related applications
Wired control data exchange	Profibus	A standard for field bus communication in automation technology
Wired control data exchange	PROFINET	The open industrial Ethernet standard of PROFIBUS and PROFINET International (PI) for automation
Wired control data exchange	Common Industrial Protocol (CIP)	An industrial protocol for industrial automation applications, which is supported by the Open DeviceNet Vendors Association (ODVA)
Wired control data exchange	CC-Link	High-speed field network able to simultaneously handle both control and information data
Wired communications	Powerlink	A deterministic real-time protocol for standard Ethernet
Wired control data exchange	EtherCAT	An open real-time Ethernet network originally developed by BECKOFF, which sets new standards for real-time performance and topology flexibility
Safety protocol	PROFIsafe	Safety extension of Profibus and PROFINET
Safety protocol	CIP-safety	Safety extension of CIP
Safety protocol	CC-Link safety	Safety extension of CC-Link
Safety protocol	Powerlink Safety	Safety extension of Powerlink
Safety protocol	TwinSAFE	Safety extension of EtherCAT
Functional security for TC 57 series	IEC 62351	Standard that handles the security of TC 57 series of protocols including IEC 60870-5 series, IEC 60870-6 series, IEC 61850 series, IEC 61970 series, and IEC 61968 series
Functional safety	IEC 61508	Functional safety of electrical/electronic/ programmable electronic safety-related systems

standards in Smart Grids are presented. Section 7.4 analyzes the security mechanisms of wireless Smart Grids. The security/safety mechanisms of wired Smart Grids are analyzed in Section 7.5. Section 7.6 discusses the key concerns and open issues of these security/safety mechanisms. Finally, Section 7.7 concludes this chapter.

7.2 Threats and Vulnerabilities of Smart Grids

7.2.1 Network Vulnerabilities

The use of wired and wireless network technologies not only provides benefits for Smart Grids but also introduces the threats of network attacks. Attacks may either be initiated by persons outside the plant or by insiders. Some security features are required to deal with the vulnerabilities, including confidentiality, integrity, availability, authentication, authorization, and nonrepudiability. The following are some common types of attacks are the following:

- **Denial of Service (DoS) attack and Distributed denial of service (DDoS) attack**: This attempts to make a machine or network resource unavailable to its intended users. Although the means to carry out, motives for, and targets of a DoS attack may vary, it generally consists of the efforts of one or more people to temporarily or indefinitely interrupt or suspend services of a host connected to the network. The goal of the attacker is to decrease the availability of the system for its intended purpose.
- **Eavesdropping**: The goal of the attacker is to violate the confidentiality of the communication, for example, by sniffing packets on the local area network (LAN) or by intercepting wireless transmissions.
- **Man-in-the-middle attack**: In a man-in-the-middle attack, the attacker acts toward both end points of the communication as if the attacker was the expected, legitimate partner. In addition to confidentiality violations, this also allows modifying the exchanged messages (integrity). Via man-in-the-middle attacks, weaknesses in the implementation or usage of certain key exchange and authentication protocols can be exploited to gain control even over encrypted sessions.
- **Virus**: A virus-based attack manipulates a legitimate user to bypass authentication and access control mechanisms in order to execute the malicious code injected by the attacker. In practice, virus attacks are often untargeted and spread among vulnerable systems and users. Virus attacks often directly or indirectly decrease the availability of infected systems by consuming excessive amounts of processing power or network bandwidth.

7.2.2 Errors of Communications

In order to prevent any damage to persons and machines, it is paramount that data in safety-sensitive areas of machines and plants are transmitted in time and in their entirety. Failures can occur for various reasons, for example, if data packets are routed to the wrong recipient, or are delayed at a gateway owing to traffic overload. Adverse conditions may also lead to erroneous transfer sequences for the packets or cause incorrect data insertions. Lastly, electromagnetic interference also threatens the integrity of information transmissions. In bus-based safety systems, performance free from defects must be ensured by the protocol, which must enable cyclic checks of the network segments that are relevant for safety, as well as checks of the devices involved. In the case of an interruption in communication or an incomplete data transmission, it is to initiate a safe shutdown of the machine or plant.

For functional safety fieldbus, The IEC SC65C/WG12 committee has developed the IEC 61784-3 Industrial communication networks-Profiles-Part 3 [1]. This standard defines the communication errors in Smart Grids. Related communication errors include corruption, unintended repetition, incorrect sequence, loss, unacceptable delay, insertion, masquerading, and addressing.

7.3   Communication Network Standards of Smart Grids

7.3.1   Wireless Network Standards

With the development of wireless communication technologies, some wireless communication standards are proposed for Smart Grids. The process automation and manufacturing industries currently have two independent and competing standards specifically designed for wireless field instruments, each supported by different industry players [2]. These two standards are WirelessHART and ISA 100.11a. In September 2007, the Highway Addressable Remote Transducer (HART) Communication Foundation (HCF) released the HART Field Communication Protocol Specification, Revision 7.0, which included the definition of a wireless interface to field devices, referred to as WirelessHART [3]. In April 2010, WirelessHart was approved by the International Electrotechnical Commission (IEC) unanimously, making it the first wireless international standard as IEC 62591 [4]. Parallel to the HCF's development of WirelessHART, the International Society of Automation (ISA) initiated works on a family of standards defining wireless systems for industrial automation and control applications. The first standard to emerge was ISA100.11a [5], which was ratified as an ISA standard in September 2009. As a matter of fact, both WirelessHART and ISA100.11a aim to provide secure and reliable wireless communication for noncritical monitoring and control applications.

7.3.2   Wired Network Standards and Their Safety Extensions

There are several communication standards and their safety extensions for wired Smart Grids.

PROFIsafe is one of four safety protocols described in the IEC 61784-3 standard [6]. The PROFIsafe system is an extension of the Profibus and PROFINET system. The safe controller and the safe bus devices communicate with one another via the PROFIsafe protocol, which superposes the standard Profibus or PROFINET protocol and contains the safe input and output data as well as data-security information.

The Common Industrial Protocol (CIP) is an industrial protocol for industrial automation applications [7]. It is supported by the Open DeviceNet Vendors Association (ODVA). CIP provides a unified communication architecture throughout the manufacturing enterprise. The CIP is used in EtherNet/Internet Protocol (IP), DeviceNet, CompoNet, and ControlNet.

CC-Link [8] is the high-speed field network able to simultaneously handle both control and information data. With the high communication speed of 10 Mbps, CC-Link can achieve the maximum transmission distance of 100 m and connect to 64 stations. CC-Link Safety is compatible with the standard CC-Link.

Ethernet Powerlink [9] is a deterministic real-time protocol for standard Ethernet. It is an open protocol managed by the Ethernet POWERLINK Standardization Group (EPSG). It was introduced by the Austrian Automation Company B&R in 2001. This protocol has nothing to do with power distribution via Ethernet cabling or power over Ethernet (PoE), power line communication, or Bang and Olufsens PowerLink cable. The Powerlink Safety protocol basically features three principal characteristics: its definition of data transfer, its upper-level configuration services, and most notably, its encapsulation of data that is relevant to safety (i.e., CRC Checksum) into a flexible telegram format.

EtherCAT is the open real-time Ethernet network originally developed by BECK-OFF [10]. EtherCAT sets new standards for real-time performance and topology flexibility. On the basis of EtherCAT, TwinSAFE [11] from BECKOFF provides a consistent hardware and software technology for achieving integrated and simplified utilization, ranging from safe input and output terminals and safe miniature controllers for the Bus Terminal system to the AX5000 Servo Drives.

7.4 Wireless Network Security Mechanisms in the Smart Grids

7.4.1 An Overview of Security Mechanisms in the Wireless Standardized Smart Grid

WirelessHART and ISA100.11a apply security protection through payload encryption and message authentication for both single-hop (hop-by-hop) messages and end-to-end messages. For both standards, the single-hop protection takes place on the Data Link Layer (DLL), while end-to-end message protection is handled by the Network Layer (NL) in WirelessHART and transport layer (TL) in ISA100.11a. The DLL security defends against attackers who are outside of the system, while NL/TL security defends against attackers who may be on the network path between the source and destination. The basic stack models and security measures of WirelessHART and ISA100.11a are shown in Figure 7.1.

7.4.2 Device Joining

7.4.2.1 State Transitions of Device Join

In wireless Smart Grids, joining is an important process by which a network device is authenticated and allowed to participate in the network. Each device must perform

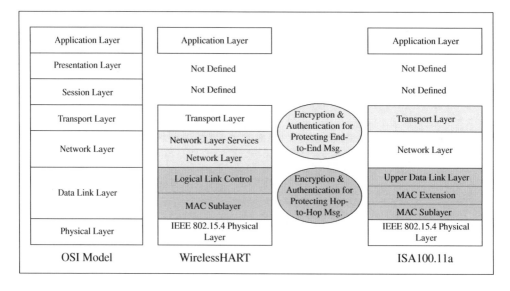

Figure 7.1   Basic framework of WirelessHART and ISA100.11a security

membership enrollment before the join process, which is based on out-of-band (OOB) mechanisms. A secure channel can be established between the unjoined device and the system manager based on the asymmetric key of the system manager. The individual information of the unjoined device (e.g., EUI-64) with its digital signature, which is generated by a certificate authority (CA), can be transferred to the system manager through the established secure channel. After membership enrollment, related join operations can be implemented.

   In the wireless Smart Grids, there are multiple paths (and state transitions) available for an unjoined device to be provisioned and ultimately to join a Smart Grid. These paths are illustrated via the state transition diagram in Figure 7.2. In practical systems, one of the modes can be chosen for realizing the join process.

- **Symmetric key join**: The path of this mode is $S0 \rightarrow S1 \rightarrow S2.1 \rightarrow S3 \rightarrow S4$. In this mode, OOB or preinstall mechanisms are used to provide an unjoined device with the target control network join key and network information. Here, the join key works as a password to join the target control network. Further symmetric key agreement can be performed on the basis of the join key.
- **Asymmetric key join**: The path of this mode is $S0 \rightarrow S1 \rightarrow S2.2 \rightarrow S3 \rightarrow S4$. An unjoined device can get asymmetric key certificates based on OOB or preinstall mechanisms. Symmetric keys for further communications can be generated on the basis of an asymmetric key.
- **No security join process**: The path of this mode is $S0 \rightarrow S1 \rightarrow S5$. The no-security option uses a global key for transfer of join keys. The global key is a well-known key whose value is static and published.

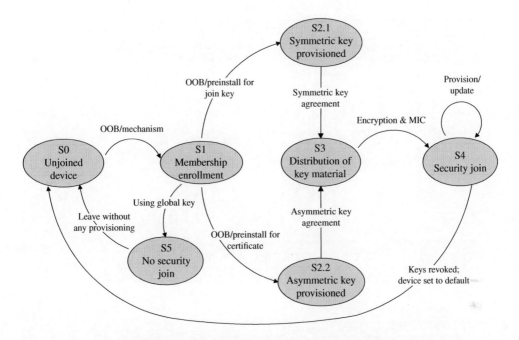

Figure 7.2   State-transition diagram of the join process

In ISA100.11a, a joining device can join the target network with one of the following security options: symmetric key, asymmetric key, and no security. In contrast to ISA100.11a, there are only two options for the join process in WirelessHART, which are a symmetric key and an asymmetric key.

7.4.2.2   Key Mode Involved in Device Joining

During the device join process, four types of keys are used in the security architecture in WirelessHART and ISA100.11a:

Asymmetric Keys are used to perform membership enrollment. **Network Keys and Data Link Keys** are shared by all network devices and used by existing devices in the network to generate Medium Access Control (MAC) Message Integrity Code (MICs) in WirelessHART and ISA100.11a, respectively. **Join Keys** are unique to each network device and are used during the joining process to authenticate the joining device with the system manager. **Symmetric session keys for further communications** are generated by the system manager and are unique for each end-to-end connection between two network devices.

In order to describe the process more clearly, we take the asymmetric key join mode, for example. The asymmetric key join mode of WirelessHART and ISA100.11a are shown in Figure 7.3. When a new device joins the Smart Grid, the joining device will use the asymmetric key to establish a secure channel between the unjoin node and

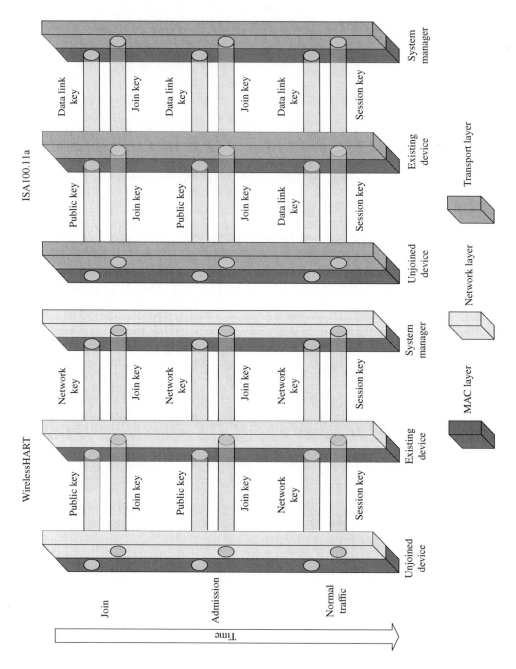

Figure 7.3 Model of asymmetric key join process in WirlessHART and ISA.100.11a

existing node in the Smart Grid. After key agreement, the MIC on the MAC layer uses the join key to generate the MIC and encrypt the join request at the network layer or the transport layer. After the joining device is authenticated, the system manager will create a session key for the device and thus establish a secure session between them. Then, the unjoined device becomes an authenticated node, the same as the existing devices. Thus, it can perform normal communication with the system manager.

7.4.3   Securing Normal Traffic

After the join process, the join device can exchange normal traffic with the existing devices in the Smart Grid. During these communications, both WirelessHART and ISA100.11a use the CCM* (Counter (CTR) with Cipher Block Chaining (CBC)-MAC (corrected)) mode together with the Advanced Encryption Standard (AES)-128 as the underlying block cipher to generate and compare the MIC [12]. CCM* is a minor variation of the CCM, CCM* includes all of the features of CCM and additionally offers encryption-only and integrity-only capabilities. This authenticated encryption algorithm is used to provide both data authentication and privacy. This section analyzes the security mechanisms of communication after the join process.

7.4.3.1   Security Operation Flow

WirelessHART and ISA100.11a networks provide the data-authentication service based on CCM* mode as show in Figure 7.4. A sender collects the field data and intends to send this data. If the layer needs to perform security protect, it will get the security level of this particular packet and construct the relevant security header. Then, on the basis of the security level, the payload may be encrypted, and the MIC may be generated and added to the end of the packet. The other layer will process a similar security operation. Finally, the packet will be sent to the MAC layer.

After receiving a data packet, the receiver analyzes the packet header to check whether this packet is to be protected or not. If this packet is not a security packet, the layer shall forward this payload with no further security interaction. Otherwise, this packet needs to go through the security check. If the security check fails, the data packet process will be stopped and the packet will be discarded. If the check operation succeeds, the payload will be decrypted and the result will also be carried to next layer. Until it reaches the last layer, it will use a similar security operation to process this packet and get the field data.

7.4.3.2   Encryption and Data Integrity

The CCM* algorithm is designed to provide both data authentication and privacy. The device encrypts the packet using the symmetric key algorithm and sends it. The device that receives the packet will decrypt the packet by the symmetric key algorithm and forward the data to the higher layer. In the MAC layer, the Payload Data Unit (PDU)

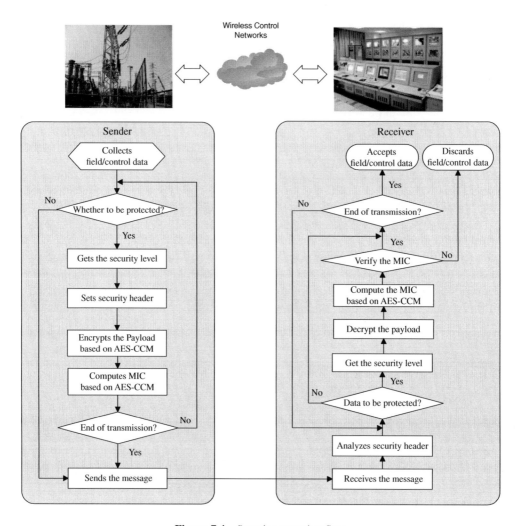

Figure 7.4 Security operation flow

is authenticated with the network key/data link key; on the network/transport layer, the packet is authenticated and encrypted by the session key.

7.4.3.3 Structure of Nonce

A nonce used in CCM* mode stands for a number only used once to differentiate the ciphertexts for the same plaintext. It is often a random or pseudorandom number or time-variant including a suitably granular time stamp in its value. By checking the reuse of nonces or limiting the accepting time range when time stamp is used, it can also ensure that old communications cannot be reused in replay attacks.

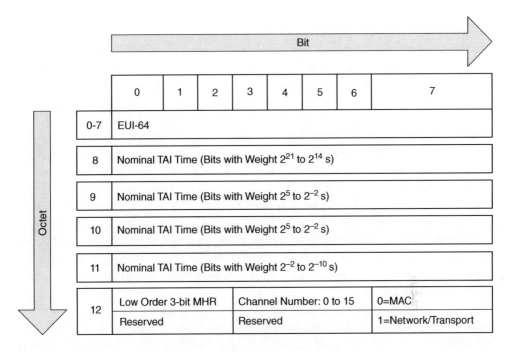

Figure 7.5   Structure of nonce

On the basis of IEEE 802.15.4, the nonce can be constructed based on the structure in Figure 7.5. A 13-octet nonce is required for the CCM* engine. The nonce shall be constructed as the concatenation from first (leftmost) to last (rightmost) octets of data fields The EUI-64 (64-bit extended unique identifier) shall be used as an array of eight octets and the truncated International Atomic Time (TAI) and counter (CTR) as an array of four octets. The TAI time shall be least significant 32 bits of the TAI time in units of 2^{-10} s, as described in Figure 7.5.

The last octet shall be constructed as follows. Bit 7 shall be zero for the MAC layer, and be one for the network layer (in WirelessHART) or transport layer (in ISA100.11a). For the MAC layer, Bits $6-3$ (4 bits) shall indicate the radio channel of transmission, in a range of $0-15$, corresponding to IEEE Std 802.15.4 channel numbers $11-26$, in the same order. Bits $2-0$ shall be copied from the corresponding low-order 3 bits of the sequence number of the MAC header (MHR). On the other hand, Bits $1-6$ are reserved for the network layer or transport layer.

7.5   Wired Network Security/Safety Mechanisms in the Smart Grid

In this section, we give a detailed analysis of the main communication standards and their safety extensions in wired Smart Grids.

7.5.1   An Overview of Security Technologies in the Wired Smart Grid

The basic framework of related standards is shown in Figure 7.6. On the one hand, the wired communication standards can provide the basic security for Smart Grids, which is based on existing Information Technology (IT) security protocols, such as MACsec, IPsec, and Transport Layer Security and Secure Socket Layer (TLS/SSL) [13–15]. On the other hand, most of the safety standards can be regarded as the safety extensions of related communication standards in Smart Grids.

Basically, the existing security standards include three principal characteristics. (i) They use black channels that can provide upper-level safety configuration services based on standard transmission schemes; (ii) they encapsulate data that is relevant to safety (i.e., Cyclic Redundancy Check (CRC), time stamp, etc.) into a flexible telegram format in the safety layer; and (iii) with the security mechanisms of communication standards, they complement each other. These communication standards can be performed on the basis of the underlying security infrastructure, such as MACsec, IPsec, and TLS/SSL. These security infrastructures use related security measures, such as AES encryption.

In this section, in order to explain the implementation principle of related security standards, we analyze PROFIsafe and CIP-Safety in detail. In addition, we briefly introduce the basic principle of CC-Link Safety, Powerlink Safety, and TwinSAFE.

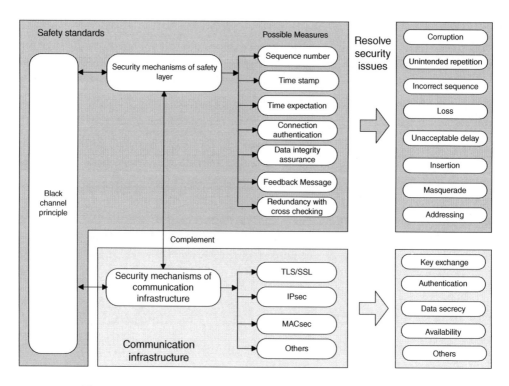

Figure 7.6   Basic framework of safe standards of wired control networks

Table 7.2 The security mechanisms of safety standards in wired Smart Grids

Safety standard	Corresponding communication standard	Basic network/ bus	Security of communication infrastructure	Principle of safety extension	Security of safety standard
PROFIsafe	Profibus	RS485/fiber optics/ MBP	MACsec	Black channel	CRC, time monitoring
	PROFINET	Ethernet	MACsec, IPsec		
CIP-safety	Common Industrial Protocol (CIP)	EtherNet/IP	MACsec, IPsec	Black channel	CRC, time stamp
		DeviceNet	MACsec, IPsec, TLS/SSL		
		CompoNet	MACsec, IPsec, TLS/SSL		
		ControlNet	MACsec, IPsec, TLS/SSL		
CC-Link safety	CC-Link	RS-485	MACsec	Black channel	Time stamp, connection ID, CRC
Powerlink safety	Powerlink	Ethernet	MACsec	Black channel	Time stamp, time monitoring, identification tag
TwinSAFE	EtherCAT	Ethernet	MACsec	Black channel	CRC, time stamp, time monitoring, sequence number

The security mechanisms of safety standards for wired Smart Grids are given in Table 7.2.

7.5.2 Basic Security Mechanisms of Communication Infrastructure

Communication standards can provide the basic security for wired Smart Grids. In order to analyze the security mechanism clearly, in this section we take PROFIBUS/ PROFINET as an example for explanation.

7.5.2.1 Security Data Tunneling

The safety concepts of PROFIsafe and the data-security concepts of PROFIBUS/ PROFINET complement each other [16]. Data-security tunneling can be performed on the basis of IPsec Virtual Private Network (VPN). By integrating PROFINET (Ethernet), the protection against unauthorized access to PROFIsafe islands is of special concern. For this purpose, the entire network is structured into subsegments which provide only a single point of access. This access shall be secured by a security gate (PROFINET security device) employing proven security measures for this purpose, at least including VPN and a firewall. As mentioned before, the PROFIsafe system is an extension of the Profibus and PROFINET system. PROFINET IO uses the protocol set of IPsec, which is defined by related Internet Engineering Task Force (IETF) standards [17].

7.5.2.2   Authentication of Security Gates (Devices) and VPN Clients

In this section, we still take PROFINET as the example for explanation. By integrating PROFINET (Ethernet), the protection against unauthorized access to PROFIsafe islands is of special concern. For this purpose, the entire network is structured into subsegments which provide only a single point of access. This access shall be secured by a security gate (PROFINET security device). During the key exchange between two partner security gates and/or clients, authentication must be performed. A user/password security token is used for authentication of the clients. Certificates according to X.509 can also be used for authentication of the security gates.

7.5.2.3   Encryption Algorithm

Encryption algorithms according to AES with a CBC-MAC mode shall be used. In order to ensure compatibility with other IPsec implementations, triple Data Encryption Standard (3DES) can be used for encryption. Usage of simple DES is not permitted.

7.5.3   Principles of Safety Extensions

Most of the safety standards can be regarded as the safety extensions of related communication standards in Smart Grids. The safety concepts of safety standards and the data-security concepts of corresponding communication standards complement each other [18]. In other words, the security infrastructure of control buses/network underlies the data-security foundation, which can provide the basic security for the high-level safety layer of safety standards.

   Most safety standards of wired Smart Grids use the black channel principle for the transmission of safe data via a standard network. The safe transmission function comprises all measures to deterministically discover all possible faults and hazards that could be infiltrated by the black channel, or to keep the residual error probability under a certain limit. Using the black channel, related safety schemes perform safe communication by using (i) a standard transmission system and (ii) an additional safety

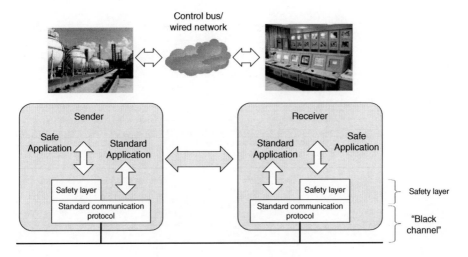

Figure 7.7   Black channel principle

transmission protocol on top of the standard transmission system. The black channel principle is shown in Figure 7.7.

7.5.4   Security Measures of Safety Extension

This section analyzes the principle of encapsulated safety data at the safety layer. In order to describe the security measures clearly, we take some typical mechanisms in the existing safety standards as examples for the analysis in this section.

7.5.4.1   Message Structure Including Container

Safe data, consisting of purely safety-related user data and the protocol overhead required for protection, are transmitted via standard control networks together with data that are not safety related. Here, we take CIP Safety, for example, to explain the principle of encapsulated safety data at the safety layer.

Figure 7.8 shows the telegram setup of a "Master Data Telegram (MDT) Data Field" within the scope of an Ethernet frame, which contains a configurable data container for real-time data of each device. The real-time data of a device are again divided into standard and safety data. The safety data are CIP Safety telegrams either in the short format (2 bytes) or in the long format (up to 250 bytes).

Note that not all the devices in the Smart Grids include the safety data container. In other works, only some devices perform CIP Safety communication, which depend on their security requirements. The devices without CIP Safety capability only send and receive data on the basis of standard CIP connections.

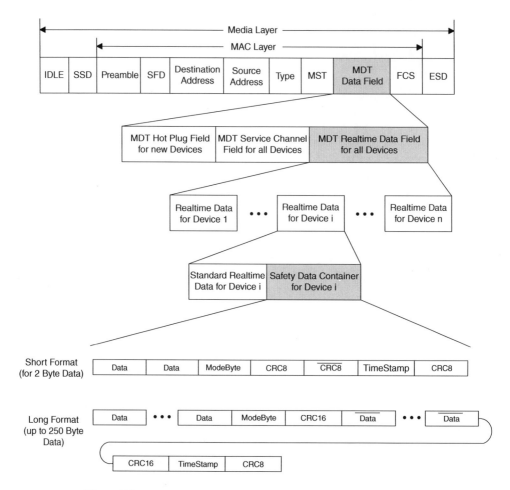

Figure 7.8 Message structure of encapsulated safety data in CIP Safety

7.5.4.2 Cyclic Redundancy Codes Checksum for Integrity Check

Cyclic Redundancy Codes (CRCs) are used in most safety standards for integrity checks. For example, PROFIsafe uses several different CRCs to protect the integrity of safety-related messages; therefore, CRCs with different numbers appear to differentiate them. The safety-related IO data of a safe node are collected in the safety PDU, and the data type coding corresponds to PROFINET IO. One safety container corresponds to one subslot in PROFINET IO. When the safety parameters have been transferred to the safe device, the safe host, the safe device/module produces a 2-byte CRC1 signature [19] over the safety parameters.

The CRC1 signature, safe IO data, status, or control byte and the corresponding consecutive number are used to produce the CRC2 signature as illustrated in Figure 7.9. The CRC1 signature provides the initial value for CRC2 calculation that is transferred

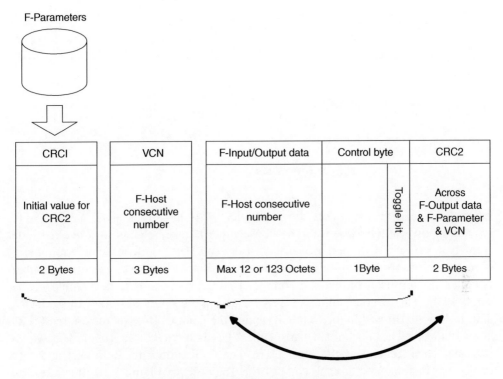

Figure 7.9 CRC2 generation in PROFIsafe

cyclically, thus limiting the CRC calculation for each cyclic PROFIsafe container to CRC2. In Figure 7.9, the symbol "F" is used throughout the PROFIsafe to identify the "fail-safe" function component introduced. The F-Parameters contain information for the PROFIsafe layer to adjust its behavior to particular customer needs and to double-check the correctness of assignments. F-Input and F-Output denote the input and output data of a PROFIsafe device, respectively.

7.5.4.3 Consecutive Number for Delay Control

The consecutive number is used as a measure to deal with some of the possible communication errors. It is also used to monitor the propagation delay between transmission and reception. Each message is equipped with a consecutive number, which is used by the recipient for monitoring the life of the sender and the communication link. Both communication partners continuously check whether the other partner manages to update the consecutive number before a defined watchdog time has elapsed.

The consecutive number check was carried out on different versions of the PROFIsafe model, considering input and output slave configurations with different

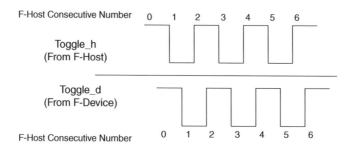

Figure 7.10   Virtual consecutive number

ranges of consecutive numbers. For example, a 24-bit CTR is used in PROFIsafe for consecutive numbering, thus the consecutive number counts in a cyclic mode from 1 … FF FF FF wrapping over to 1 at the end [19]. The consecutive number 0 is reserved for error conditions and synchronization. Here, the consecutive number is called the Virtual Consecutive Number (VCN), because it is not visible in the safety PDU. The mechanism uses CTRs located in the safety host and safety device and the Toggle Bit within the Status Byte and the Control Byte increment the CTRs synchronously. The transmitted part of the VCN is reduced to a Toggle Bit, which indicates an increment of the local CTR. The CTR within the safe host and safe device are incremented at each edge of the Toggle Bits. Figure 7.10 illustrates the VCN mechanism. To verify the correctness and to synchronize the two independent CTRs, the consecutive number is included in the CRC2 calculation that is transmitted with each safety PDU.

7.6   Typical Standards of Functional Security and Safety

7.6.1   IEC 62351 Standards

7.6.1.1   Overview of IEC 62351

IEC 62351 [20–27] is a standard developed by WG15 of IEC TC 57. This was developed for handling the security of TC 57 series of protocols including IEC 60870-5 series, IEC 60870-6 series, IEC 61850 series, IEC 61970 series, and IEC 61968 series. The different security objectives include authentication of data transfer through digital signatures, ensuring only authenticated access, prevention of eavesdropping, prevention of playback and spoofing, and intrusion detection. The published IEC 62531 standard includes the following parts:

- IEC 62351-1: Introduction to the standard
- IEC 62351-2: Glossary of terms
- IEC 62351-3: Security for any profiles including TCP (Transmission Control Protocol)/IP

- IEC 62351-4: Security for any profiles including MMS
- IEC 62351-5: Security for any profiles including IEC 60870-5
- IEC 62351-6: Security for IEC 61850 profiles
- IEC 62351-7: Security through network and system management
- IEC 62351-8: Role-based access control.

7.6.1.2 Relationships between IEC 62351 Standards and Corresponding TC 57 Standards

Security standards have been developed for different profiles of the four communication protocols: IEC 60870-5, its derivatives, IEC 60870 (TASE.2, Telecontrol Application Service Element 2), and IEC 61850. In addition, security through network and system management has been addressed. These security standards must meet different security objectives for the different protocols, which vary depending upon how they are used. Some particular profiles are discussed. The different security objectives include authentication of entities through digital signatures, ensuring only authorized access, prevention of eavesdropping, prevention of playback and spoofing, and some degree of intrusion detection. For some profiles, all of these objectives are important; for others, only some are feasible given the computation constraints of certain field devices, the media speed constraints, the rapid response requirements for protective relaying, and the need to allow both secure and nonsecured devices on the same network.

Here, we give the relationships among published IEC 62351 standards from Part 1 to Part 7 and related TC 57 communication standards, which are shown in Figure 7.11.

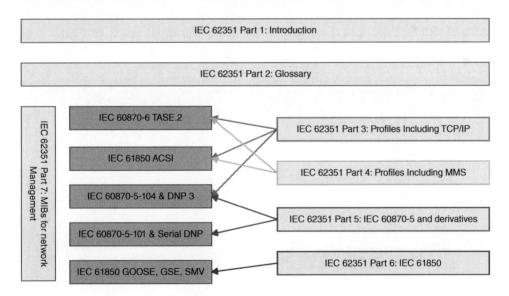

Figure 7.11 Correlation of IEC 62351 security standards to different profiles of the TC 57 standards. Reproduced with permission from International Electrotechnical Commission (IEC) (IEC 62351-1 ed.1.0 Copyright © 2007 IEC Geneva, Switzerland. www.iec.ch [20])

7.6.1.3   Security Countermeasures

Security countermeasures, as illustrated in Figure 7.12, are also a mesh of interrelated technologies and policies. Not all security countermeasures are needed or desired all of the time for all systems: this would be vast overkill and would tend to make the entire system unusable or very slow. Therefore, the first step is to identify which countermeasures are beneficial to meet which needs.

The four security requirements (confidentiality, integrity, availability, and nonrepudiation) are shown. The basic security threats are shown below each requirement. The key security services and technologies used to CTR these threats are shown in the boxes immediately below the threats. These are just examples of commonly used security measures, with arrows indicating which technologies and services participate in supporting the security measures above them. For instance, encryption is used in many

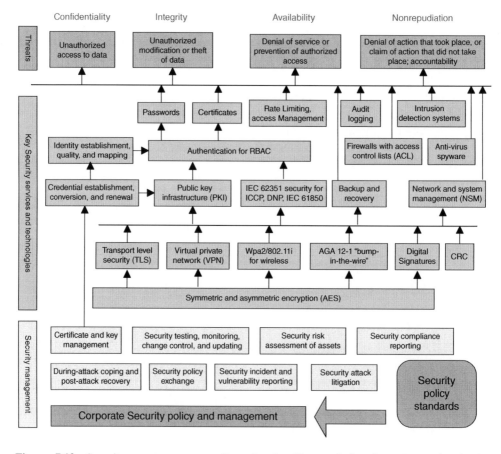

Figure 7.12 Security countermeasures. Reproduced with permission from International Electrotechnical Commission (IEC) (IEC 62351-1 ed.1.0 Copyright © 2007 IEC Geneva, Switzerland. www.iec.ch [20])

security measures, including TLS (Transport Layer Security), VPNs, wireless security, and "bump-in-the-wire" technologies. These in turn support IEC 62351 security standards and Public-Key Infrastructure (PKI), which are commonly used for authentication so that passwords and certificates can be assigned. At the bottom, below the security services and technologies, are the security management and security policies that provide the underpinning for all security measures.

7.6.2 IEC 61508 Standards

7.6.2.1 Overview of IEC 61508

IEC 61508 [28–34] is an international standard of rules applied in industry. It is entitled Functional Safety of Electrical/Electronic/Programmable Electronic (E/E/PE) or Electrical/Electronic/Programmable Electronic Safety-related Systems (E/E/PES).

IEC 61508 is intended to be a basic functional safety standard applicable to all kinds of industries. It defines functional safety as "part of the overall safety relating to the EUC (Equipment Under Control) and the EUC control system which depends on the correct functioning of the E/E/PE safety-related systems, other technology safety-related systems, and external risk reduction facilities."

The standard covers the complete safety life cycle, and may need interpretation to develop sector-specific standards. It has its origins in the process control industry. The published IEC 61508 consists of the following parts, under the general title Functional Safety of E/E/PES:

- IEC 61508-1: General requirements
- IEC 61508-2: Requirements for E/E/PES
- IEC 61508-3: Software requirements
- IEC 61508-4: Definitions and abbreviations
- IEC 61508-5: Examples of methods for the determination of the safety integrity levels (SILs)
- IEC 61508-6: Guidelines on the application of IEC 61508-2 and IEC 61508-3
- IEC 61508-7: Overview of techniques and measures.

7.6.2.2 Overall Safety Life Cycle Requirements

In order to deal in a systematic manner with all the activities necessary to achieve the required SIL for the E/E/PES, this standard adopts an overall safety life cycle as the technical framework.

The safety functions necessary to ensure the required functional safety for each determined hazard shall be specified. This shall constitute the specification for the overall safety functions requirements.

The necessary risk reduction shall be determined for each determined hazardous event. The necessary risk reduction may be determined in a quantitative and/or qualitative manner.

Where failures of the EUC control system place a demand on one or more E/E/PE or other technology safety-related systems and/or external risk-reduction facilities, and where the intention is not to designate the EUC control system as a safety-related system, the following requirements shall apply:

- The dangerous failure rate claimed for the EUC control system shall be supported by data acquired through one of the following:
  - actual operating experience of the EUC control system in a similar application;
  - a reliability analysis carried out to a recognized procedure;
  - an industry database of reliability of generic equipment.
- Dangerous failure rate that can claim lower than 10^{-5} dangerous failures per hour.
- All reasonably foreseeable dangerous failure modes of the EUC control system shall be determined and taken into account in developing the specification for the overall safety requirements.
- The EUC control system shall be separate and independent from the E/E/PE safety-related systems, other technology safety-related systems, and external risk-reduction facilities.

If the above requirements cannot be met inclusively, then the EUC control system shall be designated as a safety-related system. The SIL allocated to the EUC control system shall be based on the failure rate that is claimed for the EUC control system in accordance with the target failure measures specified in Tables 7.3 and 7.4. In such cases, the requirements in this standard, relevant to the allocated SIL, shall apply to the EUC control system.

Table 7.3 Safety levels in low-demand mode of operation

Safety level	Average probability of failure to perform its design function on demand
1	$\geq 10^{-2}$ to $< 10^{-1}$
2	$\geq 10^{-3}$ to $< 10^{-2}$
3	$\geq 10^{-4}$ to $< 10^{-3}$
4	$\geq 10^{-5}$ to $< 10^{-4}$

Table 7.4 Safety levels in high-demand or continuous mode of operation

Safety level	Average probability of failure to perform its design function on demand
1	$\geq 10^{-6}$ to $< 10^{-5}$
2	$\geq 10^{-7}$ to $< 10^{-6}$
3	$\geq 10^{-8}$ to $< 10^{-7}$
4	$\geq 10^{-9}$ to $< 10^{-8}$

The safety integrity requirements, in terms of the necessary risk reduction, shall be specified for each safety function. This shall constitute the specification for the overall safety integrity requirements.

7.6.2.3 Safety Requirements Allocation

The designated safety-related systems that are to be used to achieve the required functional safety shall be specified. The necessary risk reduction may be achieved by

- external risk reduction facilities;
- E/E/PE safety-related systems;
- other technology safety-related systems.

The safety integrity requirements for each safety function shall be qualified to indicate whether each target safety integrity parameter is either

- the average probability of failure to perform its design function on demand (for a low-demand mode of operation); or
- the probability of a dangerous failure per hour (for a high-demand or continuous mode of operation).

7.7 Discussion

7.7.1 Safety versus Security

Safety is protection against random incidents that are unwanted. Security is protection against intended incidents that happen due to a result of a deliberate and planned act.

On the basis of the aforementioned analysis, both wireless and wired communication standards can provide the basic security for Smart Grids, including authentication, secrecy, and integrity check. These measures can deal with the network attacks.

In order to provide a set of safety services in Smart Grids, related safety standards for wired Smart Grids have been proposed. The safety devices usually must realize more secure and reliable services than normal devices. Thus, more security measures are added into the safety standards for reliable communications. These safety standards can be regarded as the safety extensions of communication standards. The safety data can be transmitted on the basis of these safety standards.

At present, there are only safety extension standards for wired communication standards, but no safety extensions for wireless communication standards.

7.7.2 Security Level

Although both wireless and wired standards of Smart Grids implement a number of security mechanisms to ensure the integrity of the network, some possible security weaknesses have been identified.

For wired standards and their safety extensions, MACsec, IPsec, and TLS/SSL can be used to provide the security for frames, packets, and segments, respectively. These technologies are the security foundation for the wired Smart Grids. For WirelessHART, the security measures are mandatory. However, in ISA100.11a, many security features are defined as optional. Considering that security schemes consume additional processing time, memory, and power, having mandatory security features means that devices may not require strict security policies that cannot disable them to achieve benefits such as extended battery life. However, the added flexibility of the optional security features in ISA100.11a might be a security threat in itself and an issue when it comes to interoperability. Vendors might not choose to implement the full-security suite, and different vendors might choose to implement different parts of the security features. Also, signals from one of the ISA100.11a vendors indicate that their first generation of ISA100.11a devices will not implement any of the optional security features. According to the security levels, ISA100.11a has a series of levels ranging from default "authentication only with MAC layer MIC (MMIC) of 32 bits" to maximum "encryption and MMIC of 128 bits." However, WirelessHART has only one level of 32-bit MIC to protect data integrity that is mentioned on the session level.

7.7.3 Safety Level

In IEC 61508, SIL means Safety Integrity Level and constitutes a rating of the failure probability of a system based on IEC/EN 61508. The categories run from SIL level 1–4, with the probability of failure decreasing as the level rating increases. SIL 3 corresponds to a probability of failure of $10^{-7}-10^{-8}$ failures per hour. The responsible IEC commission once established the general rule that the bus of a safety system must not be involved with more than one percent of all failures.

All the wired communication standards can meet the requirements of IEC 61508 [35] SIL 3. Some safety standards not only fulfill the SIL 3, but also satisfy other requirements. The CIP Safety concept has been approved by TÜV Rheinland for adoption in IEC 61508 SIL 3 and EN954-1 Cat. 4 applications. CC-Link Safety is a network with high reliability in data transmission suitable for safety applications that require compliance with IEC61508 SIL3 and EN954-1/ISO13849-1 Category 4. For Powerlink, the quality of these measures will fulfill the requirements of SIL 3 (and within specific architectures also SIL 4). Also, for Powerlink safety, there is even the potential in this system to satisfy reliability and availability demands of category SIL 4 according to IEC 61508. In practice, Powerlink Safety may cause no more than 10^{-9} errors per hour. In other words, there is no more than one fault in about every 115 000 years.

7.7.4 Open Issues

7.7.4.1 Implementation Efficiency

First, how to enhance the software and hardware implementation is very important. For example, experiences from a practical effort to build a WirelessHART protocol stack have shown that performing AES calculations in software on embedded platforms is too time consuming to meet the 10-ms time-slot requirements of WirelessHART. To fulfill the requirements, it is suggested to use an AES hardware accelerator. Many variants of the CBC-MAC can also be used to enhance performance efficiency [36–38].

7.7.4.2 Security/Safety Breach

There are many security or safety breaches in wireless and wired Smart Grids. CCM* based on CBC-MAC and CTR are used in both WirelessHART and ISA100.11a. In addition, AES with CBC-MAC mode is used in wired standardized Smart Grids. However, attacks can be generated on CBC-MAC without the message length. If the length of the message is not included in the CBC-MAC integrity check, some attacks can be applied. For example, if the t_1 is known as the MAC of message m_1, then the MAC t_1 can be forged on the basis of $(m_1, m_1 t_1)$, although the key is unknown to the attacker. For another example, if some padding bits are added into the last block of CBC-MAC, for example, the form of (message|padding), the MAC can be maintained as the MAC of the original message without the padding bits. Although these attacks are not strong, some practical attacks might be applicable under certain conditions. In addition, attacks can be generated on the CTR only. One vulnerability is that an adversary can modify the plaintext. Another is that the plaintext can be revealed if a weak integrity check, such as CRC or a padding pattern check, is used. When CRC is used with CTR, the same attack as the Chopchop attacks on Wired Equivalent Privacy (WEP) and Temporal Key Integrity Protocol (TKIP), can be applied. When a padding pattern is checked, adversaries can know the last byte of the modified ciphertext by seeing whether the ciphertext has been accepted by the receiver.

In addition, some well-known attacks can also destroy the CCM* in the Smart Grid such as Birthday paradox attacks. In order to protect MACs against Birthday paradox attacks, a unique identifier can be added to each message to randomize the MACs. CTR mode encryption uses an arbitrary number (CTR) that increments with each 128-bit block of data encrypted. The CTR is first encrypted with AES, and the output is XORed with a 128-bit plaintext block to produce a ciphertext block. All CTRs used are unique and all the AES CTR mode encryption or decryption may be performed in parallel or precomputed in advance for speed gain.

Furthermore, the attacks can occur on the safety data in the safety extension standards. The analysis in [39] show that it is possible to attack PROFIsafe and change the safety-related process data without any of the safety measures in the protocol detecting the attack. By getting one safety container, and using brute force to compute all valid combinations of CRC1 and VCN that generates the same CRC2 as in the received message, a set of possible CRC1 can be obtained. With the knowledge that the CRC1 is static over the session lifetime, the remaining combinations can be reduced to the CRC1 that is in use. This has to be done as an iterative process that terminates when the correct CRC1 has been found. The remaining challenge is to find the actual VCN very quickly for all received safety containers. The VCN will increase monotonically at a rate depending on the bus period time, host, and device period time executing the safety layer. If the attacking application is fast and can receive all safety containers, the VCN would not update for each and every frame received, thus relaxing the computational efforts to derive the VCN in "real-time."

7.7.4.3 Others

Besides the aforementioned open issues, there are some other topics. For example, most existing security schemes in wireless networks are essentially the specific aspect of traditional computer security, which cannot provide enough security. In other words, security in wireless networks has traditionally been considered to be an issue to be addressed at the higher layers of a network. As a matter of fact, a physical-layer channel is the essential difference between wireless networks and wired networks. Physical-layer parameters can be considered to design the security for wireless networks.

The use of latest information technologies provides the benefits for power system control and operation, but also introduces the potential for cybersecurity vulnerabilities. A security scheme for existing standards must be proposed on the basis of these new information technologies. For example, Web services architecture is platform independent and interoperable. Thus, web services technologies are introduced into WirelessHART [40]. PROFINET also consists of the functionalities of web integration. There are two kinds of web integrations. One is making system configuration pages with html/xml. Another is the support of http/https and SSL/TLS/DTLS (Datagram Transport Layer Security). In the distributed environment in the Smart Grid, different data providers have their access rules linked to their data items, and trusted data integration services may fetch data from different providers and send clients responding to documents compliant to their own integration schema. The security schemes for these web services communication is also an important topic. In addition, as a new software interface specification and application framework based on web service, Object Linking and Embedding (OLE) for Process Control (OPC) Unified Architecture (OPC UA) can provide a cohesive, reliable crossplatform framework for Smart Grids. The OPC UA security model and mapping is presented in part 2 and part 6 of the series of OPC UA standards. However, it just gives the basic model and concepts and does not discuss the network security of OPC UA application. The existing OPC UA security model cannot meet the requirement of configuring the security strategy flexibly for adapting to a different application [41, 42]. Security for OPC UA-based web service integration still needs to be studied more deeply.

7.8 Conclusion

The analysis of security/safety mechanisms of standardized Smart Grid networks is given in this chapter. WirelessHART and ISA100.11a are analyzed and compared as the wireless data exchange standards. Then, standards of Profibus, PROFINET, CIP, CC-Link, Powerlink, and EtherCAT, which are commonly used in the control systems of Smart Grids, are introduced as the wired standards for the security analysis. Next, the safety extensions of the wired communication standards, PROFIsafe, CIP-Safety, CC-Link Safety, Powerlink Safety, and TwinSAFE, were analyzed focusing on their safety mechanisms. After that, functional security and safety standards, IEC 62351 and IEC 61508, were analyzed regarding their security and safety schemes. Finally, discussions about related security and safety technologies and some important open issues were given. On the basis of the existing communication standards and security extensions, engineers can choose the methods and techniques that are most appropriate for their projects. On the other hand, we indicate that there are still many open research topics in this area.

References

[1] IEC (2007) IEC 61784-3. *Industrial Communication Networks – Profiles – Part 3: Functional Safety Fieldbuses – General Rules and Profile Definitions*, International Electrotechnical Commission.

[2] Petersen, S. and Carlsen, S. (2011) WirelessHART versus ISA100.11a. *The Format War Hits the Factory Floor*, **5** (4), 23–34.

[3] HART Communication Foundation (2007) *HART Field Communication Protocol Specification*, Revision 7.0.

[4] IEC (2010) IEC 62591. *Industrial Communication Networks – Wireless Communication Network and Communication Profiles – WirelessHART*.

[5] ISA (2009) ISA-100.11a-2009 Standard*Wireless Systems for Industrial Automation: Process Control and Related Applications*.

[6] Sirkka, L. and Jamsa, J. (2007) Future trends in process automation. *Annual Reviews in Control*, **31** (2), 211–220.

[7] ODVA *CIP Common Specification*, ODVA, www.odva.org (accessed 6 December 2013).

[8] CC-Link www.cc-link.org (accessed 6 December 2013).

[9] Powerlink www.ethernet-powerlink.org (accessed 6 December 2013).

[10] EtherCAT www.ethercat.org (accessed 6 December 2013).

[11] BECKOFF *TwinSAFE – Open and Scalable Safety Technology* www.BECKOFF.com/twinsafe (accessed 6 December 2013).

[12] Lennvall, T., Svensson, S., and Hekland, F. (2008) A comparison of WirelessHART and Zig-Bee for industrial applications. *Proceeding of the IEEE International Workshop Factory Communication Systems*, pp. 85–88.

[13] Wahid, K.F. (2010) Rethinking the link security approach to manage large scale ethernet network. *Proceeding of the 17th IEEE Workshop on Local and Metropolitan Area Networks (LANMAN)*.

[14] Dzung, D., Naedele, M., Von Hoff, T. and Crevatin, M. (2005) Security for industrial communication systems. *Proceedings of the IEEE*, **93** (6), 1152–1177.

[15] Treytl, A., Sauter, T., and Schwaiger, C. (2004) Security measures for industrial fieldbus systems – state of the art and solutions for IP-based approaches. *Proceeding of the IEEE International Workshop on Factory Communication Systems*.

[16] PROFIBUS (2005) *PROFIBUS Guideline: PROFINET Security Guideline*, 1.0(7.002).

[17] RFCs www.rfc-editor.org (accessed 6 December 2013).

[18] PROFIsafe (2007) *PROFIBUS Guideline: PROFIsafe – Environmental Requirements, Version 2.5, Order No. 2.232*.

[19] IEC (2007) IEC 61784-3-3. *Industrial Communication Networks - Profiles – Part 3–3: Functional Safety Fieldbuses-Additional Specifications for CPF 3*, International Electrotechnical Commission.

[20] IEC (2007) IEC 62351-1. *Power Systems Management and Associated Information Exchange – Data and Communications Security- Part 1: Introduction to the Security Issues*, International Electrotechnical Commission.

[21] IEC (2008) IEC 62351-2. *Power Systems Management and Associated Information Exchange – Data and Communications Security- Part 2: Glossary of Terms*, International Electrotechnical Commission.

[22] IEC (2007) IEC 62351-3. *Power Systems Management and Associated Information Exchange – Data and Communications Security- Part 3: Security for Any Profiles Including TCP/IP*, International Electrotechnical Commission.

[23] IEC (2007) IEC 62351-4. *Power Systems Management and Associated Information Exchange – Data and Communications Security- Part 4: Security for Any Profiles Including MMS*, International Electrotechnical Commission.

[24] IEC (2009) IEC 62351-5. *Power Systems Management and Associated Information Exchange – Data and Communications Security- Part 5: Security for IEC 60870-5 and Derivatives*, International Electrotechnical Commission.

[25] IEC (2007) IEC 62351-6. *Power Systems Management and Associated Information Exchange – Data and Communications Security-Part 6: Security for IEC 61850 Profiles*, International Electrotechnical Commission.

[26] IEC (2010) IEC 62351-7. *Power Systems Management and Associated Information Exchange – Data and Communications Security, Part 7: Network and System Management (NSM) Data Object Models*, International Electrotechnical Commission.

[27] IEC (2011)IEC 62351-8. Power Systems Management and Associated Information Exchange – Data and Communications Security- Part 8: Role-Based Access Control, International Electrotechnical Commission.

[28] IEC (2010) IEC 61508-1. *Functional Safety of Electrical/Electronic/Programmable Electronic Safety-Related Systems-Part 1 ed2.0: General Requirements*, International Electrotechnical Commission.

[29] IEC (2010) IEC 61508-2. *Functional Safety of Electrical/Electronic/Programmable Electronic Safety-Related Systems-Part 2 ed2.0: Requirements for Electrical/Electronic/Programmable Electronic Safety-Related Systems*, International Electrotechnical Commission.

[30] IEC (2010) IEC 61508-3. *Functional Safety of Electrical/Electronic/Programmable Electronic Safety-Related Systems-Part 3 ed2.0: Software Requirements*, International Electrotechnical Commission.

[31] IEC (2010) IEC 61508-4. *Functional Safety of Electrical/Electronic/Programmable Electronic Safety-Related Systems-Part 4 ed2.0: Definitions and Abbreviations*, International Electrotechnical Commission.

[32] IEC (2010) IEC 61508-5. *Functional Safety of Electrical/Electronic/Programmable Electronic Safety-Related Systems-Part 5 ed2.0: Examples of Methods for the Determination of Safety Integrity Levels*, International Electrotechnical Commission.

[33] IEC (2010) IEC 61508-6. *Functional Safety of Electrical/Electronic/Programmable Electronic Safety-Related Systems-Part 6 ed2.0: Guidelines on the Application of IEC 61508-2 and IEC 61508-3*, International Electrotechnical Commission.

[34] IEC (2010) IEC 61508-7. *Functional Safety of Electrical/Electronic/Programmable Electronic Safety-Related Systems-Part 7 ed2.0: Overview of Techniques and Measures*, International Electrotechnical Commission.

[35] IEC (1998) IEC 61508. *Functional Safety of Electrical/Electronic/Programmable Electronic Safety-Related Systems - Part 1: General Requirements*, International Electrotechnical Commission.

[36] Vaudenay, S. (2002) Security flaws induced by CBC padding. *Proceeding of the International Conference Theory and Applications of Cryptographic Techniques (EUROCRYPT 2002)*.

[37] Krawczyk, H. (2001) The order of encryption and authentication for protecting communications (or: how secure is SSL?). *Proceedings 21st Annual International Cryptology Conference Advances in Cryptology, CRYPTO*, pp. 310–331.

[38] Black, J. and Rogaway, P. (2001) A suggestion for handling arbitrary-length messages with the CBC MAC. *Proceeding of the NIST Second Modes of Operation Workshop*.

[39] Akerberg, J. and Bjorkman, M. (2009) Exploring network security in PROFIsafe. *Proceeding of the International Conference on Computer Safety, Reliability and Security (SAFECOMP 2009)*.

[40] Raza, S. (2010) *Secure Communication in WirelessHART and its Integration with Legacy HART*. Technical Report, Swedish Institute of Computer Science.

[41] IEC. (2007) IEC 62541-2. *OPC Unified Architecture Specification-Part 2: Security Model*, International Electrotechnical Commission.

[42] IEC. (2007) IEC 62541-6. *OPC Unified Architecture Specification-Part 6: Mapping*, International Electrotechnical Commission.

8

Interoperability

8.1 Introduction

Smart Grid technologies are developed to make the current electrical grid an intelligent grid. The current electrical grid is a complex system of power-generation plants, transmission lines, substations, transformers, feeders, and other components. These components are connected together to generate and distribute electricity to the customer end. The present grid is more than a hundred years old. When it was first introduced in the early 1900s, it was considered to be one of the wonders of engineering. However, since then no major changes have been made in the infrastructure of the grid. The current increased use of electricity in various industries and applications globally has raised the demand for electric power. It has made the electric power grid a critical component of a nation's infrastructure and development. However, on considering the current electric grid, it is obvious that the grid still relies on old design features, which were intended to meet the electricity demand of the twentieth century.

Interoperability is the key phenomenon in making the current grid smarter so that it meets the demands of the twenty-first century. It is the capability of exchange of information among Smart Grid devices in order to achieve efficiency, reliability, and network conformance. It enables two devices from different vendors to work together. In fact, interoperability does not have one formally established definition. The most concrete definition of interoperability is mentioned in [1] as *"the capability of two or more networks, systems, devices, applications, or components to exchange and readily use information securely, effectively, and with little or no inconvenience to the user."* There are many elements in interoperable equipment that coordinate and work together technically to perform useful work. This explanation provides a solid starting point for the consideration of interoperability in the Smart Grid's standards development process.

In the Smart Grid, when different equipment are lined up technically to perform useful tasks, they will have to face various interoperability challenges. In order to

Smart Grid Standards: Specifications, Requirements, and Technologies, First Edition. Takuro Sato,
Daniel M. Kammen, Bin Duan, Martin Macuha, Zhenyu Zhou, Jun Wu, Muhammad Tariq and Solomon Abebe Asfaw.
© 2015 John Wiley & Sons, Ltd. Published 2015 by John Wiley & Sons, Ltd.

address these challenges collectively, we need to look into the role of various industrial standards developed by various Standard Developing Organizations (SDOs), the interoperability agreements, technical specifications, technological transfer, and the different phases of the development in industrial architecture.

8.1.1 Interoperability and Interchangeability

The key difference between the interoperability and interchangeability is the degree of openness. According to the National Institute of Standards and Technology (NIST) the definition of interchangeability, it is an extreme degree of interoperability characterized by a similarity sometimes termed "plug and play." In general, interchangeability is the extreme end-state of interoperability and openness, where its components can be freely replaced without any change in the functionality. This replacement does not require any major additional configuration. Once achieved, it will be an important milestone for Smart Grid technologies because a state of interchangeability provides more choices to both end users and grid managers and operators to select from. A system needs to be designed such that it should be able to integrate new and the older versions of manufactured devices.

8.1.2 The Challenges of Network Interoperability

According to a recent case study [2], the most significant barrier in the way of achieving widespread adoption of Smart Grid technologies will be interoperability. Interoperability needs the stakeholder's and vendor's cooperation across all states and territories globally during the development and implementation process of Smart Grid technologies. In interoperable systems, two or more devices, which may be manufactured by different vendors, have to work together to perform useful tasks. Owing to the delicate nature of interoperable systems, appropriate consideration is required for concrete solutions of different Smart Grid applications. In contrast to human communications, where there is an opportunity for interpretation, interoperable devices only respond to exact meanings, particularly when these devices are manufactured by different vendors. The communication, application execution, data management, and security should all be well understood by the interoperating devices in order to perform the required task [3]. Interoperability among different devices whether located in the same vicinity or in remote locations, can be achieved by connecting them through networks as shown in Figure 8.1. Interoperability over a network only deals with the required amount of data and operates according to the set of given instructions. It should be noted that to physically connect two interoperable devices there is only one dimension of the interoperability. Another level of complexity is added to the system when the devices are connected via a wired or wireless network in such a way that they should conform to all the networking devices. When the two network devices initiate an exchange of messages, they

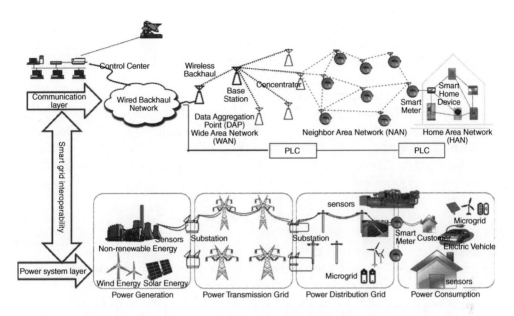

Figure 8.1 Achieving smart grid interoperability through diverse networks

must recognize and understand the language of network routing and the messages must be properly addressed in order to reach the final destination. Although it looks simple for machines, this kind of processing is complicated as there are various complex processes involved. These complexities should be covered and addressed in the developing standards. Furthermore, developing various standards in the field of networking is not simple as it encompasses the management of all network resources as well as data exchange among various network devices. To address this and similar kinds of issues, various SDOs have developed interoperability standards. These standards deal with small, medium, and large-scale network operations and architectures.

8.1.3 Adding Application Interoperability

The "Smart Grid" is a general term that specifies the integration of a number of different technological architectures in such a way that the grid looks like a single grid. These architectures, once integrated, can be used for various Smart Grid applications, such as Advanced Metering Infrastructures (AMIs), distribution management systems, grid-level storage, home automation, cybersecurity, network communications, and Plug-in Electric Vehicles (PEVs). These technologies once performed together, integrate renewable resources, enhance the efficiency of electricity transmission, minimize management operational costs, restore power interruptions or breakdown early, improve overall security, and provide more control at the user end.

There are various levels of interoperability involved among a set of devices in the Smart Grid. We have discussed that the interoperable devices among different type of devices are fundamentally based on the exchange of information (messages). However, exchanging messages and transferring data from a source device via the network to arrive at the destination only achieves a very fundamental level of network/communications interoperability. The main interoperability challenge occurs at the destination device where it should be able to read the message and act according to the instructions provided. This needs a different level of standards for message semantics/syntax. Application interoperability is accomplished when a correct action is performed by the destination device. It needs common language semantics/syntax, which the applications should recognize and understand.

In most systems, a complex or higher level of interoperability is required for system equipment to get "Plug-and-Play" type of interoperability. In higher-level interoperability, a variety of interoperable functions behind the scenes are performed by interoperable devices. These higher levels of interoperability contain a lot of details and are important to many systems administration processes. Therefore, various types of agreement among the vendors take place, in the form of interoperable standards [3].

8.2   Interoperability Standards

To achieve interoperability among devices that are developed by different vendors, various standards have been devised by different SDOs, such as International Electrotechnical Commission (IEC), Institute of Electrical and Electronics Engineers (IEEE), and NIST. Usually, extensive efforts are needed to develop a standard, because it requires a formal process which should arrive at an agreement through contributed documents, technical meetings, feasibility reports, thorough discussions of user groups, and formal voting on an agenda. Although the standards are important ingredients, they are not enough on their own to achieve interoperability among the devices. Most of the interoperable standards are usually based on compromises, multiple options, and may even leave undefined items sometimes, which are not conclusive during meetings and discussions. This is not an exact indication of the standards process, rather it is recognition of the reality of reaching a proper agreement. It should also be noted that once developed, these standards are good for providing a measure of stability for vendors of interoperable devices; however, they require to be amplified. For the said purpose, various kinds of user groups developed. User groups usually play the role of developing a synergy with the formal standard. They focus on resolving the conflicting details and often get involved with conformance testing.

Interoperability is still an ongoing topic of discussion. Some interoperable standards have already been developed and some are at the developing stage. The IEEE, NIST, IEC, and various other organizations have designed various roadmaps, initiatives and projects for building new and updated interoperable standards [4].

There are various standards that are considered as interoperable standards or standards used in different components of interoperable systems. Table 8.1 summarizes these standards, which are basically recognized and recommended by NIST. These standards contain the initial 16 specifications, along with nine standards that NIST added after reviewing and evaluating the feedback it received. On the basis of the feedback received and further evaluation of these standards, which were added to the list at the start, it was decided to move some of the current standards on the extended list to the second list because they need further analysis and assessment.

8.3 NIST-Identified List of Standards to Be Reviewed

In order to update the list of interoperable standards, the NIST and the Electric Power Research Institute (EPRI), conducted a workshop in 2009 where they identified further standards to be considered as a part of the Smart Grid interoperability. The workshop's main focus was on analyzing and enhancing use cases, locating key interfaces, determining Smart Grid interoperability requirements, and identifying additional standards for consideration. During the workshop, many of the use cases discussed referenced standards in addition to those summarized in Table 8.1. As a result of workshop recommendations, some standards were identified to be included in the interoperability standards that are not included in Table 8.1, which were consolidated in the "EPRI's *Report to NIST on the Smart Grid Interoperability Standards Roadmap*" [5]. These are

- **ANSI-based standards**: ANSI C12.22-2008/IEEE P1703/MC1222, ANSI C12.23, ANSI C12.24.Global Positioning System (GPS) Standard Positioning Service (SPS) Signal Specification.
- **HomePlug-based standards**: HomePlug AV and HomePlug C&C are home automation standards, which are explained in detail in Chapter 5.
- **IEEE-based standards**: IEEE 61400-25, IEEE P1901, IEEE 802 Family, IEEE2030 IEEE C37.2-2008, IEEE C37.111-1999, IEEE C37.232, IEEE 1159.3, IEEE 1379-2000.
- **ITU-based standards**: ITU Recommendation G.9960 (G.hn).
- **ISO/IEC based standards**: ISO/IEC 8824 Abstract Syntax Notation (ASN).1, ISO/IEC 12139-1, ISO/IEC15045, IEC 62056, IEC PAS 62559, ISO/IEC 15067−3.ISA SP100.NIST SP 500-267.Z-Wave. (Z-Wave is a home and building automation standard [6, 7, 17], which has been explained in detail in Chapter 5.)

Table 8.1   List of NIST-identified standards

Standard name	Main focus	Descriptions
ANSI/ASHRAE 135-2008/ISO 16484-5 BACnet	It is related to home and building automation and control networks	BACnet is basically a home and building automation standard which defines the information model and messages for building system communications at the consumer end. It integrates a range of networking technologies to achieve scalability from very small systems to multibuilding operations that span wide geographic areas using IP protocols
ANSI C12 Suite	This set of standards offers the industry a comprehensive protocol **suite** to transfer the newly revised data standard. The standards are open for public use in different applications	
ANSI C12.1		ANSI C12.1 is related to performance and safety-type tests for revenue meters
ANSI C12.18/IEEE P1701/MC1218		ANSI C12.18 is related to the protocol and optical interface for measurement devices
ANSI C12.19/MC1219		ANSI C12.19 is related to revenue metering end-device tables
ANSI C12.20		ANSI C12.20 is related to revenue metering-type test and accuracy specification
ANSI C12.21/IEEE P1702/MC1221		ANSI C12.21 is related to the measurement device through which data is transported via telephone networks
ANSI/CEA 709 and CEA 852.1 LON Protocol Suite	At the consumer end, these standards serve as the facility interface and are related to the Price, Demand Response (DR), and Energy Usage Priority Area Plans (PAP). Different types of PAPs will be explained in Section 8.7. These standards are supported by the LonMark international users group. Currently, these standards are used extensively. These standards are mostly mature	ANSI/CEA 709 and CEA 852.1 LON Protocol Suite is a general purpose Local Area Network (LAN) protocol. It is used for a variety of applications such as electric meters, street lighting, and in home building automation

(continued overleaf)

Table 8.1   (*continued*)

Standard name	Main focus	Descriptions
ANSI/CEA 709.1-B-2002 Control Network Protocol Specification		ANSI/CEA 709.1-B-2002 is a specific physical layer protocol designed for use with ANSI/CEA 709.1-B-2002
ANSI/CEA 709.2-A R-2006 Control Network Power Line (PL) Channel Specification		ANSI/CEA 709.2-A R-2006 is an explicit physical layer protocol designed for use with ANSI/CEA 709.1-B-2002
ANSI/CEA 709.3 R-2004 Free-Topology Twisted-Pair Channel Specification		ANSI/CEA 709.3 R-2004 is a physical layer protocol designed for use with ANSI/CEA 709.1-B-2002
ANSI/CEA-709.4: 1999 Fiber-Optic Channel Specification		ANSI/CEA-709.4:1999 protocol offers a way by using a User Datagram Protocol (UDP) to tunnel local operating network messages through an IP, thus defining a way to produce larger internetworks
DNP3	The standard is about substation and feeder device automation.	Distributed Network Protocol 3 (DNP3) is an open, mature, commonly implemented specification that is developed and supported by a group of vendors, utilities, and users. The use of this protocol is recommended by the IEEE. Some work is underway to make this standard into the IEEE standard. The DNP3 standard is related to substation and feeder device automation. In addition it is used in the communications between different control centers and substations
IEC 60870-6/ TASE.2	IEC 60870-6/ TASE.2 standard defines how to send messages among control centers of different utilities	IEC 60870-6/TASE.2 is an open, and mature standard for sending messages among control centers. It is widely implemented with compliance testing. It is part of the IEC 60870 suite included in PAP14
IEC 61850 Suite	IEC 61850 Suite defines communications within transmission and distribution	IEC 61850 Suite is an open standard that is being adopted in North America. It was developed for field device communications within substations

Table 8.1 (*continued*)

Standard name	Main focus	Descriptions
IEC 61968/61970 Suites	IEC 61968/61970 families of standards define information exchanged among control center systems using common information models	IEC 61968/61970 are open standards that are widely implemented and maintained by an SDO with support from a users group. They are part of PAPs relating to integration with IEC 61850 and Multispeak
IEEE C37.118	IEEE C37.118 standard defines Phasor Measurement Unit (PMU) performance specifications and communications	IEEE C37.118 is an open standard, widely developed and maintained by an SDO. It includes some requirements for communications and measurement. Currently, it is being updated by the IEEE Power System Relaying Committee (PSRC) Relaying Communications Subcommittee Working Group H11
IEEE P2030	This is a long-term project of the IEEE whose initiative is to address Smart Grid interoperability	The IEEE P2030 [15] and the IEEE 1547 [16] series of standards are both sponsored by (SCC21) [6] The standard is a guide for Interoperability in Smart Grid technologies, such as energy technologies and Information Technology (IT) operation with the electric power system, and user end applications
IEEE 1547 Suite	IEEE 1547 family of standards and protocols defines publications and drafts. It provides insight into systems integration and grid infrastructure, physical and electrical interconnections between utility and distributed energy generation, and storage	IEEE P1547 standard addresses demand/response planned island systems (e.g., microgrids)
IEEE 1547.1	The standards and protocols are open, having significant implementation for the parts covering physical/electrical connections. The parts of standards that describe messages are not as widely deployed as the parts that specify the physical interconnections	IEEE 1547.1 provides the conformance test procedures
IEEE 1547.2		IEEE 1547.2 publication is an application guide to 1547
IEEE 1547.3		IEEE 1547.3 publication is a guide to demand/response monitoring, information exchange, and control

(*continued overleaf*)

Table 8.1 (*continued*)

Standard name	Main focus	Descriptions
Open Geospatial Consortium Geography Markup Language (GML)	GML standard is used for exchanging of location-based (geographical) information that addresses geographical data requirements for many Smart Grid applications	GML is an open standard, which is in compliance with ISO 19118 for the transport and storage of geographic information modeled according to the conceptual modeling framework used in the ISO 19100 series of International Standards. GML is widely used along with supporting open-source software. It is also used in disaster management, home/building, and equipment-location information bases
ZigBee/HomePlug Smart Energy Profile 2.0	It is developed for Home Area Network (HAN) device communications and information model	Zigbee Energy profile 2.0 is a home and building automation standard developed by Zigbee Alliance and HomePlug. It is still under development, but expected to be technology independent and useful for many Smart Grid applications. It is explained in detail in Chapter 5
OpenHAN	OpenHAN is a specification for HAN to connect to a utility an advanced metering system, including device communication, measurement, and control	OpenHAN is a specification for home and building automation, which is developed by a user group, UCAIug, that contains a checklist of requirements that enables utilities to compare the many available HANs. OpenHAN is also explained in detail in Chapter 5
AEIC Guidelines v2.0	AEIC Guidelines v2.0 consists of framework and testing criteria for vendors and utilities who desire to implement standards-based AMI (StandardAMI) as the choice for AMI solutions	AEIC Guidelines v2.0 was created in order to assist utilities in specifying implementations of ANSI C12.19 typical metering and AMI devices. The aim was to limit the possible options chosen when implementing the ANSI C12 standards. In doing so, it is expected that the interoperability will be improved
IEC 62351 Parts 1–8	IEC 62351 Parts 1–8 is a family of open standard, developed and maintained by IEC power systems management and associated information exchange. This family of standards defines information security for power system control operations.	

Table 8.1   (*continued*)

Standard name	Main focus	Descriptions
IEC 62351-1		Part 1 is about an introduction to the standard
IEC 62351-2		Part 2 is about the key terms used in the IEC 62351 series
IEC 62351-3		Part 3 is about security for any profiles including Transport Control Protocol (TCP)/Internet Protocol (IP)
IEC 62351-4		Part 4 is about security for profiles that include Manufacturing Messaging Specifications (MMS)
IEC 62351-5		Part 5 is about security for any profiles including IEC 60870-5 (e.g., DNP3 derivative)
IEC 62351-6		Part 6 specifies security standards for the IEC TC 57 communication protocols
IEC 62351-7		Part 7 defines security through network and system management
IEC 62351-8		Part 8 deals with role-based access control for power system management
IEEE 1686-2007	The IEEE 1686-2007 is a standard that defines the functions and features to be provided in substation intelligent electronic devices (IEDs) to accommodate critical infrastructure protection programs	The IEEE 1686-2007 is also an open standard. The standard deals with the security related to the access, operation, configuration, firmware revision. This standard also specifies data retrieval from an IED. However, features such as communications for the purpose of power system protection and encryption for the secure transmission are not addressed in this standard
NERC CIP 002-009	NERC CIP 002-009 family of standards cover physical and cybersecurity standards for the bulk power system	NERC CIP 002-009 is mandatory for the bulk electric system. It was revised by NERC recently
NIST Special Publication (SP) 800-53, NIST SP 800-82	This family of standards contains cybersecurity standards and guidelines for federal information systems, which include all those for the bulk power system	This family of standards is open source, which is developed by NIST. SP800-53 describes security measures required for government standards of the United States. SP800-82 is in the completion process. It specifies security especially for industrial control systems, such as the electric grid

8.4 NIST Interoperability

The NIST has been assigned the task of designing a framework and a roadmap that includes protocols and model standards for information management. The main purpose is to achieve interoperability of Smart Grid devices and systems under the Energy Independence and Security Act (EISA) of 2007 [3]. Due to the rapid growth in Smart Grid technologies and applications, there is an urgent need to develop protocols, standards, guidelines, and specifications for the Smart Grid interoperability. The deployment of various Smart Grid technologies in diverse applications is already underway through various means, including smart sensor nodes on power distribution lines, smart meters in homes, and widely detached sources of renewable and alternative energy such as solar Photovoltaic (PV), Concentrated Solar Power (CSP), wind, biomass, and geothermal such as hydrothermal. With the US Department of Energy (DOE) Smart Grid Investment Grants and other incentives, such as loan guarantees for renewable energy production projects, the advancement in the Smart Grid is expected to be further accelerated in the near future. It is obvious that without the formal standards' recommendation, it is highly unlikely that the systems that are developed will be long lasting. Rather, it is highly expected that such developed or implemented systems with considerable investment may become obsolete prematurely or be implemented without proper procedures that are important to guarantee security. Therefore, formal standards development is necessary to get concrete solutions and capital return of the investment.

EISA basically delegates the development of the Smart Grid as a US national policy goal. It specifies that the interoperability framework must be neutral in terms of technology. In addition, the framework must also be very flexible and uniform. The law also specifies that the framework should not only have room for centralized power generation and distribution resources but should also facilitate the incorporation of new and innovative Smart Grid technologies, such as distributed renewable energy resources, PEVs, AMI, and smart energy storage.

NIST developed a three-phase plan to speed up the identification of a basic set of interoperability standards. It has to establish a vigorous framework for the long-term development of the many additional standards that will be required, and for establishing an infrastructure for conformance testing and certification.

8.5 Conceptual Reference Model for the Smart Grid

Smart Grid technologies are usually not simple, rather they are a combination of many complex systems. It requires a common understanding to make use of the major building blocks of Smart Grid technologies. In addition, it requires that all these technologies must be openly shared with each other. "NIST *Framework and Roadmap for Smart Grid Interoperability Standards, Release 1.0*," [3], has been published as a result of the first phase of the NIST developed three-phase plan. This framework is a very high-level Conceptual Reference Model (CRM) for the Smart Grid technologies. It specifies a total of 75 existing standards to be the important standards

Table 8.2 Domain and actors in the Smart Grid's CRM

Domain	Actors
Customers	There are three types of customers. These are Residential (home users) Commercial users Industrial users Customers are the end users who use electricity. In addition, it is also possible that they generate electricity (through renewable energy technologies), store the energy, and manage the overall system
Markets	They are operators and participants in the power/electricity market
Service providers	They can be either companies or operators who provide services to the electricity market
Operations	They are the managers who operate and control the electricity distribution and transmission
Bulk power generation	They deal with the electricity generation in bulk. Their tasks also include storage of electricity at a later stage
Power transmission	They deal with the transmission of electricity through power lines
Power distribution	They deal with the distributors who deal in providing electricity to the customers or take energy from the customers (in the case of renewable generation)

of interoperability. It rectifies 15 harmonization issues and high-priority gaps, which need revision of some of the existing standards or development new standards, and works out action plans by which selected Standards Setting Organizations (SSOs) will tackle these gaps. To help secure the Smart Grid objectives, it will work out the plan to identify requirements and standards.

CRM is a tool that can be used to analyze a variety of use cases and to identify interfaces for which interoperability standards are needed. CRM basically facilitates in developing strategies for cybersecurity. For example, the NIST Smart Grid CRM identifies seven different areas of Smart Grid technologies, namely, power generation, transmission, distribution, markets, operations, service provider, and end users (customers). The CRM specifies interfaces among various domains and actors. It also includes applications requiring exchanges of information, for which interoperability standards are needed [8–10]. Table 8.2 identifies different domains and actors of the Smart Grid's CRM.

8.6 Different Priority Areas Identified for Standardization

Considering the variety of applications of the Smart Grid, it is confirmed that the Smart Grid will require hundreds of interoperable standards, specifications, and requirements for rapid replacement of existing power grid with the Smart Grid. Owing to the urgent demand in a Smart Grid application, some proposed standards should be given priority as they will be required more immediately than others. In order to prioritize some areas that need urgent response, NIST chose to initially focus on the standards needed, which are identified in the Federal Energy Regulatory Commission

(FERC) Policy Statement [11] in addition to some areas identified by NIST. These priority areas are given as follows:

- Wide-Area Situational Awareness (WASA)
- Demand Response (DR) and Consumer Energy Efficiency
- Energy Storage (Smart Energy Storage)
- Electric Transportation (PEV)
- Cybersecurity
- Network Communications
- AMI
- Distribution Grid Management (DGM).

NIST has conducted numerous workshops to prioritize various action plans. As a result of the workshops, NIST found that many potentially useful standards would be required thorough revision for further enhancement before they could actually be deployed to fulfill the Smart Grid requirements. In addition, some stakeholders recognized gaps for which new standards must be developed as the existing standards are not appropriate to tackle these gaps. To date, around 70 such gaps have been identified. NIST selected 15 gaps out of 70 existing gaps for which a solution is urgently needed to support one or more of the Smart Grid priority areas.

As mentioned earlier, owing to the diverse nature of Smart Grid technologies and their applications, it will ultimately require hundreds of standards, some of which will be more urgently needed than others. To prioritize its work, NIST chose to focus on eight main areas. Among these areas or functionalities, cybersecurity and network communications are highly prioritized, as these two aspects will play an important role in the ongoing projects and also in Smart Grid technologies and its future deployments.

8.6.1   Wide-Area Situational Awareness

The goals of WASA are to recognize and ultimately optimize the management of power-network components, behavior, and performance. In addition, it is a sound approach to foresee, prevent, or respond to problems before disruptions such as power fluctuation, outage, or complete blackout can arise. Here, the main aim is to monitor and display the components of the power system and performance across interconnections over geographically divided regions in real time. Various Smart Grid technologies and standards related to WASA have been explained in Chapter 3.

8.6.2   Demand Response and Consumer Energy Efficiency

DR is important for keeping the balance of electric power between supply and demand in an optimized way. In particular, it is very important in distributed power systems where the sources of electricity generation may be more than one. DR devises mechanisms and incentives for utilities, businesses, industrial, and residential customers in order to minimize the energy usage during peak hours or when power is unstable and fluctuating. Complete details about various Smart Grid technologies and standards related to DR have been provided in Chapter 5.

8.6.3 Smart Energy Storage

Smart energy storage means the storing of energy either directly or indirectly in a smart way in contrast to the existing storage mechanism. Significant bulk energy storage technology is available today in the form of pumped hydroelectric storage technology. Smart energy storage also deals with the capability of storing energy from distributed source, which will ultimately benefit the entire grid, from power generation to the customer/end user. Recently, in October 2014, a real breakthrough has been made in the field of Smart Energy Storage. Researchers from Singapore at the Nanyang Technology University (NTU) have developed a Li-ion battery, which is capable of 20 years of deep discharges. It is more than 10 times that of existing Li-ion batteries. Instead of charging batteries in hours, the NTU researchers claim that the new battery design can be charged very quickly, in fact, 70% of the battery will be charged in just two minutes [18]. For more details about smart energy storage, please refer to Chapter 4 of this book.

8.6.4 Electric Transportation

Electric transportation refers to the ability of automobiles to run on electrical energy. It deals with the large-scale integration of PEVs. Once fully developed, electric transportation will possibly minimize the dependence of developed countries (e.g., United States, Japan, Europe) and developing countries (e.g., China and India) on foreign oil. In doing so, it will dramatically reduce these nations' dependence on hydrocarbons. Therefore, the overall demand of oil and the ever-increasing prices will be controlled and more importantly, the ever-growing threat to the environment will be reduced considerably. For more details about electric transportation, please refer to Chapter 4 of this book.

8.6.5 Cybersecurity

In terms of security there is a key difference between the existing power grids and the Smart Grid. In the existing power grids, availability is the primary requirement, that is, keeping the lights on. In the present power grid, the key security requirements are Availability and Integrity, not Confidentiality (AIC), whereas in the Smart Grid, the key security requirements will be Confidentiality, Integrity, not Availability (CIA).

The purpose of cybersecurity is to ensure the security of the electronic information and communication systems and the control systems necessary for the management, operation, and protection of the Smart Grid's energy, Information and Communication Technologies (ICT), and various other infrastructures. Various cyber security features, their respective functions and methodologies are specified in Table 8.3. Cybersecurity in the Smart Grid systems and components is a very important issue and a main concern.

Table 8.3 Cybersecurity features, their respective functions, and methodologies

Cyber security features	Function(s)	Methodologies
Confidentiality	To protect a system from unauthorized users	For confidentiality encryption, key management, public/private key Infrastructure, data separation are required
Integrity	Prevents the system from unauthorized alteration of data Provides detection and notification mechanism	For data integrity, time stamping, digital signatures, message integrity safeguards, are the key parameters
Availability	Provides accessibility of information/system when required	Protection from attack, unauthorized users and resistance to failures, are features of the availability
Identification	Identifies individuals/entities to enter the system	For identification, unique user name, distinctive ID, and strong passwords are the key
Authentication	Verifies the claimed identity of users	For authentication, it requires, secure tokens, smart cards, and single sign-on
Authorization	Identifies users who have been authorized to use the system	It requires certificates and the attribute use
Access Control	Role-based access of users to systems and services provided by systems	There should be a strong passwords and access control should be role-based
Non-repudiation	To prove that a system did take part in the exchange of data	Digital signatures, time stamping, and certificate authority are the methodologies for non-repudiation

In order to achieve this goal, it requires incorporating security at the architectural level. For Smart Grid systems security, a NIST-led Cyber Security Coordination Task Group consisting of almost 300 participants from the private and public sectors is leading the development of a cybersecurity strategy and cybersecurity requirements. The task group is identifying use cases with cybersecurity parameters. It includes assessing risks, vulnerabilities, threats, and impacts. In addition, it deals with privacy impact assessment, assessing relevant standards, and specifying the relevant R&D topics. The achievements of the NIST-led Cyber Security Coordination task group are summarized in [12, 13]. Cybersecurity issues, standards, and specifications are explained in details in Chapter 7.

8.6.6 Network Communications

Smart Grid systems will not be complete without use of some particular communication network. In Smart Grid technologies and applications, various domains and subdomains will use a variety of public and private communication networks, which can be either wired or wireless. Since there are a variety of networking environments, the identification of performance metrics and core operational requirements of different applications are very critical for Smart Grid technologies to become successful.

Various network and communication standards that are related to Smart Grid systems have already been explained in detail in Chapters 5 and 6.

8.6.7   Advanced Metering Infrastructure

At present, utilities are focusing on developing AMI or smart metering to implement residential DR and to serve as the chief mechanism for implementing dynamic pricing. AMI is very critical in keeping record of energy provided to the grid, which is produced from renewable energy sources or energy taken from the electrical grid. It consists of the communication hardware and software and associated system and data-management software that creates a two-way communication between advanced meters and utility business systems. AMI helps in collection and distribution of information to and from the customers and other parties, such as the competitive retail suppliers or the utility itself. It provides customers with the facility of getting real-time pricing of electricity based on demands, peak hours and discounted hours (fewer loads). For more details about AMI, please refer to Chapter 5 of this book.

8.6.8   Distribution Grid Management

It deals with maximizing the performance of grid components such as feeders, transformers, and other components of networked power distribution systems. As Smart Grid technologies' capabilities are improving (such as advancement in AMI and DR), and as large numbers of distributed energy resources and PEVs are deployed, the automation of distribution systems becomes necessary for efficient and reliable operation of the overall electric power system. The anticipated benefits of DGM include increased reliability, system efficiency, reductions in peak loads, overall security of the system, and improved capabilities for managing distributed sources of various renewable technologies. For more details about issues related to DGM, please refer to Chapter 3 of this book.

8.7   Priority Action Plans

NIST organized a smart grid workshop on August 3-4, 2009, by inviting more than 20 SDOs and user groups. The main aim of the workshop was to identify the initial set of priorities for the development and improvement of standards to build an interoperable smart grid through Priority Action Plans (PAP).

The main criteria for inclusion on the initial list were [3] the following:

1. Propinquity of need
2. Importance to high-priority smart grid functionalities
3. Availability of existing standards to respond to the need
4. The extent and stage of the deployment of affected technologies

Table 8.4 provides details about nineteen PAPs, related smart grid applications, standards, and their developmental progress.

Table 8.4 Priority Action Plans

PAP No	Targeted Area	Related standards	Developmental progress
PAP 00	Advanced Metering Infrastructure(AMI) Upgradation of meters	NEMA SG-AMI	Remote meter upgradeability Completed already
PAP 01	Role of Internet Protocol (IP) in the Smart Grid	Informational Internet Engineering Task Force (IETF) Read for Comments (RFC)	IETF Smart Grid IP RFC-Completed already
PAP 02	Wireless communications for the smart Grid	IEEE 802.x, 3GPP, 3GPP2	Wireless Guidelines Report (NISTIR)
PAP 03	Common price communication model	OASIS EMIX, ZigBee SEP 2	In the read for comments /public comments period at OASIS to seek more participation of utilities
PAP 04	Common scheduling mechanism	OASIS WS-Calendar	Out of final public comment period at OASIS
PAP 05	Standard meter data profiles	AEIC V2.0 Meter Guidelines (addressing use of ANSI C12)	AEIC guideline completed. Addressing technical issues raised by different meter manufacturers
PAP 06	Common semantic model for meter data tables	ANSI C12.19-2008, MultiSpeak V4, IEC 61968-9	Addressing scope to ensure alignment with NIST and PAP team objectives
PAP 07	Electric storage Interconnection guidelines	IEEE 1547 61850-7-420, ZigBee SEP 2	Completed in the 4th quarter of 2012
PAP 08	CIM for Distribution Grid Management (DGM)	IEC 61850-7-420, IEC 61968-3-9, IEC 61968-13,14, MultiSpeak V4, IEEE 1547	Developing requirements affecting IEEE 1547 and IEC 61850-7-420
PAP 09	Standard Demand/Response and DER signals	NAESB WEQ015, OASIS EMIX, OpenADR, ZigBee SEP 2	Seeking additional utility stakeholder participation
PAP 10	Standard energy usage information	OpenADE, ZigBee SEP 2, IEC 61968-9, ASHRAE SPC 201P	Model complete in February 2011. PAP closed in the first quarter of 2011
PAP 11	Common object models for electric transportation	ZigBee SEP 2, SAE J1772, SAE J2836/1-3 , SAE J2847/1-3, ISO/IEC 15118-1,3, SAE J2931, IEEE P2030-2, IEC 62196	SAE standards completed Last standard has been completed in the second quadrant of 2011
PAP 12	IEC 61850 Objects/DNP3 Mapping	IEC 61850-80-5, Mapping DNP to IEC 61850, DNP3 (IEEE 1815)	SDOs like IEEE and IEC are currently working on the development of standards

Table 8.4   (*continued*)

PAP No	Targeted Area	Related standards	Developmental progress
PAP 13	Time synchronization, IEC 61850 Objects/IEEE C37.118 Harmonization	IEC 61850-90-5, IEEE C37.118, IEEE C37.238, Mapping IEEE C37.118 to IEC 61850, IEC 61968-9	Drafting standards in the committees of SDOs while requirements have been completed
PAP 14	Transmission and distribution power systems model mapping	IEC 61968-3, MultiSpeak V4	SDOs are dealing with use cases and requirements
PAP 15	Harmonize Power Line Carrier (PLC) standards for smart home	DNP3 (IEEE 1815), HomePlug AV, HomePlug C&C, IEEE P1901 and P1901.2, ISO/IEC 12139-1, G.9960 (G.hn/PHY), G.9961 (G.hn/DLL), G.9972 (G.cx), G.hnem, ISO/IEC 14908-3, ISO/IEC 14543, EN 50065-1	Harmonizing PLC standards for home appliances' communication (i.e., home automation) Work on broadband coexistence has completed Working on narrowband is in progress
PAP 16	Wind plant communications	IEC 61400-25	Use cases, requirements, and guideline development completed by PAP team in the last quarter of 2011
PAP 17	Facility smart grid information standard	New Facility Smart Grid Information Standard ASHRAE SPC 201P	Completed in the 3rd quarter of 2011
PAP 18	SEP 1.x to SEP 2 transition and coexistence	SEP 1.x, 2	Completed in the 3rd quarter of 2011

8.8   Different Layers of Interoperability

For large, integrated, and complex systems, a single layer of interoperability is not sufficient. It requires multiple layers of interoperability, from a plug or wireless connection to compatible processes and procedures for participating in distributed business transactions. In developing the CRM, which is described previously, the high-level categorization approach developed by GWAC is considered [14]. In fact, a different layer of interoperability is realized there. Different layers of the Smart Grid's interoperability in different categories, such as organizational, informational, and technical, and subcategories are specified in Figure 8.2.

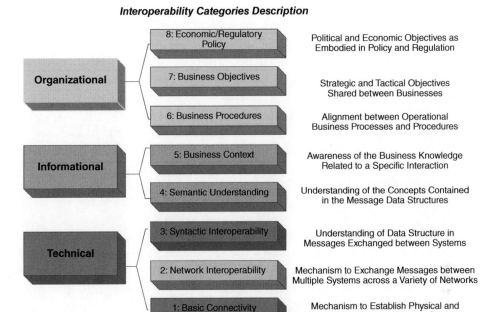

Interoperability Categories Description

	8: Economic/Regulatory Policy	Political and Economic Objectives as Embodied in Policy and Regulation
Organizational	7: Business Objectives	Strategic and Tactical Objectives Shared between Businesses
	6: Business Procedures	Alignment between Operational Business Processes and Procedures
Informational	5: Business Context	Awareness of the Business Knowledge Related to a Specific Interaction
	4: Semantic Understanding	Understanding of the Concepts Contained in the Message Data Structures
Technical	3: Syntactic Interoperability	Understanding of Data Structure in Messages Exchanged between Systems
	2: Network Interoperability	Mechanism to Exchange Messages between Multiple Systems across a Variety of Networks
	1: Basic Connectivity	Mechanism to Establish Physical and Logical Connections between Systems

Figure 8.2 Different layers of the Smart Grid's interoperability. Reproduced from the "GridWise® Interoperability Context-Setting Framework", March 2008, page 5, with permission from the GridWise Architecture Council. The "GridWise® Interoperability Context-Setting Framework" is a work of the GridWise Architecture Council

8.9 Conclusion

Interoperability has an important role to play in the Smart Grid's future application development. For the said purpose, various organizations have developed standards, protocols, guidelines, and specifications for interoperable devices. For a designer, it is very important to identify and select suitable interoperable standards, protocols, guidelines, or specifications for a particular application of the Smart Grid. By evaluating candidate standards and its utilization in various applications, new needs and priority areas will be identified, and as a result new technologies will emerge. For example, the NIST task group has concentrated on eight priority areas for the first phase of its standard development and coordination efforts. In addition, the NIST framework report has identified a list of anticipated Smart Grid benefits that is not included in the current definition of the Smart Grid but they can be achieved once Smart Grid technologies become fully functional.

To include new standards, NIST has also recommended various principles put forward by the World Trade Organization's Committee on Technical Barriers to Trade. These include transparency in standards development process, openness of the standardizing body to all interested parties, maintaining neutrality and consensus in the process of standards development, relevance, and effectiveness in responding to regulatory and market needs, as well as scientific and technological developments. In addition, some other principles were also put forward to be considered in devising new standards.

The IEEE 1547 and P2030 interoperability projects are going to address various priority areas identified by the NIST. In future, both of these IEEE projects will be addressing the additional NIST recommendations by either expansions of existing standards or new standards' projects. The priority areas include but not limited to energy storage systems, DGM standard requirements, technical management of distributed energy resources, static and mobile electric storage, and electric transportation/vehicles.

To summarize, interoperability is a key feature in the Smart Grid's applications. Every Smart Grid system requires an established interoperability standard to connect different devices. Connecting different devices manufactured by different vendors without an established interoperability standard will most probably lead to a system that will likely be unproductive and inefficient.

References

[1] U.S. DOE Office of Electricity Delivery and Energy Reliability *Smart Grid Investment Grant Program.* Funding Opportunity Number: DE-FOA-0000058, December 2009. DOESGIGQuestions@HQ.DOE.GOV.

[2] Kominers, P. (2012) *Interoperability Case Study: The Smart Grid*, http://papers.ssrn.com/sol3/papers.cfm?abstract_id=2031113 (accessed November 17, 2014).

[3] NIST (2010) *Framework and Roadmap for Smart Grid Interoperability Standards Release 1.0*, http://www.nist.gov/public_affairs/releases/smartgrid_interoperability.pdf (accessed 17 March 2012)

[4] Hughes, J. (2008) *Interoperability 101-The Basics of an Interoperable Grid*, SmartGridNews.com (Nov. 26, 2008), Grid Modernization Initiatives.

[5] GridWise Architecture Council (2008) *GridWise Interoperability Context-Setting Framework*.

[6] IEEE Standards Coordinating Committee 21, http://grouper.ieee.org/groups/scc21/index.html (accessed November 17, 2014).

[7] Zensys, A.S. (2006) *Z-Wave Protocol Overview*, Copenhagen.

[8] Pacific Northwest National Laboratory, Department of Energy, USA (2003) *Gridwise Architecture Tenets and Illustrations*.

[9] U. S. Department of Energy, Office of Electricity Delivery and Energy Reliability *Recovery Act Financial Assistance, Funding Opportunity Announcement*, Smart Grid Investment Grant Program Funding Opportunity Number: DE-FOA-0000058, June 2009.

[10] GridWise Architecture Council (2005) *Interoperability Path Forward Whitepaper*.

[11] Federal Energy Regulatory Commission, (2009) *Smart Grid Policy*, 128 FERC 61,060, Smart Grid Section 1301.

[12] (2007) *Energy Independence and Security Act Public Law No: 110–140* Title XIII, Sec. 1301.

[13] NISTIR (2009) *Smart Grid Cyber Security Strategy and Requirements* DRAFT NISTIR 7628.

[14] Electric Power Research Institute (EPRI) (2009) *Report to NIST on the Smart Grid Interoperability Standards Roadmap*.

[15] IEEE *IEEE P2030 Draft Guide for Smart Grid Interoperability of Energy Technology and Information Technology Operation with the Electric Power System (EPS), and End-Use Applications and Loads*, http://grouper.ieee.org/groups/scc21/dr_shared/2030/ (accessed November 17, 2014).

[16] IEEE *1547 Series of Standards*, http://grouper.ieee.org/groups/scc21/dr_shared/ (accessed November 17, 2014).

[17] Z-Wave Alliance (2012) www.Z-Wave.com/modules/iaCM-ZW-PR/readMore.php?id=577765376 (accessed November 17, 2014).

[18] IEEE Spectrum (2014) http://spectrum.ieee.org/nanoclast/semiconductors/nanotechnology/nanotubebased-liion-batteries-can-charge-to-near-maximum-in-two-minutes (accessed online November 17, 2014).

9

Integration of Variable Renewable Resources

9.1 Introduction

The opportunity to reduce greenhouse gas (GHG) emissions from electricity generating systems has become much more viable and economic as technological and market-based strategies to deploy intermittent renewable sources, such as wind and solar power, have increased along with continuing improvement in baseload renewable energy sources, including geothermal power, biomass energy, and waste-to-energy systems. The contribution of these variable resources to the electricity generation mix has significantly increased over the past decades throughout the world. New total capacity of installed wind and solar energy systems over the year 2011 was 40 and 29 GW, respectively [1]. Estimates show that this trend will continue in the coming years. However, as system size increases, their negative impact on the power grid and their ability to serve the intended purpose decreases.

The present electric power grid is designed to generate electricity as economically as possible and distribute it to the customers in a manner that maintains its quality and reliability [2–5]. Generally, the electricity demand pattern follows the daily and seasonal cycle of human activities. The demand is high during daytime when industrial activity is high (particularly in summer afternoons, when air-conditioning is called for), and in the early evening when domestic heating (particularly in winter), cooking, and lighting come into operation. On the other hand, the demand in the late evenings and the early mornings when people are asleep is lower. In this kind of situation, the dictates of economy, and the requirement of balancing load and generation – which is a physical constraint independent of market structure – do not allow the continuous operation of generators at a fixed power level for all 24 hours of the day.

Adding the output of large intermittent energy sources, such as wind and solar, is expected to cause a technical challenge of many dimensions [6–15]. The most

Smart Grid Standards: Specifications, Requirements, and Technologies, First Edition. Takuro Sato,
Daniel M. Kammen, Bin Duan, Martin Macuha, Zhenyu Zhou, Jun Wu, Muhammad Tariq and Solomon Abebe Asfaw.
© 2015 John Wiley & Sons, Ltd. Published 2015 by John Wiley & Sons, Ltd.

important is related to the required generator-ramping capabilities in order to balance demand and generation at all times. But it is possible that grid systems can be restructured in anticipation of electricity produced by large intermittent renewable energy generating systems. However, even if we could achieve a proficient power system restructuring, it is possible that – depending on the type of energy generating resources and the grid type – the mismatch between the demand time and electricity generation of these renewable sources could further limit its ability to play its role in reducing GHG emissions unless we use energy-storage mechanisms [13–15]. Recent studies have found that depending on geographic locations, many potential paths of decarbonizing the present grid exist [9, 16, 17]. Such a future grid differs from the present grid in many ways. One such difference is the level of the contribution of intermittent renewable resources. This chapter is devoted to examining the grid interoperability challenges and possible solutions at different levels of intermittent renewable penetration.

9.2 Challenges of Grid Integration of Intermittent Renewable Systems

9.2.1 Operation of a Conventional Electric Power System

Electricity demand varies depending on the national business cycle, residential activity, and weather. A daily demand curve is usually divided into three layers: base, intermediate, and peak load, as shown in Figure 9.1. The base load is that part of the load curve that varies little during the day. The intermediate layer is that part of the load that varies throughout the day in a roughly predictable manner. The peak load layer is the highest part of the variable load, which varies in a less predictable manner. The pattern of the daily demand profile and the corresponding characteristics shown in Figure 9.1 vary from day to day. The trend over weekends and holidays is different from that on weekdays. The typical daily pattern can also vary with season.

Unlike some consumer goods, electricity supply requires that generation and consumption should always be as closely balanced as possible. Thus, the demand pattern is one of the factors that dictates how electric power systems are operated [2–5]. The baseload layer in Figure 9.1 is met by baseload generating units. These generating units can be ramped up to meet some of the demands above the baseline and ramped down to reduce output when the demand is unusually low. The intermediate generating units, which are also known as "load-following" or "cycling" units, meet most of the varying daily demand. These can quickly change their output in response to the change in demand. Such load-following units can also provide an online spinning reserve capacity, that is, unused capacities that are immediately available in response to unexpected increases in demand or forced outages of some units. Peaking units, which are quick-start units that can go from shutdown to full load in a few minutes, supply the highest daily loads. Such units have low "capacity factors" (i.e., during the year in which they operate) and are also used as offline spinning reserves.

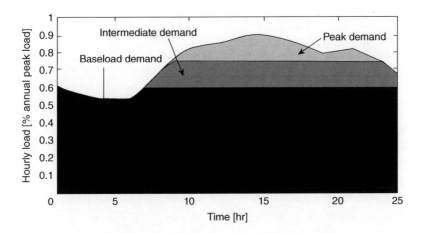

Figure 9.1   Illustrative daily load curve

In "dispatching" the available generating units, dispatchers deal with two contradicting problems. These are meeting the varying demands at lowest possible cost and maintaining reliability of the power system as well as the quality of the service they provide. Maintaining reliability requires the ability to make sure that sufficient generation is available to meet the demand at all times, while maintaining power quality requires delivering electricity at the standard consumption voltage and frequency.

The economically optimal dispatching of a system of power plants dictates that units with lowest cost per kilowatt hour be put online first [2–5]. Depending on the particular utility, power plants with low operation costs, such as nuclear, hydropower, coal, geothermal baseload units are dispatched almost continuously throughout the year except at times of planned or forced outages. Combined-cycle units using expensive natural gas are used for supplying intermediate loads. The very expensive and less-efficient combustion turbines are used as peaking units. Variable renewable energy generating systems do not fall into any of these three categories but can be used as available to help meet demands.

As indicated above, power system dispatching is not a simple function of economical factors alone. It is also dependent on operational constraints associated with each of the various generator types employed, load-curve characteristics, system operation risks (i.e., the possibility that the system may fail to meet the required load), and so on [2–16, 18]. Preparing sufficient spinning reserve can minimize the operational risk. The spinning-reserve requirement is a complex function of system load, unit sizes, component failure rates, available water for hydropower plants (if relevant), startup time of additional generating units, and so on.

The particular set of committed units operated under any given load level can differ according to circumstances. This shows that the corresponding ramping rate and the ramping range also varies on the basis of the actual scheduling. In the presence of intermittent renewable energy sources, the instantaneous grid ramp rate and ramp range could be limiting factors on the grid penetration of these resources [9, 13–15, 19, 20].

9.2.2   Impact of Adding Intermittent Renewable Systems to the Power Grid

The electrical output of wind and solar technologies substantially vary from time-to-time. The variability depends on geographic location, weather, time of the day, season, and various topographic conditions. Figure 9.2 presents the hourly profile of simulated output of wind and solar technologies, and the corresponding load profile for one week in California. The figure shows that both solar and wind output vary significantly. For example, on a clear day (days five and six) solar generation slowly increases from the morning to noon when it starts its gradual decrease until the evening. On a (semi-) cloudy day (days one to four and six), moving clouds may add more variability to its generation by temporarily reducing/blocking the solar irradiance that the solar generators would have received. Its generation also has seasonal characteristics (Figure 9.3). In general, in the northern hemisphere, solar generation peaks during summer, while reaching its lowest generation during winter seasons. Unlike solar, wind does not have a defined diurnal output profile but for most geographic places it is reported to reach its peak generation during the evening time. However, wind tends to demonstrate some sort of periodic output, that is, a few continuous days of good generation (which depends on season) after/before days of

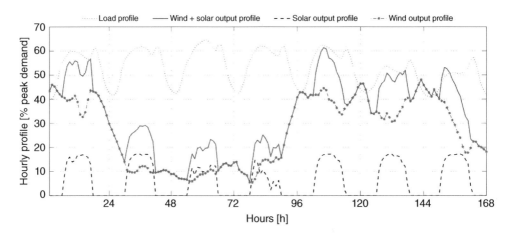

Figure 9.2   Simulated hourly profiles of wind, solar, wind–solar hybrid system output, and the corresponding load profile for one spring week in California. The total capacity of wind–solar hybrid in this simulation was 40.9 GW (11.1 GW solar and 29.8 GW wind). The system was distributed throughout California and would have supplied the maximum possible energy (38.3% of the annual demand) that California could have obtained from intermittent renewable in the year 2011 without energy dumping. Note that achieving the same penetration with a solar or wind technologies alone would have required some energy dumping or storage or both

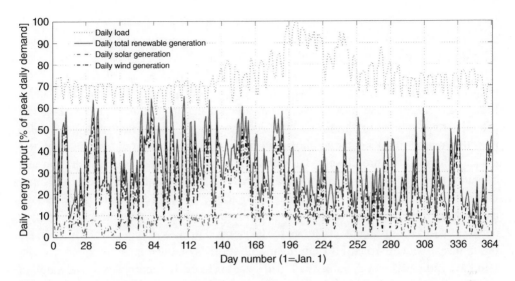

Figure 9.3 Daily simulated output of wind, solar, and their hybrid – for systems given in Figure 9.2 – and daily demand of California for the year 2011

poor generation. This may be evident from Figure 9.2 and the spiky nature of its daily output, as shown in Figure 9.3. Seasonally speaking, wind output peaks, depending on location, during springtime. In addition, as shown in Figure 9.2, the hybrid system output profile appears to have smother output that almost fits better to the load profile as compared to wind or solar technologies.

The variability of wind and solar generators output introduces new challenges to electric system operation and design. A key performance characteristic of generators in general is their degree of dispatchability, specifically their ability to satisfy opera- tors' need to control its output over a range of specified output levels. Wind and solar photovoltaic (PV) have little dispatchability – the output from these sources can be reduced, but not increased, on demand. An additional challenge is the uncertainty in the output profile of these resources, with wind and solar having limited predictability over various time scales.

If the energy input to the grid from intermittent sources is sufficiently small, it can be treated as a negative contribution to the overall load and does not cause any par- ticular control problems. However, integration of larger intermittent energy sources is undergoing serious scrutiny by utilities due to the likelihood of enhanced uncertainties in the grid system. Ideally, the grid system should be able to manage instantaneous changes in input from large intermittent sources of energy [21–25].

Several studies have reported the impact of adding different sizes of intermittent energy sources into the grid system [9–26]. Many of these studies have revealed an increase in the effective cost of wind energy due to the increased cost incurred to counter imbalances caused by variability of the source [8, 25, 26]. In general, the various impacts on a power system caused by large intermittent energy sources are categorized as short-term (minute to hours) and long-term (years) effects. Typical short-term effects involve voltage and frequency management, whereas typical long-term effects would include component degradation.

In fact, the major short-term effects are voltage and frequency management but questions of grid instability resulted owing to disturbances, transmission and distribution losses, and energy losses due to overgeneration are also important [7, 8, 22–24]. The principal challenge of integrating variable energy sources to the electricity grid is also maintaining the frequency and voltage within their prescribed limits. For a stable grid system, these limits are closely determined by the balance between generation and demand, the balance being maintained by having some plants increase or decrease their output. For large, interconnected, grid systems, this challenge is minimal as long as the intermittent energy source is not too large. However, the variability of such sources causes a significant challenge for an isolated grid. A number of studies carried out to investigate the role of different variable sources on the conventional power system have indicated that the ability of grid systems to cycle rapidly limits the value of such sources to conventional power systems [9–15]. It has been found that an annual contribution of up to about 20% of the total demand is possible from intermittent sources if a modest amount of energy dumping from those sources is allowed.

The addition of large variable energy sources affects how the conventional grid system should be operated, because special measures are needed in order to control instabilities [9–15, 19, 20]. These measures may include increasing the number of available reserve units and/or increasing the frequency of shutdown and start-up cycles of the existing load-following units. In the short term, this leads to increased costs for variable energy due to the resulting increase in maintenance and operation costs of the conventional plants. It may also lead to more frequent forced outages. In the long run, the excessive shutdown and start-up of such plants will significantly decrease their lifetime [26–28, 43] requiring further capital expenditure to replace them. The alternative is to employ more units to allow operation of the power system on a shift basis, which also increases the effective cost of a variable energy source.

In addition to plant aging, another important long-term consideration, which affects grid reliability, is the importance of long-term planning. Specifically, because of the intermittent nature of grid input from solar and wind, as the grid system grows to incorporate ever more of these sources it is of vital importance for there to be a concomitant growth of conventional generating capacity of a type that can maintain grid stability under all possible demand conditions. To put it another way: solar and wind are not dependable resources for capacity. Therefore, the conventional grid system must be designed in a manner that will enable it to meet anticipated future peak demands.

When we go to a very high penetration of intermittent renewable systems, an additional challenge emerges. This is the ability to match the seasonal and diurnal profile of our resources to the local demand. In the future grid, when intermittent renewables make a significant part of the generating capacity, a small increase in the variable generating capacity could come at the cost of dumping some of its excess energy, especially in spring season, when its output exceeds the demand that it could supply [9–15].

9.3   Transitioning to Highly Renewable Electricity Grid

9.3.1   Planning Studies

Recently, many planning studies assessing the possibility of transforming the present carbon-intensive power grid to a system that emits at most 20% of the 1990 GHG emission by 2050 have been released. These studies differ, inter alia, in their research approaches/tools as well as the geographic areas they cover. Consequently, detailed comparison is difficult. Here, we will briefly outline the research approaches and the most important findings related to the role of variable generators in the future grid.

9.3.1.1   Decarbonizing Scenarios for the Western Electricity Coordinating Council (WECC)

This study was performed using the SWITCH model that has been developed by researchers in our Renewable and Appropriate Energy Laboratory at the University of California – Berkeley [16–18]. SWITCH is a capacity expansion and dispatch model developed to study policy options for decarbonizing the power sector in the entire geographic extent of the Western Electricity Coordinating Council (WECC). The model is a mixed-integer linear program whose objective function is to minimize the cost of meeting electricity demand between the present day and some future time, say the year 2030. Figure 9.4 shows the diagram summarizing the model's input, optimization, and output, while Figure 9.5 provides the objective function together with necessary information. It has high geographic resolution that enables it to assess the optimal deployment of low carbon generating resources by representing the entire WECC region as 50 subregions referred to as load areas.

In order to capture elements of the day-to-day operation, a switch investment model employs sampling techniques based on historical real-time demand but due to runtime the highest sample hour stands at 144 h per year, which is composed of (12 month/year) × (2 days/month) × (6 h/day). The day with median demand and peak demand was sampled from each historical month. The peak demand day is assigned a weight of one day per month, while the median day gets a weight of the total days per month minus one. The model uses hourly dispatch similar to the investment optimization to check that the capacity expansion model result properly balances electricity demand and supply throughout the year of interest.

SWITCH has an outstanding feature that enables the assessment of the impact of continuing the present-day business as usual power sector policy versus an alternative path on future emissions from the sector. The model uses a carbon price adder to reach a restructured and less carbon-intensive grid by some future time.

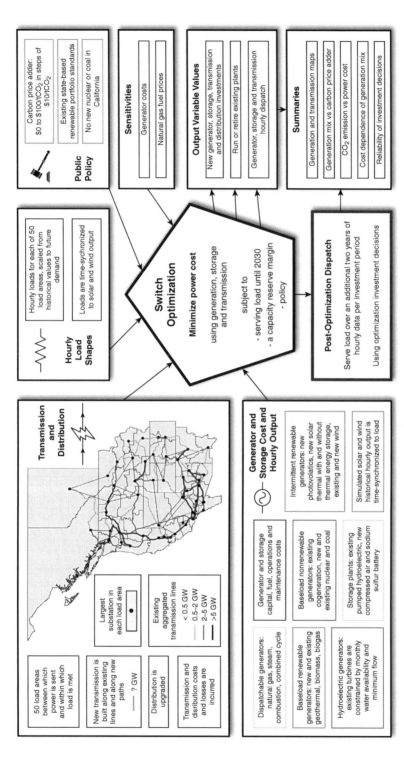

Figure 9.4 Diagram of data inputs, optimization, and output of the SWITCH model. Reprinted from Energy Policy, Vol. 43, April 2012, James Nelson, Josiah Johnston, Ana Mileva, Matthias Fripp, Ian Hoffman, Autumn Petros-Good, Christian Blanco, Daniel M. Kammen, "High-resolution modeling of the western North American power system demonstrates low-cost and low-carbon futures", pp. 436–447, Copyright © 2012, with permissions from Elsevier [16].

		Objective function: minimize the total cost of meeting load	
Generation and storage	Capital	$\sum_{g,i} G_{g,i} \cdot c_{g,i}$	The capital cost incurred for installing a generator at plant g in investment period i is calculated as the generator size in MW $G_{g,i}$ multiplied by the cost of that type of generator in \$2007/MW $C_{g,i}$
	Fixed O&M	$+ (ep_g + \sum_{g,i} G_{g,i}) \cdot x_{g,i}$	The fixed operation and maintenance costs paid for plant g in investment period i are calculated as the total generation capacity of the plant in MW (the pre-existing capacity ep_g at plant g plus the total capacity $G_{g,i}$ installed through investment period i) multiplied by the recurring fixed costs associated with that type of generator in \$2007/MW $x_{g,i}$
	Variable	$+ \sum_{g,i} O_{g,t} \cdot (m_{g,t} + f_{g,t} + c_{g,t}) \cdot hs_t$	The variable costs paid for plant g operating in study hour t are calculated as the power output in MWh $O_{g,t}$ multiplied by the sum of the variable costs associated with that type of generator in \$2007/MWh. The variable costs include per MWh maintenance costs $m_{g,t}$, fuel costs $f_{g,t}$, and carbon costs $c_{g,t}$, and are weighted by the number of hours each study hour represents, hs_t.
Transmission		$+ \sum_{a,a',i} T_{a,a',i} \cdot l_{a,a'} \cdot t_{a,a',i}$	The cost of building or upgrading transmission lines between two load areas a and a' in investment period i is calculated as the product of the rated transfer capacity of the new lines in MW $T_{a,a',i}$, the length of the new line $l_{a,a'}$, and the regionally adjusted per-km cost of building new transmission in \$2007/MW. km, $t_{a,a',i}$. Transmission can only be built between load areas that are adjacent to each other or that are already connected.
Distribution		$+ \sum_{a,i} d_{a,i}$	The cost of upgrading local transmission and distribution within a load area a in investment period i is calculated as the cost of building and maintaining the upgrade in \$2007/MW $d_{a,i}$.
Sunk		$+ s$	Sunk costs include ongoing capital payments incurred during the study period for existing plants, existing transmission networks, and existing distribution networks. The sunk costs do not affect the optimization decision variables, but are taken into account when calculating the cost of power at the end of the optimization.

Figure 9.5 SWITCH objective together with the necessary description. Reprinted from Energy Policy, Vol. 43, April 2012, James Nelson, Josiah Johnston, Ana Mileva, Matthias Fripp, Ian Hoffman, Autumn Petros-Good, Christian Blanco, Daniel M. Kammen, "High-resolution modeling of the western North American power system demonstrates low-cost and low-carbon futures", pp. 436–447, Copyright © 2012, with permissions from Elsevier [16].

It also models uncertainties related to generator cost/technology improvement using a scenario-based approach, which includes changing the projected generator capital cost within a feasible range, or expanding/reducing the set of generator mix from which the model can choose. Table 9.1 summarizes some of the scenarios that have been examined. The model also examines the impact of electrification of vehicles and heating, and that of energy-efficiency measures by utilizing and adjusting load profiles.

SWITCH results show that multiple paths to a decarbonized future grid exist. The corresponding sets of installed generators by year 2050 are given in Figure 9.6.

Table 9.1 Potential electricity system scenarios for 2050

Scenario	Load profile	California load in 2050 (TWh/yr)	Total WECC load in 2050 (TWh/yr)	Carbon cap (% reduction from 1990 emission levels) (%)	Extra capital cost declination relative to base case (%/yr)	Generators included or excluded
Frozen, no carbon cap	Frozen efficiency	395	1368	N/A	N/A	Biomass solid CCS excluded
Frozen efficiency with carbon cap	Frozen efficiency	395	1368	80	N/A	Biomass solid CCS excluded
Base case	Base case	424	1310	80	N/A	Biomass solid CCS excluded
Inexpensive nuclear	Base case	424	1310	80	Nuclear: −2	Biomass solid CCS excluded
Inexpensive CCS	Base case	424	1310	80	CCS: −1.5	Biomass solid CCS excluded
No CCS	Base case	424	1310	80	N/A	All CCS excluded
Inexpensive solar and wind	Base case	424	1310	80	Solar and wind: −1	Biomass solid CCS excluded
Expensive photo-voltaics	Base case	424	1310	80	Photovoltaics: +1.5	Biomass solid CCS excluded
Biomass solid CCS	Base case	424	1310	100	N/A	Biomass solid CCS included
Extra electrifica-tion	Extra electrifica-tion	484	1478	80	N/A	Biomass solid CCS excluded

Reproduced with permissions from [18].

Figure 9.6 shows that the required generator capacity by 2050 increases significantly under all circumstances that reduce carbon emission to 20% below the 1990 level as compared to the no carbon cap scenario. The major reason for such a significant difference in the installed capacity was the level of installed wind and solar technologies. Comparing each scenario also reveals that the type of generators being built vary with the amount of wind and solar technologies. When large inflexible generators such as coal with carbon capture and sequestration (Coal-CCS) gets built, the amount of solar

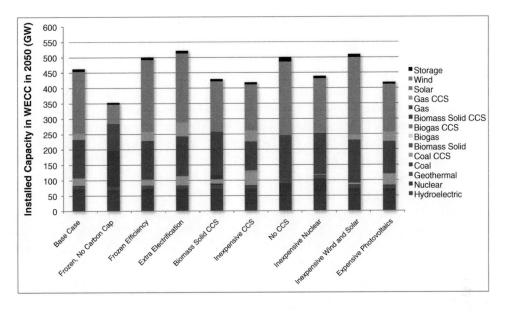

Figure 9.6 Generation and storage capacity installed throughout WECC in 2050 by scenarios. Source [18].

and wind technology decreases, and by contrast, when Coal-CCS technologies are not available, it builds more solar and wind technologies together with more flexible gas-firing units plus some storages that could provide the flexibility for the required balancing services. The study shows that depending on scenarios, wind and solar was found to supply up to 50% of the electricity demand by 2050.

9.3.1.2   Renewable Electricity Future by NREL

The National Renewable Energy Lab (NREL) has recently released a comprehensive report on the renewable energy future of the US grid. This study investigated the potential mix of different renewable resources – which comprises wind, PV, concentrating solar power (CSP), hydropower, geothermal, and biomass – requirements to meet various levels of renewable penetration – from 30% to 90% of the annual demand – by 2050 [9]. The study was performed using NREL's Regional Energy Deployment System (ReEDS) model and ABB GridView model. The former was used to analyze the potential to meet the US electricity demand over the coming decades using the geographically diverse renewable resources while the later model is used to test the hourly operation of the US grid under various 2050 scenarios. ReEDS is represented by 17 time slices per year and as a result uses statistical treatment to assess the impact of variable wind and solar resources on capacity planning and dispatch. The study focuses on exploring how by the year 2050, 80% of US electricity demand can be met by renewable generation. Special attention was given to the 80% generation level because it was determined to be adequate to achieve the targeted reduction of GHG gas emission of the electricity sector to 20% of the 1990 level.

The study shows that multiple paths to achieving the 80% renewable generation by the year 2050 exist. The renewable generation comes from various renewable technologies that have different degrees of dispatchability. Depending on the scenarios at 80% renewable generation level, energy from wind and solar technologies accounts for almost 50% of the total generation. This study shows that compared to the baseline scenario, which assumes the use of the present-day conventional generators by 2050, more diverse technologies emerge as renewable penetration increases. Other technologies, such as energy storage, become increasingly important as variable renewable technologies such as wind and solar increases.

According to their hourly simulation performed using GridView with nearly 80% of energy from renewables – including when 50% is from variable generators – the demand and supply of electricity was balanced for all regions [9]. These studies have identified some features of the future grid as compared to the existing one. Here we emphasize three important lessons related to integrating large amounts of intermittent renewable energy into the power grid. These are as follows:

1. **Increased need for grid flexibility**: The study shows that an increased use of variable renewable energy resources requires the capability to ramp the output of the conventional technologies output in accordance with the nature of the incoming renewable generations. Real-time balance of the demand and supply was achieved using the grid flexibility coming from the use of energy storage and demand-side technologies (such as interruptible load), the increased transmission interconnection, and increased dispatch flexibility on the part of conventional plants including fossil-fuel generators.
2. **Increased operational challenge**: In contrast to the fossil-fuel-dominated grid where the main concern is meeting the peak demand hours (e.g., summer afternoons), a grid with high renewable was shown to face additional challenges when the abundance in renewable generation coming during low demand hours (e.g., spring evenings) forces most of the thermal generations to cycle down to their minimum generation levels. This adds to the list of uncertainties that comes with high penetration of intermittent renewable generations.
3. **Energy curtailments**: On the basis of the hourly dispatch analysis for the 80% by 2050 scenarios, 8–10% of the renewable energy will be curtailed. This would reduce the economic value of the power plants.

9.3.1.3   Road Map to 2050 – the Study of the European Grid

Following the announcement of the targeted reduction of CO_2 emission by the year 2050 to 80% below the 1990 level set by the leaders of the European Union and the G8, European Climate Foundation (ECF) has performed a fact-based assessment of this policy, which was an instrument for the initial set of technical and policy proposals known by the term "Road Map to 2050" [19]. The study investigated the potential carbon abatement across various economic sectors of the European countries (EU-27 plus Norway and Switzerland). The study shows that to achieve the targeted reduction by the year 2050, the emission in the electric sector should be reduced by as much as 95% below the 1990 level. The study considered three main pathways that are perceived as representative of prevailing views of the potential power mix evolution in order to investigate the nature of the potential 2050 European decarbonized grid. The difference between the three pathways being variation in the relative contribution of their three classes of the low/zero carbon generation technologies, that is, fossil fuel with CCS, nuclear, and renewable energy sources (which includes wind and solar technologies, geothermal, hydropower, and biomass), to the electricity supply by the year 2050. The share of renewable resources in the three pathways are 40%, 60%, and 80%, the remaining supply being assumed to be shared equally by nuclear and fossil-fuel with CCS. The demand supply balance and other security measures were assessed using a generation dispatch model. Similar to the previous findings, this study shows that variable generators such as wind and solar make up a significant part of the future grid. Unlike the above two studies, this study does not use any capacity expansion model. However, it used back-casting methods (a method of working backward from 2050 to the present) to identify the time at which the current path should be altered to reach the future target. This study also indicated that the need for storage increases as the share of the intermittent renewable system increases.

9.4 Very High Penetration and Grid-Scale Storage

The foregoing long-term planning studies were performed under significant economic and technical uncertainties. In designing a grid that accommodates large intermittent renewable energy systems, we deal with the challenge of handling significant fluctuation of intermittent renewable energy and the possibility of reaching an improved matching of their varying output profile to the local demand profile throughout the year. In the following sections, we will discuss the essence of reaching an improved matching of the varying renewable systems output to the demand profile using two studies, namely, the grid matching study to the Israeli grid and the result of our own storage design and dispatch model for an interconnected grid.

9.4.1 Grid-Matching Analysis – Case of the Israeli Grid

A series of papers detailing a grid-matching possibilities to an island Israeli grid was published by Solomon et al. [10–15]. The study aspires to identify possible ways of incorporating larger and larger intermittent renewable energy systems into an electricity grid. This study was performed using simulated hourly wind and solar technologies output using metrological data for the year 2006 and the corresponding historical load data for Israeli grid. Various mechanisms that can increase grid matching of intermittent renewable systems were examined using a simple mathematical algorithm that was developed for this purpose by researchers at Ben-Gurion University of the Negev. The study assumes that some percent of the peak demand is provided by a nonflexible baseload power plant, while the remaining part of the load, which is also considered amenable to be supplied by variable renewable systems output, was assumed to be supplied by more flexible generators. Grid flexibility was, therefore, defined as

$$\text{ff} = 1 - \frac{G_{\text{min}}}{G_{\text{max}}}$$

where G_{min} and G_{max} are the minimum generation levels to which the grid could be ramped and the maximum generation, respectively. These studies have shown that grid penetrations of intermittent renewable systems is limited by the grid flexibility and its ability to match to the local demand profile. Increasing grid flexibility, that is, replacing the baseload power plants by more flexible generators, could enhance the capacity of the grid to accommodate more energy from intermittent renewable resources. Even under an ideal grid flexibility of ff = 1 (no baseload), grid penetration achieved by wind or solar technology alone barely reaches 20% depending on the type of generating technology. Other mechanisms, excluding storage, to increase grid penetration were shown to be allowing some energy dumping and the use of more diverse resources. As regards the role of energy dumping, the study shows that for any technology type or their mixture, grid penetration increases significantly for smaller energy dumping. For a modest 5% energy dumping, grid penetration of about 27%, 32%, and 32% of the annual demand could be achieved using stand-alone static-flat plate PV, Concentrating Photovoltaic (CPV), and wind technologies, respectively. Any further increase in energy dumping appears to help little. The other mechanism that was found to result in a significant energy penetration was the use of a mix of wind and solar technologies. The study shows that any mix of the two technologies achieves better energy penetration than any one of them as a stand-alone. This is because of the nature of complementarities of the output of these technologies (e.g., solar output peaks during midday while wind peaks in the evening), which increases their demand matching. At 5% energy dumping, wind CPV hybrid was shown to reach as high as 46% of the annual demand. This is about the maximum penetration that can be achieved for an ideally flexible Israeli grid without storage. Because of the observed smoothing effect of the two resources, on top of its ability to increase grid penetration, the use of a wind–solar mix could also reduce the grid ramping requirement as compared to the stand-alone use of both technologies.

Now, let us see the case of storage design requirement and its role in grid penetration of PV technology. Instead of putting a priori limits on the energy and power capacity of the storage, their model calculates those parameters based on the interaction between the electricity grid and solar PV output as the PV system size increases. Figure 9.7 shows the dependence of PV penetration on both power capacity and energy capacity of storage. Figure 9.7a shows that depending on grid flexibility, penetration shows a

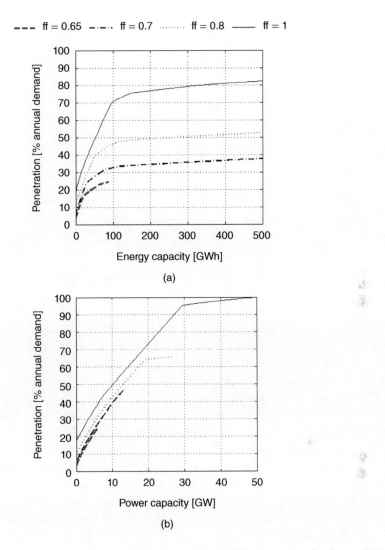

Figure 9.7 Dependence of penetration on (a) power capacity and (b) energy capacity. Source [12]

sharply increasing trend for smaller storage energy capacity. The rate of penetration starts to level off as we increase storage energy capacity in order to accommodate more surplus PV energy. This shows that increasing energy capacity far beyond the turning point in Figure 9.7a becomes an increasingly poor strategy because a small increase in penetration requires a large increase in the storage energy capacity. In other words, the storage usefulness decreases as we further increase the storage energy capacity beyond the turning point. Figure 9.8 presents how the storage "usefulness index (UI)," that is, an identifier created by taking the ratio of energy delivered by storage in a year to the energy capacity of storage, varies with its energy capacity. The figure shows that "UI" initially increases with energy capacity until it reaches some peak where it starts

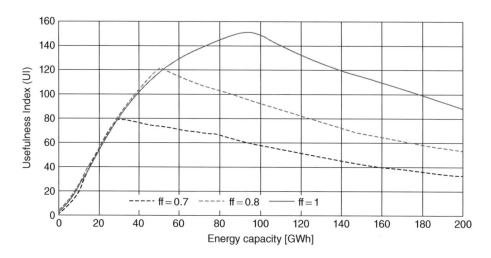

Figure 9.8 Dependence of usefulness index (UI) on energy capacity of storage. Source [11]

to decrease. For all grid flexibility, "UI" peaks where the change in slope observed in Figure 9.7b occurs. The study shows that building significantly larger storage than the one that corresponds to peak UI leads to a reduced benefit of storage. For ff = 1, storage that corresponds to peak UI is 94.5 GWh.

The corresponding high penetration for ff = 1 was approximately 70% of the annual demand. However, the study reported that penetration could be increased significantly by increasing the storage use by changing the dispatch. In order to do that, one has to allow energy dumping when the storage power and energy capacities are exceeded. Using storage with energy capacity of 94.5 GWh (which is about 70% the daily average demand of Israel) and the corresponding power capacity, PV grid penetration of about 86% has been achieved by allowing 20% energy dumping. At the same time, a proper dispatch strategy would have reduced the conventional power plant capacity by at least 3 GW below the total 10.5 GW capacity that Israel has operated that year. In general, the study concludes that designing the storage according to the seasonal and diurnal matching of intermittent renewable output profile to the load profile carries a significant role in achieving very high penetration and high system performance.

9.4.2 Storage Design and Dispatch – Case of Interconnected Grid

In order to assess the possibility of reaching a very high penetration in an interconnected grid we have developed a noneconomic mathematical model (based on a linear program), which was also instrumental in assessing the potential impact of storage design and dispatch as we increase the capacity of variable generators. The model was designed in a way that enables optimizing the renewable penetration while minimizing the corresponding storage energy and power capacity requirements. To ensure hourly demand and supply balance throughout the year, the model also builds the minimum conventional backup capacity required to meet the demand according to the circumstances. For simplicity, we assume that the backup represents a set of quick-start and fast-ramping generators in each load area.

The study was performed using a one-year hourly load data of the State of California and hourly simulated output of various wind and solar technologies. For this study, we took the data for the year 2011 from the SWITCH model database. We also divide the state of California into 12 load areas as in the SWITCH model [16]. Power exchanges between load areas were treated as a transportation model with the maximum transfer between load areas constrained not to exceed the total thermal capacity of the transmission lines connecting them. To force the model to build the renewable generators, we constrain that at a given condition some amount of renewable system gets built. From the set of technologies distributed throughout the state, the model then builds technologies that optimize the energy penetration of intermittent renewable systems.

Now let us see how grid penetration of renewable systems and the corresponding energy storage requirement relate to one another as we increase the total variable system size. Figure 9.9 shows how grid penetration of energy from variable resources varies with network energy capacity. The term Network Energy Capacity refers to the sum of total storage energy capacity to be built in the entire network. The two curves in this figure correspond to two storage design models that we constructed for the purpose of comparison, that is, one allows the storage to supply its stored energy to the network (SET model) while the other limits the use of the stored energy only in

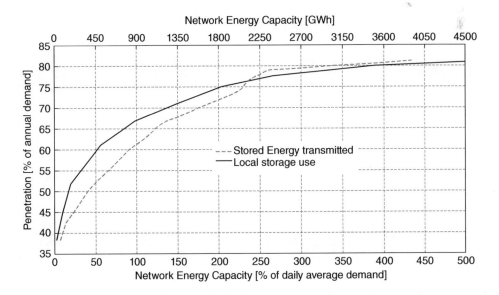

Figure 9.9 Dependence of penetration on network energy capacity

the load area (SEUL model) where it is stored. The figure shows that, under both conditions, grid penetration sharply increases for smaller storage. As in the case of the Israeli grid, the rate of penetration gradually decreases as the network energy capacity increases to comply with increasing generations. This figure also shows that, for smaller network energy capacity, penetration increases at a faster rate when stored energy is used locally than when it is transmitted. This phenomenon occurs because when we allow transmission of the stored energy, the model builds larger storage to reach almost the same grid penetration but achieves a slight reduction of the conventional backup capacity needed. This can be seen from the close correspondence between the lower conventional backup capacities for most of the renewable system sizes given in Figure 9.10 and the relatively higher network energy capacity of the storage under the SET model. Figure 9.10 also indicates that total conventional backup capacity decreases as the network energy capacity of the storage increases. However, Figure 9.11 indicates that an increase in network energy capacity beyond the turning region, as in the case of the above Israeli study, is a poor strategy. Figure 9.11 presents storage UI for both SET and SEUL versions. The figure shows that UI initially increases before it starts declining. Similar trends were shown above for the Israeli grid. However, the UI trend in the present study is not as smooth as in the above case, especially for the SET model that demonstrated other smaller peaks. This should be expected because in the present study, the model can build systems from

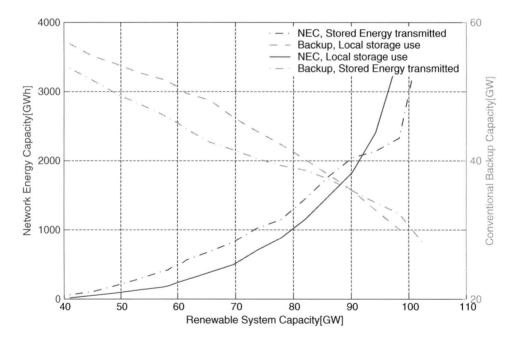

Figure 9.10 The network energy capacity of storage and the corresponding conventional backup capacity needed to meet the hourly load at each renewable system size

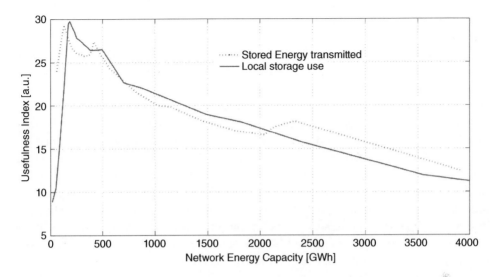

Figure 9.11   Storage UI versus network energy capacity

a large geographic domain as well as various wind (onshore and offshore) and solar (e.g., Static-PV, concentrating solar (thermal) power, tracking-PV, etc.) technologies as compared to the single-site PV resource in the Israeli study. Despite this behavior, the UI curve together with the corresponding Figure 9.9, is the only tool that simplifies approximating the maximum threshold on storage energy capacity requirement. Here, we would like to note that the difference in the UI curves for the two models in the present study may be a result of the SET model's tendency to reduce the conventional backup need more than its counterpart.

This study and the previous one show that increasing energy storage beyond the turning region to avoid energy dumping is a bad policy. These findings suggest that, depending on grid type and local resources, the seasonal and diurnal matching capability of intermittent renewable system output and demand profile effectively establishes the maximum threshold on the required energy storage capacity (which may also depend on storage technology). Figure 9.10 and the previous section show that the maximum of this threshold is of the order of daily average demand.

The nature of the UI in the SET model suggests that its peak could be arbitrary; however, we keep using its result for the purpose of comparison. As we will see later, the SEUL model captures the impact of seasonal and diurnal interaction of intermittent renewable system output and load profile better than the SET model.

Figure 9.11 also reveals that under all conditions the UI remains very low. The only mechanism that could increase grid penetration now allows energy dumping when the energy capacity and power capacity of the storage is exceeded. Unfortunately, in the present market no energy dumping is economically valuable. But, as shown in the following, energy curtailment could be one of the potential ways of increasing grid penetration of intermittent renewables. Moreover, it is shown that it carries a significant advantage of reducing the conventional backup capacity. To demonstrate this we selected two network energy capacities, one defined at the peak UI for the SEUL model (which is 184 GWh) and the other one corresponding to the second peak of the SET model (which is 411 GWh). The first peak of the SET model occurs at a smaller storage size and was ignored as a result. Figure 9.12 presents the manner of grid penetration and the required conventional backup capacity versus the total energy loss (the sum of energy loss due to storage efficiency and curtailed excess energy). Figure 9.12 shows that grid penetration of renewables sharply increases for smaller total energy loss, reaching a penetration of about 85% of the annual demand at 20% total energy loss. On the contrary, the conventional backup capacity requirement significantly decreases as we increase energy loss. When total energy loss reaches 20% of the total renewable generation, the conventional backup capacity requirement has decreased approximately to 35 and 33 GW for the SET and SEUL models, respectively. This is very significant because the indicated backup capacity was sufficient to meet the year-round hourly demand, including the 59 GW peak demand hour plus the 5.3% distribution loss.

Figure 9.12 also reveals that the 418 GWh storage shows a small advantage over the 186 GWh in increasing grid penetration. Our subsequent analyses confirms that proper storage design should be achieved in any grid, as in the case of the Israeli grid, in order to achieve higher penetration in an efficient way. As a result, we conclude that the largest storage need for California, under the present study, is approximately 186 GWh (as suggested by the SEUL model) if grid penetration is taken as the only measure. Even though from the figure, the 418 GWh storage has shown a significant advantage in reducing the conventional backup capacity need, its energy capacity is more than double the 186 GWh, while the power capacity is larger by at least 2 GW.

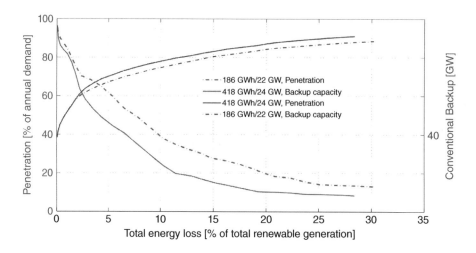

Figure 9.12   The dependence of penetration and backup capacity on the level of energy dumping

This shows that designing storage based on the seasonal and diurnal interaction of the intermittent renewable output and the load profile significantly matters in achieving very high penetration and an efficient system performance.

In order to see how much energy dumping increases storage use, we present Figure 9.13. The figure reveals two important findings. The first is that the storage delivered a smaller amount of the energy to the grid (only 10% of the renewable energy at the highest condition) but, as discussed above, it played a significant role of an enabler for intermittent renewable energy resources. As a result of this, the storage capacity was significantly smaller, that is, 22% of the daily average demand (taking 186 GWh as a reference). Recall that the corresponding capacity for the Israeli grid was about 72% of the annual demand and storage delivers a larger share of the energy from PV [11]. The second is that the energy delivered by storage sharply increases when we start allowing energy dumping but the trend reverses its direction after reaching its peak at about 15% of the total loss. A similar trend was also reported for the Israeli study [11]. By comparison, storage delivered a significant amount of the energy from intermittent renewable resources in the case of the Israeli grid. This is because that study, on top of being an island grid (one load area system), considers the case of PV technology only. In the presence of diverse resources and large interconnectivity, the complementarities of the resources and the power exchange potential reduce storage requirement. This possibility was also proposed in the study of the Israeli grid [15].

At this point, it may be instructive to in how an economic model could measure the role of storage. The foregoing discussion suggests that to reach an efficient grid in a lowest-cost way that allows any level of high grid penetration with storage, our

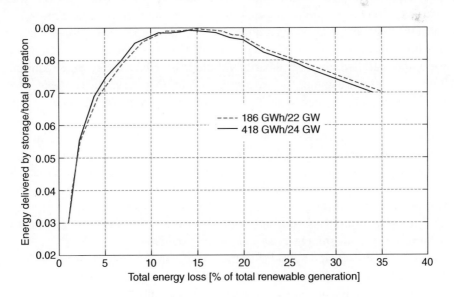

Figure 9.13 Share of the renewable energy delivered by storage

models should have the capability to measure many physical and policy dimensions. In the following, we will briefly discuss the most important criteria:

1. **Flexibility of storage design and dispatch**: Unlike conventional generators, which convert other forms of energy to electrical energy, energy storage stores electrical energy in whatever form (depending on storage technology) and converts that back to electrical energy when needed. This process carries time dynamics because storage cannot deliver energy if it did not store. Nor can it store more than its energy capacity at a given time; more importantly, when it stores excess energy generated by variable generators and delivers it at later hours. The time dynamics become more important because (i) the benefit of storage in performing the task depends on temporal matching between demand and the intermittent renewable system output and (ii) the amount of energy that it can deliver/store depends both on its power capacity and energy capacity. Therefore, the model should have the flexibility to capture the required storage design. Together with this comes the importance of dispatch flexibility. The foregoing storage studies show that the storage energy capacity requirements depend on how they are dispatched from day to day. To find the optimal design, one has to capture the optimal dispatch performance under certain operational policy. But as shown above, the present operational policies appear to undermine the value of storage.

2. **Capturing complementarities of renewable resources**: Wind and solar technologies output have the tendency to complement one another. Consequently, at a given condition, a large wind–solar hybrid system can achieve higher penetration than any one of them as a stand-alone [15]. In addition, as shown in Figure 9.2, the hybrid system output profile appears to have a smother output that almost fits better to the load profile as compared to wind or solar output. This property may help in mitigating the cost incurred due to variability of one of them. Most importantly, we found that their complementarities could also reduce the storage requirement. Table 9.2 presents a comparison of storage requirement to reach a comparable penetration level. The table presents the result of the runs performed to assess how the complementarities of wind and solar technology affect storage requirements at a certain level of penetration by artificially setting a solar (or wind) technology in a particular scenario to a constant fraction of the total variable capacity. Table 9.3 reveals that storage required to reach almost the same level of penetration is lower for a hybrid wind–solar system than either one of them as a standalone. We present two tables to demonstrate this phenomenon at different levels of penetration as storage requirement may also depend on whether the point of discussion is above or below the corresponding turning point of that scenario.

   These tables show that a hybrid system requires one of the lowest storage levels to achieve almost the same penetration. This may be because of the improved demand matching capability of the hybrid system. This multiple benefit of complementarities can be measured only if (i) we have the ability to capture the year-round compatibility of these resources to the load profile and (ii) we are able to make proper valuation of ramping cost in our models.

3. **Flexible operational policy**: The present power market does not allow any kind of energy curtailment. In the future grid where spring excess generation becomes common, we may need to have a policy that separates excess renewable generation time from undergeneration time. Unlike the present-day grid, in which excess energy is very low, the massive excess generation of the future grid may require us to design the market in a way that motivates safe curtailing or storage service or both to maintain grid stability rather than keeping online conventional spinning reserves. The above study also shows that the implementation of storage and some curtailment significantly reduces the conventional backup capacity requirement. As a result, developing a model that can test such operational policy scenarios could help in examining the cost of dumped energy plus storage system versus avoided investment in the conventional capacity that would have been built under storage without dumping. Alternatively it could be useful to investigate the cost of energy under a largely storage and renewable energy system that can dump some energy versus an equally decarbonized grid with present-day operational policy.

4. **Complementarities of an grid operation**: High penetration of an intermittent renewable system requires the ability of conventional backup system to substitute for the shortcomings of the intermittent renewable system plus storage technologies. This will require that we have a significant number of units that can be online at short notice and have the capability to do as many on/off cycles as necessary. This operational complementarity requires that the models have the ability to build and operate power plants that have this capability, constrained by their number of on/off cycles, minimum up/down time, ramp rate and range if any, and so on. Even if we have a baseload unit that should continuously be online, variable renewables' ability to supply substantial part of the remaining load depends on our ability to transition to such a kind of operational strategy.

Table 9.2 Network storage requirement to reach a grid penetration of about 52.2 ($\pm$0.5)% of annual demand at different level of wind–solar capacity mix. The hybrid system is defined as a fraction of capacity, even though molecular weight of wind is not equivalent to that of solar. Each run was made using the SEUL model

Total capacity (GW)	Penetration (% of annual demand)	Network energy capacity (GWh)	Network power capacity (GW)	WIND (% of total capacity)
65.5	52.6	4730.6	44.4	100
58.2	51.9	424.4	35.8	75
58.2	52.2	232.2	20.6	65
58.2	52.2	154.2	18.1	50
58.2	52.1	153.8	18.4	45
63.0	52.0	152.4	16.0	25
67.6	51.9	322.3	29.0	0

Table 9.3 Network storage requirement to reach a grid penetration of about 38 ($\pm$0.5)% of annual demand at different levels of wind–solar capacity mix. The hybrid system is defined as a fraction of the total capacity, even though a molecular weight of wind is not equivalent to that of solar. Each run was made using the SEUL model

Total capacity (GW)	Penetration (% of annual demand)	Network energy capacity (GWh)	Network power capacity (GW)	WIND (% of total capacity)
41	37.6	303.8	22.3	100
41	38.3	15.5	3.7	77
43	37.8	14.0	3.3	50
45	37.9	30.5	5.6	25
47	38.3	93.6	12.8	0

9.5   List of Standards Related to Integration of Renewable Resources

Here, we present standards that are relevant to grid integration of renewable resources. Note that these standards and several other closely related standards are presented in the previous chapters.

Standard name	Short description
IEC 61850 suite	Substation automation
IEC 61400-25-1	Communications for monitoring and control of wind power plants
Underwriters laboratories	Standard for safety inverters, converters, controllers, and interconnection system equipment for use with distributed energy resources
IEEE 1547 suite	Standards for interconnecting distributed resources with electric power grid

The IEC 61850 is discussed in detail in Chapter 3. It defines communications within transmission and distribution substations for automation and protection [29–33, 44, 45]. It is also being extended to cover communications beyond the substation to integration of distributed resources and between substations. As a result, the IEC 61400-25-1 is considered part of the IEC 61850.

The family of standards included in the "IEEE 1547 Suite" deals with physical and electrical interconnections between utilities, distributed generation (DG), and storage [34–41]. This standard is under revision to include energy storage interconnections and further details on this effort can be found in Chapter 4. It supplements, Underwriters Laboratories (UL) 1741, "Standard for Safety Inverters, Converters, Controllers, and Interconnection System Equipment for Use With Distributed Energy Resources" protocol when it is applied to utility interactive devices [42]. The IEEE 1547 Suite is also discussed in Chapter 4 together with other standards relevant to energy storage technology.

Several new approaches related to variable renewable integration, such as the use of storage, energy curtailment, forecasting tools, and accuracy are currently under deliberation. There is a good possibility that some sort of new criteria or even standards may be developed in the future.

9.6 Conclusion and Recommendations

Many studies have shown the possibility of future grid that accommodate energy from large intermittent renewable resources. The inclusion of large intermittent renewable systems significantly changes the nature of power system composition as compared to the present one. Technologies such as energy storage are expected to play a significant role in mitigating the impact of their variability and at the same time add some capacity to the future generation need. However, we have also seen that the storage need depends on the diversity of the local resources and their matching capability to the local demand profile. Depending on these factors, among other things, we have seen that the maximum storage energy capacity need is of the order of the daily average demand.

We have also seen that the performance of different technologies/measures that may help increase the compatibility of the renewable resources to the electricity grid depends on the nature of the operational policy that we implement. In transitioning to a future renewable energy grid, it is important that we have the tools that may help us evaluate diverse physical and operational policy scenarios of the future grid. This is important because the performance of some technologies may depend on the future operational rules as it also depends on technological advancements and their cost. For example, depending on the level of the energy curtailment that we allow, storage technology was shown to increase the grid penetration of variable renewable resources, while reducing the required conventional backup capacity need for meeting the hourly demand throughout the year. At the same time, complementarities between solar and wind technologies were shown to have the capacity to increase grid penetration and result in a smoothed out composite output that matches to the load profile better than either of the solar/wind technologies alone. In addition, it was shown that this also reduces the required storage to reach the same penetration. This indicates that designing an efficient and lowest-cost grid requires the capability to bring together those values and measure how they can be used gainfully in the future market.

References

[1] BP www.bp.com/extendedsectiongenericarticle.do?categoryId=9041560&contentId=7075261 (accessed 9 November 2012).
[2] Wood, A.J. and Wollenberg, B.F. (1984) Power Generation, Operation and Control, John Wiley & Sons, Inc., New York.
[3] Kirby, B. and Milligan, M. (2005) A Method and Case Study for Estimating the Ramping Capability of a Control Area or Balancing Authority and Implications for Moderate or High Wind Penetration, Wind Power, Denver, CO.
[4] Ter-Gazarian, A. (1994) Energy Storage for Power Systems, Peter Peregrinus Ltd, London.

[5] Chowdhury BH, Rahman S. A review of recent advances in economical dispatch. *IEEE Transactions in Power Systems.* **5**(4)(1990):1248–1259.

[6] Archer C, Jacobson M. Supplying baseload power and reducing transmission requirements by interconnecting wind farms. *Journal of Applied Meteorology and Climatology.* **46**(11)(2007):1701–1717.

[7] Gouveia EM, Matos MA. Evaluating operational risk in a power system with a large amount of wind power. *Electric Power Systems Research* **79**(5)(2009):734. doi: 10.1016/j.epsr.2008.10.006

[8] Holttinen, H. (2008) Estimating the impacts of wind power on power systems - summary of IEA wind collaboration. *Environmental Research Letters*, (3), 025001, (pp6).

[9] NREL *Renewable Electricity Futures Study*, www.nrel.gov/analysis/re_futures/ (accessed 9 November 2012).

[10] Solomon, Abebe Asfaw, Faiman, D. and Meron, G. (2012) The role of conventional power plants in a grid fed mainly by PV and storage and the largest shadow capacity requirement. *Energy Policy*, **48**, 479–486.

[11] Solomon, Abebe Asfaw, Faiman, D. and Meron, G. (2012) Appropriate storage for high-penetration grid-connected photovoltaic plants. *Energy Policy*, **40**, 335–344.

[12] Solomon, Abebe Asfaw, Faiman, D. and Meron, G. (2010) Properties and uses of storage for enhancing the grid penetration of very large scale photovoltaic systems. *Energy Policy*, **38**, 5208–5222.

[13] Solomon, Abebe Asfaw, Faiman, D. and Meron, G. (2010) The effects on grid matching and ramping requirements, of single and distributed PV systems employing various fixed and sun-tracking technologies. *Energy Policy*, **38**, 5469–5481.

[14] Solomon, Abebe Asfaw, Faiman, D. and Meron, G. (2010) An energy-based evaluation of the matching possibilities of very large photovoltaic plants to the electricity grid: Israel as a case study. *Energy Policy*, **38**, 5457–5468.

[15] Solomon, Abebe Asfaw, Faiman, D. and Meron, G. (2010) Grid matching of large-scale wind energy conversion systems, alone and in tandem with large-scale photovoltaic systems: an Israeli case study. *Energy Policy*, **38**, 7070–7081.

[16] Nelson, J., Johnston, J., Mileva, A. *et al.* (2012) High-resolution modeling of western North American power system demonstrates low-cost and low-carbon futures. *Energy Policy*, **43**, 436–447.

[17] Fripp, M. (2012) Switch: a planning tool for power systems with large shares of intermittent renewable energy. *Environmental Science and Technology*, **46** (11), 6371–6378.

[18] Wei, M., Nelson, J.H., Ting, M., and Yang, C. (2012) *California's Carbon Challenge: Scenarios for Achieving 80% Emissions Reduction in 2050*, http://eaei.lbl.gov/sites/all/files/california_carbon_challenge_feb20_20131.pdf (accessed 30 January 2013), http://eaei.lbl.gov/sites/all/files/california_carbon_challenge_feb20_20131.pdf permission (accessed 5 December 2013).

[19] ECF (European Climate Foundation) (2010) *Roadmap 2050: A Practical Guide to a Prosperous, Low-Carbon Europe*, European Climate Foundation, The Hague, www.roadmap2050.eu/ (accessed 9 November 2012).

[20] Denholm, P. and Margolis, R.M. (2007) Evaluating the limits of solar photovoltaics (PV) in electric power systems utilizing energy storage and other enabling technologies. *Energy Policy*, **35** (9), 4424–4433. doi: 10.1016/j.enpol.2007.03.004

[21] Denholm, P. and Margolis, R.M. (2007) Evaluating the limits of solar photovoltaics (PV) in traditional electric power systems. *Energy Policy*, **35** (5), 2852–2861doi: 10.1016/j.enpol.2006.10.014.

[22] Moore, L.M. and Post, H.N. (2008) Five years of operating experience at a large, utility-scale photovoltaic generating plant. *Progress in Photovoltaics: Research and Applications*, **16** (3), 249–259.

[23] Dany G. (ed.) (2001) Power reserve in interconnected systems with high wind power production. *IEEE Porto Power Tech Conference, Porto, Portugal, September 10–13, 2001.*

[24] Xu, Z., Gordon, M., Lind, M. and Ostergaard, J. (2009) Towards a Danish power system with 50% wind-smart grids activities in Denmark. *IEEE Xplore.* doi: 10.1109/PES.2009.5275558

[25] Parsons, B, Ela, E., Holttinen, H. *et al.* (2008) Impacts of large amounts of wind power design and operation of power systems; results of IEA collaboration. *AWEA Windpower 2008, Houston, TX*.

[26] Katzenstein, W. and Apt, J. (2012) The cost of wind power variability. *Energy Policy*, **51**, 233–243.

[27] Ihle, J. (2003) *Coal Wind Integration Strange Bed Follows may Provide a New Supply Option: The PR&C Renewable Power Service*, PRC-3, 2003.

[28] Lefton, S. and Besuner, P. (2006) The cost of cycling coal fired power plants. *Coal Power Magazine*, Winter, 16–20.

[29] IEC IEC 61850–2. *Communication Networks and Systems in Substations - Part 2: Glossary*, International Electrotechnical Commission.

[30] IEC IEC 61850–3. (2003) *Communication Networks and Systems in Substations - Part 3: General Requirements*, International Electrotechnical Commission.

[31] IEC IEC 61850–4. (2003) *Communication Networks and Systems in Substations - Part 4: System and Project Management*, International Electrotechnical Commission.

[32] IEC IEC 61850–5. (2003) *Communication Networks and Systems in Substations - Part 5: Communication Requirements for Functions and Device Models*, International Electrotechnical Commission.

[33] IEC IEC 61850–6. (2003) *Communication Networks and Systems for Power Utility Automation - Part 6: Configuration Description Language for Communication in Electrical Substations Related to IEDs*, International Electrotechnical Commission.

[34] IEC (2003) IEEE 1547–2003. *Standard for Interconnecting Distributed Resources with the Electric Power Systems*, International Electrotechnical Commission.

[35] IEC (2005) IEEE P1547.1-2005. *Standard For Conformance Test Procedures for Equipment Interconnecting Distributed Resources With Electric Power Systems*, International Electrotechnical Commission.

[36] IEC (2008) IEEE P1547.2-2008. *Application Guide for IEEE Std. 1547 Standard for Interconnecting Distributed Resources With Electric Power Systems*, International Electrotechnical Commission.

[37] IEC (2007) IEEE P1547.3-2007. *Guide for Monitoring, Information Exchange, and Control of Distributed Resources Interconnected With Electric Power Systems*, International Electrotechnical Commission.

[38] IEC IEEE P1547.4. (2011) *Draft Guide for Design, Operation, and Integration of Distributed Resource Island Systems with Electric Power Systems*, International Electrotechnical Commission.

[39] IEC IEEE P1547.5. *Draft Technical Guidelines for Interconnection of Electric Power Sources Greater than 10 MVA to the Power Transmission Grid*, International Electrotechnical Commission. (in process of development)

[40] IEC IEEE P1547.6. (2011) *Draft Recommended Practice for Interconnecting Distributed Resources With Electric Power Systems Distribution Secondary Networks*, International Electrotechnical Commission.

[41] IEC IEEE P1547.7. *Draft Guide to Conducting Distribution Impact Studies for Distributed Resource Interconnection*, International Electrotechnical Commission. (in process of development)

[42] UL UL 1471. (2010) *Inverters, Converters, Controllers and Interconnection System Equipment for Use With Distributed Energy Resources*, Underwriters Laboratories.

[43] Leyzerovich, A.S. (2008) Steam Turbines for Modern Fossil – Fuels Power Plants, The Firmont Press, Inc., Lilburn, GA.

[44] IEC IEC 61850–1. (2003) *Communication Networks and Systems in Substations – Part 1: Introduction and Overview*, International Electrotechnical Commission.

[45] IEC IEC 61400-25-1. (2006) *Communications for Monitoring and Control of Wind Power Plants – Overall Description of Principles and Models*, International Electrotechnical Commission.

10

Future of the Smart Grid

10.1   The Premise of the Smart Grid

Electricity is consumed at the time of production. The present power grid is composed of generating technologies that were invented many decades ago but are designed to supply the demand for electricity, at least in the developed world, continuously and reliably. These generators are predominantly fossil-fuel firing power plants, which are largely responsible for the polluting greenhouse gas (GHG) emissions of the electric industry. In the past decade, environmental concerns, coupled with significant technological improvements of the intermittent renewable generators, have increased interest in renewable energy technologies such as wind energy and solar energy. Together with the integration of these variable resources, new challenges to the electricity sector have emerged. Unlike the dispatchable technologies of the legacy grid that perfectly follow the varying customer demand, the new technologies bring additional variability and uncertainties. As a result, depending on their level of penetration and the grid type, utilities need to prepare the necessary resources to maintain the balance between demand and supply. As discussed in Chapter 9, utilities are also expected to confront issues such as energy dumping and limited demand matching the capability of these resources sooner or later. To increase grid matching of variable renewable resources output to the electricity grid, energy storage, which is also a relatively new technology, will have to be used.

Moreover, the traditional electric grid has a tree-like structure, in which one generator or sets of generators are connected to many customers via a high-voltage transmission line. Long-distance high-voltage transmission lines carry electricity from generators to customers, which finally connects to the loads through the branching distribution networks. However, the suitability of solar and wind technologies for distributed application – or the use of smaller electricity generators located close to loads/customer end – has prompted a significant interest in devising grid-connected microgrids. This will introduce a new level of hierarchy to the existing grid. As the prevalence of distributed generators in the electricity grid grows, reliable operation of the current grid can be compromised.

Smart Grid Standards: Specifications, Requirements, and Technologies, First Edition. Takuro Sato,
Daniel M. Kammen, Bin Duan, Martin Macuha, Zhenyu Zhou, Jun Wu, Muhammad Tariq and Solomon Abebe Asfaw.
© 2015 John Wiley & Sons, Ltd. Published 2015 by John Wiley & Sons, Ltd.

On the other hand, electricity demand, especially peak demand, grows year after year. The lack of regulations on the part of small consumers need together with the use of the same flat price per unit of electricity for this group of consumers may have encouraged customers to look for new energy services. Consequently, utilities should plan to meet the expected increase in demand without any control. Despite the environmental concern over fossil fuels being consumed, it is known that today's grid relies on a limited resource and as a consequence its sustainability is in jeopardy. In addition to the way we generate electricity, utilities are making an effort to change the way we use electricity by reducing waste. The above comment is by no means a criticism of the existing grid. Undoubtedly it is one of the most fascinating human achievements and yet, just like any other industry, the electric power sector is also making its own gradual transformation to its own better version.

Nations around the world have taken the initiative to transform the existing grid so as to address the challenges that are described above. The transformation is aspired to result in real-time bidirectional communications within a network and between networks. This information flow is expected to lead to better operational efficiency through eased fault detection and remedy, reduced load, better maintenance, demand control, and energy-efficiency measures. The efficient and reliable operation of the grid that comes as a result of this automation may bring significant benefits, especially by allowing customer participation, facilitating use of dynamic pricing, facilitating the integration of intermittent renewable energy systems, and so on. These promising features of the Smart Grid may help address some of the potential barriers/problems discussed above.

Transitioning to a Smart Grid will not be free and yet the electricity services need to be affordable. This will require making thoughtful strategic choices during the design process of the Smart Grid. Most Smart Grid programs appear to emphasize communication technologies, dynamic pricing, distribution grid, and policies related to them. However, as noted in Chapter 9, depending on geographic location, the future grid composition and its operational policy could be significantly different from the present one. In this chapter, we will argue for attentive Smart Grid design and a proper comprehensive policy framework that will guide the present initiatives toward the aspired capability of addressing multiple challenges through the Smart Grid framework.

In the following section, we will briefly describe what the Smart Grid should actually be delivering and the potential means of reaching that target. That will be followed by a discussion of the potential challenges of the Smart Grid. Finally, we will present some suggestions on the pathways of the Smart Grid in Section 10.4 before providing our conclusion in the last section of this chapter.

10.2   What the Smart Grid Should Deliver

In the foregoing section, we have discussed some of the barriers of the existing grid. The Smart Grid is anticipated to address those caveats. That means, together with being as reliable and affordable as the legacy grid, the future version should be more flexible, clean, and sustainable. In this section, we will briefly present how the Smart Grid could deliver most of the aspired qualities.

10.2.1   Clean Electricity

Decarbonizing the existing grid requires no less than abandoning our addiction to the cheap polluting fossil fuels and consequently existing fossil fuel burning power generators must gradually be replaced with the cleaner alternatives such as biomass firing generators, fossil-fuel generators with carbon capture and sequestrations, wind technologies, solar technologies, geothermal, hydropower, and possibly nuclear power [1–4]. Chapter 2 of this book provides more detailed information on the renewable electricity generating technologies. As discussed in Chapter 9 of this book, many regions have shown the presence of multiple paths to a decarbonized future grid. We would also like to note that energy storage together with curtailing limited excess variable energy could also produce another potential alternative path via increased use of intermittent renewable energy.

On the other hand, it was also found that reducing the carbon emissions of the electric industry alone could not adequately reduce the overall emissions as aspired [1, 2]. As a result, researchers have concluded that electrifying the transportation industry and some heat load would help drive the emission reduction further [1, 2]. This indicates that the electric industry should become the most dependable low-carbon energy infrastructure of the future.

10.2.2   System Flexibility

In many ways, today's grid has limited flexibility. Typical examples of its limited flexibility include the following: (i) its generation mix is defined on the basis of the anticipated traditional load following requirements, that is, no consideration of the additional variability caused by intermittent renewable generators; (ii) the amount and quality of information exchange between various parts of the network are poor and yet limited to the supply side; (iii) the power flows mostly in one direction, especially near distribution centers; and (iv) the network lacks demand control mechanism. Its inflexibility is partly what created a multidimensional barrier to grid integration of variable renewable systems.

Improved flexibility should be one of the most important promises of the Smart Grid. Various studies have shown the possible solutions to the inflexibility of the existing grid as regards the challenges of variability [1–10]. As discussed in Chapter 9, various energy-modeling studies performed for different areas of the world show that as variable renewable energy technology penetration increases the required flexibility defines the composition of the future grid. To handle the ramping requirement, the grid is expected to rely on energy storage and demand-side technologies (such as interruptible load), increased transmission interconnections, and increased dispatch flexibility on the part of conventional plants including fossil-fuel generators. Most importantly, the economics will define the dispatch of these technologies in the load-following process.

For example, at a time when the upramping requirement exceeds the plants that are already in operations, one may have to start the peaking units. On the contrary, if the downramping is larger, the excess energy needs to be dumped. But alternatively, if demand responses (DRs) participate in the load-following process, it may offer a cheaper mechanism. A broader description of some of the DR technologies can be found in Chapter 5 of this book. According to Hesser and Succar [11], employing dispatchable demand control mechanism could increase flexibility and enhance the grid penetration of intermittent renewable resources. There are various potential residential, commercial, and industrial DR resources that could be exploited for this purpose [11–13]. According to Kirby [12], loads such as water pumping, irrigation, municipal treatment facilities, thermal storage in large buildings, industrial electrolysis, aluminum smelting, electric vehicle charging, and shale-oil extraction are well suited to benefit during extreme upramps of renewable sources [12, 13]. On the contrary, it could be a good alternative to keeping generators spinning for a long time at the time of wind downramp by providing regulations and ancillary services.

On the other hand, careful design may mitigate the ramping requirement by emphasizing spatial and technological diversity of the intermittent resources [10]. Wind and solar generators have an inverse ramping behavior in the morning and evening, which could significantly reduce the load-following requirements. To take advantage of the complementarities of the two resources, other technologies, such as the conventional generators and energy storage, should have the flexibility to complement wind energy and solar energy at a time when the two resources are not sufficient to meet the load. Consequently, it is very important to examine the load-following and regulation capabilities of systems, under various circumstances, to determine its response to sudden changes that come with renewable systems variability and load. Together with that, it is important to also note that excess generation could occur very often, especially in the spring season, when demand is low [4–10]. This will lead to significant energy curtailment, depending on the renewable system size [4–10]. The above load types that are recommended for DR participation may benefit from such events. The economic benefit of these DR measures may depend on their ability to take advantage of this seasonality as it may also depend on technologies to be built in the future grid. For example, if the future grid is optimized for very high renewable penetration using appropriately sized energy storage, as discussed in Chapter 9 and in [5–7], DR may have limited value.

Maintaining the quality and reliability of the electricity service in the presence of massive distributed technology, variable generators, and DR participations makes real-time exchange of information between customers and various component of the electricity network very important. This is where massive communication infrastructures, software, and hardware, as well as control devices to be built in the Smart Grid play significant roles. Utilities will be able to obtain quality real-time information about customer preferences, renewable generation, transmission loading, and condition of generation and transmission infrastructures. Customers will also

receive information about real-time electricity cost and the condition of the generation. This may give customers the opportunity to manage their electricity bill, while
helping utilities to diffuse customers from network stress time. This communication
infrastructure will also help them to make more accurate forecasts, manage their
excess generation better, optimize their dispatch, and so forth. In addition, as opposed
to the present grid that has no capability to detect faults, the Smart Grid is expected
to detect faults and take remedial action. Such capability may be able to reduce the
worry about the customer-side generators.

10.2.3 Affordable Service

No one would expect that adding information and communication technologies, building new generating technologies, and providing better service could come at no cost.
But, as usual, we expect electricity services to be affordable.

According to various studies exploring options for decarbonizing the power grid,
building the cleaner version of today's grid costs a lot of money over a long period
[1–4]. Nevertheless, many planning studies have shown that under almost all scenarios,
the cost per unit of electricity by the year 2050 does not show a significant change,
from their base case decarbonized scenario [1–3]. Figure 10.1 presents the average
generation by technology and the cost of electricity, for various scenarios discussed in
Chapter 9, in the geographic region of the Western Electricity Coordinating Council
(WECC) of North America as estimated using the SWITCH model [1]. The figure
shows that the cost per unit of electricity varies little between the scenarios except for the
Frozen No Carbon cap case. The Frozen No Carbon cap scenario is a business as usual
extension of the present grid/policy that will consequently rely heavily on the cheap coal
resources by the year 2050. That study did not find a significant difference between
the present-day cost of electricity and the cost of electricity under the decarbonized
scenarios. Figure 10.2 presents the trend of system load and cost of electricity from
the present day to 2050 under the base-case scenario [1]. Note that these estimations
are based on the generator capital cost, their operation and maintenance cost as well as
fuel price. As a result they do not include the entire cost of building the Smart Grid.

A major report by the EPRI (Electric Power Research Institute) shows that building a Smart Grid in the United States would cost between $338 and $476 billion
over a 20-year period [14]. The study also promises a benefit-to-cost ratio of about
2.8–6.0. First, the estimate varies significantly because of the complexity of the technologies discussed together with a Smart Grid. Secondly, according to Felder [15],
the promised benefits are not guaranteed. Thirdly, the estimate does not include the
cost of generation and transmission expansions that we see in the typical planning and
decarbonizing scenario studies. It also does not include customer costs for Smart Grid
appliances. According to Felder [15] to justify large Smart Grid investments regulators must make sure that customers have various choices. In the following sections,
we will discuss some options that could help control the cost.

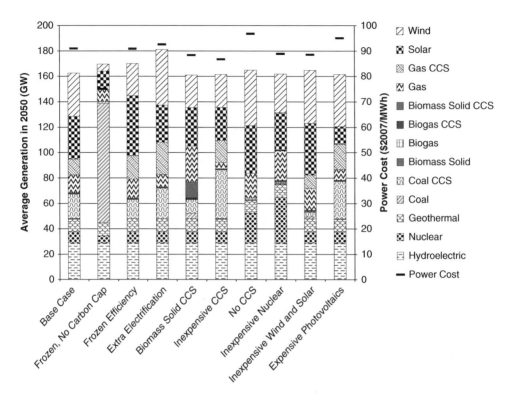

Figure 10.1 Average generation by technology and the average energy cost ($2007/MWh) for the WECC region in the year 2050 under all scenarios as studied by the SWITCH model. The details of these scenarios can be found in Chapter 9. Source [1]. Note that the cost of electricity remains almost the same except for the Frozen No Carbon cap scenario

10.2.3.1 Designing an Optimal and Lowest-Cost Systems

Designing a grid for high penetration of intermittent renewable requires the ability to value the potential benefit of spatial and technological diversity in matching the local demand with high temporal resolution. As discussed in Chapter 9, we have seen that at a very high penetration of intermittent renewable, where storage starts to play a significant role, grid design becomes even more complicated because of the following: (i) the dependence of the storage energy capacity and power capacity interlink on the seasonal and diurnal interaction of intermittent renewable output and load profile; (ii) the surprising benefits that energy dumping brings in increasing storage use, decreasing the conventional backup capacity requirement, and increasing the energy penetration of intermittent renewable resources. These studies have shown the possibility of supplying about 85% of the annual demand from intermittent renewable resources using storage with energy capacity significantly lower than daily average demand and allowing energy dumping of about 20% of the energy. This suggests that we need a different power market rule to the one that we have today. On top of that,

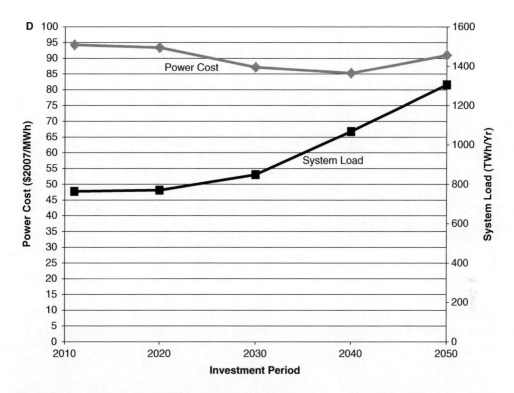

Figure 10.2 Cost of electricity ($2007/MWh) and total system load as a function of investment period for the base-case scenario of the WECC system. Source [1]

it also indicates that designing a proper grid could be one potential way of making an electricity service affordable. Unfortunately, this task appears to be very challenging, as will be discussed below.

The other cost reduction could come from the development of the Smart Grid itself, which promises system automation and modernization of the appliances. Through its automatic communication and control system, the Smart Grid could easily detect and diagnose problems, identify local outage, remotely connect or disconnect accounts, monitor voltage levels, and implement a better means of maintenance. Such advancement could reduce time consumed for the detection and remedy of network fault or local outage, reduce losses and equipment failures, increase equipments lifetime, and so on. These are some of the benefits of the Smart Grid, which might lead to a reduced cost of operation. However, one should bear in mind that it requires deployment of a massive number of sensors. The majority of these communication devices will be going into the distribution centers, which, according to recent EPRI report [14], takes up most of the cost and benefits of the Smart Grid. As suggested by Hauser and Crandall [16], it is important to design a cost-effective Smart Grid if full system automation is not economically feasible.

10.2.3.2 Demand-Side Management

In the past century, utilities were planning to provide whatever energy consumers needed, leading to a significant increase in demand for electricity year after year. But the increase in demand, especially the peak demand, has led to an increased need for generators. According to the projections by the North American Electric Reliability Corporation (NERC), peak demand in the United States (US) will grow from about 800 GW in 2009 to 900 GW in 2018 [17]. The corresponding demand for electricity is also expected to grow from approximately 4000 to 4500 TWh, respectively. Over the past few decades, utilities in the developing world have started to find ways of controlling the increase in demand through DR and efficient energy use. While DR is employed to control the growth in peak demand, energy efficiency would reduce the overall electricity consumption.

Energy Efficiency

Researchers agree that energy efficiency has a significant potential to decrease energy demand while supporting economic growth. According to a study by the American Council for an Energy Efficient Economy (ACEEE), by the year 2020, a 35% larger economy could be supported by 7% less energy if the present, near-commercial semiconductor appliances are employed [18, 19]. The future electricity consumption in the United States under three efficiency scenarios is shown in Figure 10.3. Customer transition to such appliances will lead to decreased electricity bills as they consume less for the services. Such benefits can be obtained if the Smart Grid has the capability to identify and implement the region of opportunities for efficiency [16].

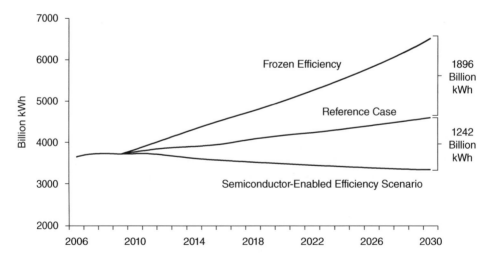

Figure 10.3 The future electricity consumption in the United States under three efficiency scenarios. Reproduced from Journal of Industrial Ecology, Vol. 14, No. 5, Oct 7. 2010, J. A. Laitner, *et al.*, "Semiconductors and Information Technologies", pp. 692–695, Copyright © 2010 by Yale University, with permissions from John Wiley and Sons [19].

Demand Response

In the Smart Grid, DR is expected to be the intrinsic method of grid operation. In the past few years, utilities in the United States reported a reduction of peak demand using incentive-based peak-load reduction DR method. According to a study performed by the FERC (Federal Energy Regulatory Commission), various DR options could reduce the US peak demand by up to 20% [11, 16, 17]. In the Smart Grid, the inherent communication devices as well as customer preference could make more flexible use of the DR measure to the benefit of both utilities and customers. DR may play various roles depending on the local generation capacity mix of the future grid. If the grid relies on a massive variable renewable resources and storage, it could be used to reduce energy loss during excess generation hours, while also driving customers away from the network stress hours that requires putting significant conventional backup capacity online. It also be a tool for increasing grid flexibility, as discussed above. However, there should be a clear metric that allows a proper sharing of the cost and benefits of the Smart Grid between customers and suppliers [15].

10.2.4   Reliable and Sustainable Electricity Grid

At least in the developed world, the existing grid supplies reliably around the clock customer demand. Our increasingly digital economy needs even more reliability and quality. According to Hauser and Crandall [16], the Smart Grid, with its invented dynamic pricing, could allow differentiating customers by level of service quality as opposed to the present broader requirements. It could also improve the sustainability of electricity service via its shift from fuel-based generating technology to renewables such as wind and solar.

10.3   Challenges of the Smart Grid

On the basis of the foregoing discussions, the Smart Grid could simply be considered a hub of heterogeneous technological and policy measures that will make the future grid more efficient, reliable, and clean. This takes, among other things, customers involvement in problem solving, building cleaner energy generation, automating operation, using some sort of DR mechanisms and energy efficiency measures, and devising proper operational and regulation policy. Owing to the inherent extensiveness of the Smart Grid concept, Hauser and Crandall [16] argue that the Smart Grid is a lot more than "Technology." In the next section, we will briefly discuss possible challenges for achieving Smart Grid goals.

10.3.1   Designing for a Broader Purpose

Transforming to smarter, cleaner, and efficient grid needs significant design work. In the foregoing discussion, we have said the grid should be flexible to (i) accommodate significant energy from variable sources; (ii) support or differentiate heterogeneous customers; (iii) allow penetration of new technologies such as electric vehicles and addition of heating loads; (iv) allow bidirectional power flows especially in distribution centers, and so on. The study given in Chapter 9 reveals that designing a grid for very high penetration of renewables opens up a new set of challenges. The challenges include the lack of enough policy tools to assume opportunity cost of energy dumping during planning, computational challenges related to designing hybrid storage systems that can perform power quality and energy services, and the difficulty to see how the year-to-year variation of demand and intermittent renewable resources affect our designing. The variability of the wind energy, solar energy, and load also requires high temporal and spatial resolution modeling. This is further complicated because of the difficulty to correctly predict the impact of expected electrification of transportation and heating loads.

On the other hand, there is no clear boundary on the level of penetration of Distributed Generations (DGs) into the network. DG technologies have the potential to transform the electricity grid to a cleaner system through residential PV (photovoltaic), wind, and biomass technologies. Their potential benefit as compared to the centralized system include their resilience, reducing transmission and distribution loss, and the capability of avoiding investment in transmissions. They are also expected to significantly affect the grid reliability and power quality. To reduce this challenge, organizing nearby DGs into microgrids is considered one solution [19]. Unfortunately, there are still several concerns, inter alia, over its control mechanism and grid connectivity. Wind and PV can also be built as centralized large power plants. These centralized power plants have the potential to produce significantly larger energy as compared to a similar size of DG. Economics suggests building the best resources that better fits the local demand profile. To see how distributed systems perform against a centralized system, we calculated the grid penetration of an intermittent renewable as shown in Figure 10.4. The figure shows the grid penetration of an intermittent renewable for two conditions, that is, 90% solar-distributed generation and 90% solar without any special preference between solar technologies. This model is based on California's 12 load-area system that was described in Section 9.3.1.1. The power transfer between load areas are constrained by the total thermal capacity of the existing transmission between load areas as discussed in that section but no storage is used. We also assume an ideally 100% flexible system that can absorb any energy by DGs or centralized variable renewable sources. The DGs are composed of the residential and commercial PV. The centralized plants include static PV, 1-axis tracking PV, solar thermal without storage, and offshore and onshore wind technologies. As shown in the figure, the grid penetration under the 90% DGs scenario achieves about 10% less penetration than that of the corresponding 90% solar case. The latter system does not build any DGs even if it could have. The main reason is because the energy generated by DG resources is significantly lower even if the PV technologies are the same. Both of scenarios increase penetration by dumping more and more energy as the system size increases. As a side note, we would like to state that, on the basis of the present market, building a central storage with a centralized renewable could lead

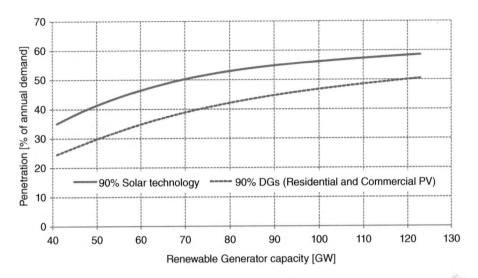

Figure 10.4 Penetration as a function of system size for two scenarios. The model increases system size starting from the reference 41 GW. Note that the model dumps more energy as system size increases, decreasing the value of the technologies

to a more-efficient and lower-cost grid over a grid that emphasizes DGs and distributed storages delivering the same energy. Furthermore, even if we ignore their impact on grid stability and the additional need for controlling devices, we can infer that DGs displace lower-polluting emissions as compared to the equivalent centralized system. But in the future, other policies related to land use and transmission construction could still favor DGs. It is, therefore, important to find the proper DG and centralized grid combination.

Furthermore, unlike the present grid in which the major application of DR is only to reduce peak demand, in the future grid, an automated DR mechanism could lead to various useful applications. Unfortunately, it is very difficult to measure the economic value of those services. What complicates the matter even more is that intermittent renewables have their own seasonal and diurnal output profiles leading to some types of interaction with the local demand because some loads, such as air conditioning and space heating, also have their own seasonal and diurnal need. This could affect their use as DR resources to enhance renewable penetration. This emphasizes the need for thoughtful decisions during our restructuring process.

10.3.2 Operational Challenges

Automation of the power grid will definitely simplify the operational challenge of the present grid. However, significant penetration of new technologies, such as intermittent renewable generators and storage, will complicate the power grid operation. Even if we have the experience to operate energy storage technologies such as pumped hydro and Compressed Air Energy Storage technologies, the present service of storage is very different from that of the future. In the present grid, storage is used to store energy from controllable generators during nighttime low-demand hours and supply the grid during the daytime peak demand hours. There is also little reason for uncertainty except due to load variation, which is also predictable with a very good accuracy. In the future grid, storage will be used to store excess energy from variable renewable generators for use at later hours, that is, when the energy from renewable generators is inadequate to meet the load or is totally unavailable. Note that storage gives a dual benefit to the future grid, that is, by storing the energy in excess of demand and by supplying the stored energy when renewable generation is short of satisfying the load.

To have a reliable supply, the charging and discharging processes of the storage device should be optimal. For such an efficient operation, we need to, inter alia, forecast renewable generations, demand profile, generation in excess of demand, storage state of charge, and also identify potential sites that curtail energy if curtailment is needed. This will require an optimal dispatch model that is capable of showing the potential operation and a set of other software to do the forecasting. The dispatch will have to be updated very often – using real-time data from the latter software or from meters that monitor the storage state of charge, and so on. This will make data handling and frequent decision making more challenging. At the present time, there are no good criteria for energy dumping or storage dispatch but several protocols may be developed in the future. Of course, these protocols and the potential composition of each grid may vary with geographic location. Such a caveat could also lead to several revisions to the present standardization efforts as we acquire more knowledge about the aspired grid. There may also be a need for localized protocols.

The other challenges are related to issues of interoperability of various intelligent devices and software. Each and every component of the Smart Grid must reach efficient communication to minimize human intervention. That requires significant efforts toward creating various standards, which is a major subject of this book, in order to make various components of the grid understand and interpret a given message in the same way, and properly respond if necessary. Further details on this and other technical challenges and possible solutions can be found in a book by Momoh [20]. As was also noted by Momoh [20], operation of the Smart Grid requires engineers and professionals with greater expertise than the present-day skilled technicians.

10.3.3 Policy Challenges

As discussed in Chapter 1 of this book, some legislation supporting the automation of the power grid is being adapted in many countries. Various large projects are also being run/planned to test Smart Grid concepts [21–23]. However, not much progress has been made in the area of clean energy policy, as is anticipated. In addition, despite studies that show many alternative paths to a cleaner power grid, there are no clear specific scenarios. In some cases, having a specific set of choices by itself is difficult as many of the studies that create those scenarios are based on various assumed technological advances and projected generator cost. This indicates that the winning path depends on the regulatory policy and/or the emergence of some disruptive technologies. This has an implication for the Smart Grid designs because the level of complexity may depend on the champion scenarios.

The above challenges may have been further complicated because of the tendency to judge/define the Smart Grid on the basis of technical performance. Even though a focus on the technical performance is not by itself a problem, the policy path for the Smart Grid should have a broader component so as to achieve the aspired reduction in polluting emissions.

10.4 Future Directions

At the present time, many large Smart Grid projects are underway throughout the world. Industries are also doing amazing jobs in developing new smart technologies and projects that will help them make an impact and gain a market [21, 22]. Governments and utilities are also making strategic decisions for Smart Grid development, in particular, and energy generation and use, in general [23]. Here, we would like to stress a few points that those strategic decisions may have to consider and address.

First, depending on the local resources and the corresponding clean energy policy, the future grid could be significantly different from the present one. It may be good to identify a few of the most probable local decarbonized scenarios and perform some heuristic and economic optimization modeling of those potential grids to identify the challenges, advantages, and possible policies that may help achieve that goal.

Secondly, for effective and lowest-cost grid automation and design, it may also be helpful to know how new technologies such as electric vehicles, DRs, and distribution generators contribute to/impact the future grid.

The above activities may help in identifying the algorithms to be developed, the best grid skeletons for the decarbonized grid, inform the type of local and network control framework, and so on. The new generator-side technology, such as wind, solar and storage systems, brings new challenges and opportunities. Our policies must be devised on the basis of a good understanding of their impacts. At present, many studies revolve around trying to accommodate the variable generators into the existing grid. Policies relying on this modeling strategy may fail to achieve very high penetration of these resources. The alternative is to try to optimize the grid for the wind and solar technologies and devise a policy pathway that would help achieve the targeted goal. In other words, we have no means of changing when the sun shines or wind blows but we could use policy measures to motivate the designing and construction of a grid that is more compatible for those resources.

10.5   Conclusion

Significant effort is being made toward the creation of the Smart Grid. Its development can transform the way we generate and use energy. However, significant challenges also exist. The challenges lie in the intricacies related to the heterogeneous Smart Grid concepts and the lack of good policy framework that could guide this effort. Owing to the role that variable generators, such as wind and solar, will be playing the future, grid composition and operation will be different from the present grid. Consequently, it is important to identify the potential characteristics of such a grid and structure the Smart Grid policy that will help transition to what we need.

References

[1] Wei, M., Nelson, J.H., Ting, M., and Yang, C. (2012) *California's Carbon Challenge: Scenarios for Achieving 80% Emissions Reduction in 2050*, http://rael.berkeley.edu/publications /californiaco2report (accessed 9 February 2013).

[2] ECF (European Climate Foundation) (2010) *Roadmap 2050: A Practical Guide to a Prosperous, Low-Carbon Europe*, European Climate Foundation, The Hague, Netherlands, http://www .roadmap2050.eu/ (accessed 9 February 2013).

[3] NREL *Renewable Electricity Futures Study*, http://www.nrel.gov/analysis/re_futures/ (accessed 9 February 2013).

[4] Nelson, J., Johnston, J., Mileva, A. *et al.* (2012) High-resolution modeling of western North American power system demonstrates low-cost and low-carbon futures. *Energy Policy*, **43**, 436–447.

[5] Solomon, Abebe Asfaw, Faiman, D. and Meron, G. (2012) The role of conventional power plants in a grid fed mainly by PV and storage and the largest shadow capacity requirement. *Energy Policy*, **48**, 479–486.

[6] Solomon, Abebe Asfaw, Faiman, D. and Meron, G. (2012) Appropriate storage for high-penetration grid-connected photovoltaic plants. *Energy Policy*, **40**, 335–344.

[7] Solomon, Abebe Asfaw, Faiman, D. and Meron, G. (2010) Properties and uses of storage for enhancing the grid penetration of very large scale photovoltaic systems. *Energy Policy*, **38**, 5208–5222.

[8] Solomon, Abebe Asfaw, Faiman, D. and Meron, G. (2010) The effects on grid matching and ramping requirements, of single and distributed PV systems employing various fixed and sun-tracking technologies. *Energy Policy*, **38**, 5469–5481.

[9] Solomon, Abebe Asfaw, Faiman, D. and Meron, G. (2010) An energy-based evaluation of the matching possibilities of very large photovoltaic plants to the electricity grid: Israel as a case study. *Energy Policy*, **38**, 5457–5468.

[10] Solomon, Abebe Asfaw, Faiman, D. and Meron, G. (2010) Grid matching of large-scale wind energy conversion systems, alone and in tandem with large-scale photovoltaic systems: an Israeli case study. *Energy Policy*, **38**, 7070–7081.

[11] Hesser, T. and Succar, S. (2012) Renewables integration through direct load control and demand response, in *Smart Grid Integrating Renewable, Distributed Generation and Energy Efficiency*, (ed F.P. Sioshansi) Academic Press, pp. 450–494.

[12] Kirby, B.J. (2007) Load response fundamentally matches power system reliability requirements. *IEEE Power Engineering Society General Meeting, June 2007*.

[13] Kirby, B. and Milligan, M. (2010) Utilizing load response for wind and solar integration and power system reliability, *Windpower 2010, Dallas, TX*.

[14] Gellings, C. (2011) *EPRI, Estimating the Costs and Benefits of the Smart Grid: A Preliminary Estimate of the Investment Requirements and the Resultant Benefits of a Fully Functioning Smart Grid*, EPRI.

[15] Felder, F. (2011) The equity implications of smart grid, in *Smart Grid Integrating Renewable, Distributed Generation and Energy Efficiency*, (ed F.P. Sioshansi) Academic Press, pp. 247–275.

[16] S.G. Hauser and K. Crandall. Smart grid is a lot more than just "technology", in *Smart Grid Integrating Renewable, Distributed Generation and Energy Efficiency*, (eds FP Sioshansi) Academic Press, pp. 109–153

[17] North American Electric Reliability Corporation (2009) *Scenario Reliability Assessment*. October 2009.

[18] Laitner, S. (2007) *Assessing the Potential of Information Technology Applications to Enable Economy-Wide Energy-Efficiency Gains*, August 17, 2007.

[19] Platt, G., Berry, A. and Cornforth, D. (2011) What role for microgrids? in *Smart Grid Integrating Renewable, Distributed Generation and Energy Efficiency*, (ed F.P. Sioshansi) Academic Press, pp. 413–449.

[20] Momoh, J. (2012) *Smart Grid Fundamentals of Design and Analysis*. IEEE Press Series on Power Engineering, John Wiley & Sons, Inc, Hoboken, NJ.

[21] www.smartcititeschallenge.org.

[22] SmartGridNews http://www.smartgridnews.com/artman/publish/Key_Players/ (accessed 5 December 2013).

[23] Global Smart Grid Federation www.globalsmartgridfederation.org (accessed 5 December 2013).

List of Standards for the Smart Grid

Smart Grid Standards: Specifications, Requirements, and Technologies, First Edition. Takuro Sato,
Daniel M. Kammen, Bin Duan, Martin Macuha, Zhenyu Zhou, Jun Wu, Muhammad Tariq and Solomon Abebe Asfaw.
© 2015 John Wiley & Sons, Ltd. Published 2015 by John Wiley & Sons, Ltd.

Type of Standard:
A: Power generation
B: Power consumption
C: Power delivery
D: Data exchange
E: Security or safety
F: Electric Storage

Application or service areas	Standard	Brief descriptions	Type	Issuer	Progress state	Issuing date	Location in this book
Renewable Energy Generation (Biomass)	BS EN 14774-1	The standard deals with the determination of moisture content (oven dry method) Part 1 explains the total moisture reference method Part 2 explains the moisture simplified method Part 3 explains the moisture in general analysis sample	A	British Standard Institution (BSI)	Published	2009	Chapter 2
Renewable Energy Generation (Biomass)	BS EN 14775	The standard is related to extracting of ash content	A	BSI	Published	2009	Chapter 2
Renewable Energy Generation (Biomass)	BS EN 14918	The standard is related to extracting of calorific value	A	BSI	Published	2009	Chapter 2
Renewable Energy Generation (Biomass)	BS EN 14961-1	The standard is related to specification and classes of bio fuel (general specifications)	A	BSI	Published	2010	Chapter 2
Renewable Energy Generation (Biomass)	BS EN 15103	The standard is related to bio fuel bulk density	A	BSI	Published	2009	Chapter 2
Renewable Energy Generation (Biomass)	BS EN 15148	The standard deals with extracting of the content of volatile matter	A	BSI	Published	2009	Chapter 2

Category	Standard	Description		Organization	Status	Reference
Renewable Energy Generation (Biomass)	BS EN 15210	The standard deals with extracting of mechanical durability of pellets and briquettes Part 1 is about the determination of the mechanical durability of pellets and briquettes Part 2 is about solid bio fuels	A	BSI	Published 15210-1 (2009) 15210-2 (2010)	Chapter 2
Renewable Energy Generation (Biomass)	CEN/TS 14588	The standard is related to various terminology, definitions, and descriptions of bio fuels	A	European Committee for Standardization (CEN)	Published 2004	Chapter 2
Renewable Energy Generation (Biomass)	CEN/TS 14778	The standard is related to bio fuel sampling Part 1 is about the methods for biomass sampling Part 2 is about methods for sampling particulate material transported in lorries	A	CEN	Published 2005	Chapter 2
Renewable Energy Generation (Biomass)	CEN/TS 14779	The standard is related to bio fuel sampling (methods for preparing sampling plans and sampling certificates)	A	CEN	Published 2005	Chapter 2
Renewable Energy Generation (Biomass)	CEN/TS 14780	The standard is related to the methods for sample preparation	A	CEN	Published 2006	Chapter 2
Renewable Energy Generation (Biomass)	CEN/TS 15104	The standard deals with the determination of total content of carbon, hydrogen, and nitrogen (instrumental methods)	A	CEN	Published 2005	Chapter 2

Application or service areas	Standard	Brief descriptions	Type	Issuer	Progress state	Issuing date	Location in this book
Renewable Energy Generation (Biomass)	CEN/TS 15105	The standard deals with the determination of the water soluble content of chloride, sodium, and potassium	A	CEN	Published	2005	Chapter 2
Renewable Energy Generation (Biomass)	CEN/TS 15149	The standard deals with the methods for the determination of particle size distribution Part 1 is related to the oscillating screen method using sieve apertures of 3.15 mm and above Part 2 is basically related to the vibrating screen method using sieve apertures of 3.15 mm and below Part 3 is related to the particle density of bio fuels	A	CEN	Published	2006	Chapter 2
Renewable Energy Generation (Biomass)	CEN/TS 15210	It is related to the determination of mechanical durability	A	CEN	Published	2005	Chapter 2
Renewable Energy Generation (Biomass)	CEN/TS 15234	This standard is related to quality of bio fuels	A	CEN	Published	2006	Chapter 2
Renewable Energy Generation (Biomass)	CEN/TS 15289	The standard is related to the total content of sulfur and chlorine that are used in bio fuels	A	CEN	Published	2006	Chapter 2
Renewable Energy Generation (Biomass)	CEN/TS 15290	The standard deals with the determination of major elements of bio fuel	A	CEN	Published	2006	Chapter 2
Renewable Energy Generation (Biomass)	CEN/TS 15296	The standard is related to the analysis of different bases in bio fuels	A	CEN	Published	2006	Chapter 2

Category	Standard	Description		Organization	Status	Year	Chapter
Renewable Energy Generation (Biomass)	CEN/TS 15297	The standard is related to the determination of minor elements in bio fuel	A	CEN	Published	2006	Chapter 2
Renewable Energy Generation (Biomass)	CEN/TS 15370-1	The standard is related to the determination of ash melting behavior of bio fuel	A	CEN	Published	2006	Chapter 2
Renewable Energy Generation (Fuel Cell)	IEC 60079-29	Part 1 of IEC 60079 is related to explosive atmospheres, gas detectors, that is, performance requirements of detectors for flammable gases Part 2 deals with detectors selection, installation, use, and maintenance of detectors for flammable gases and oxygen	A	International Electro-technical Commission (IEC)	Published	2007	Chapter 2
Renewable Energy Generation (Fuel Cell)	IEC/TS 62282	Part 1 of IEC 62282 specifies fuel cell terminologies Part 2 specifies different fuel cell modules	A	IEC	Published	62282-1 (2010) 62282-2 (2012)	Chapter 2
Renewable Energy Generation (Fuel Cell)	IEC 62282-3-100	It is related to safety of stationary fuel cell power systems	A	IEC	Published	2012	Chapter 2
Renewable Energy Generation (Fuel Cell)	IEC 62282-3-200	It mainly deals with how to measure the performance of stationary fuel cell power systems designed for residential, commercial, and agriculture systems	A	IEC	Published	2011	Chapter 2
Renewable Energy Generation (Fuel Cell)	IEC 62282-3-3	It is related to installation of stationary fuel cell power systems	A	IEC	Published	2007	Chapter 2

Application or service areas	Standard	Brief descriptions	Type	Issuer	Progress state	Issuing date	Location in this book
Renewable Energy Generation (Fuel Cell)	IEC 62282-5-1	It is related to safety of portable fuel cell appliances	A	IEC	Published	2007	Chapter 2
Renewable Energy Generation (Fuel Cell)	IEC 62282-6-100	It is related to safety of micro fuel cell power systems	A	IEC	Published	2010	Chapter 2
Renewable Energy Generation (Fuel Cell)	IEC/PAS 62282-6-150	It is related to the safety of water reactive (UN Division 4.3) compounds in indirect PEM fuel cells	A	IEC	Published	2011	Chapter 2
Renewable Energy Generation (Fuel Cell)	IEC 62282-6-200	It is related to the performance of micro fuel cell power systems	A	IEC	Published	2007	Chapter 2
Renewable Energy Generation (Fuel Cell)	IEC 62282-6-300	It is related to micro fuel cell power system's fuel cartridge interchangeability	A	IEC	Published	2009	Chapter 2
Renewable Energy Generation (Fuel Cell)	IEC 62282-7-1	It is related to single cell test method for polymer electrolyte fuel cells	A	IEC	Published	2010	Chapter 2
Renewable Energy Generation (Fuel Cell)	ISO 23273	It is related to fuel cell road vehicles safety specifications Part 1 is related to vehicle functional safety Part 2 is related to the protection against hydrogen hazards for vehicles fueled with compressed hydrogen gas Part 3 is related to protection of persons against electric shock	A	International Organization for Standardization (ISO)	Published	2006	Chapter 2

Category	Standard	Description		Organization	Status	Chapter
Renewable Energy Generation (Fuel Cell)	ISO 23828	It is related to fuel cell road vehicle's energy consumption measurement. Part 1 deals with vehicles fueled with compressed hydrogen gas)	A	ISO	Published 2008	Chapter 2
Renewable Energy Generation (Fuel Cell)	ISO/TR 11954	It deals with fuel cell based vehicle's maximum speed measurement	A	ISO	Published 2008	Chapter 2
Renewable Energy Generation (Fuel Cell)	ISO 6469-1	It deals with safety specifications of electrically propelled road vehicles; Part 1 deals with on-board rechargeable energy storage systems (RESSs); Part 2 of this standard is related to vehicle operational safety means and protection against failures; Part 3 specifies methods for protection of persons against electric shock	A	ISO	Published 6469-1 (2009) 6469-2 (2009) 6469-3 (2011)	Chapter 2
Renewable Energy Generation (Fuel Cell)	ISO/TR 8713	It deals with vocabulary of electrically propelled road vehicles	A	ISO	Published 2012	Chapter 2
Renewable Energy Generation (Fuel Cell)	ISO 13985	It deals with liquid hydrogen in land vehicle fuel tanks	A	ISO	Published 2006	Chapter 2
Renewable Energy Generation (Fuel Cell)	ISO/TR 14687-2	It deals with the product specification (Part 2 – PEM fuel cell applications for road vehicles)	A	ISO	Published 2012	Chapter 2

Application or service areas	Standard	Brief descriptions	Type	Issuer	Progress state	Issuing date	Location in this book
Renewable Energy Generation (Fuel Cell)	ISO/PAS 15594	It deals with the airport hydrogen fueling facility operation	A	ISO	Published	2004	Chapter 2
Renewable Energy Generation (Fuel Cell)	ISO 17268	It deals with the compressed hydrogen surface vehicle (refueling connection devices)	A	ISO	Published	2006	Chapter 2
Renewable Energy Generation (Fuel Cell)	ISO/TS 15869	It deals with the gaseous hydrogen blends and hydrogen fuels (land vehicles fuel tanks)	A	ISO	Published	2009	Chapter 2
Renewable Energy Generation (Fuel Cell)	ISO TR 15916	It deals with the basic considerations for the safety of hydrogen systems	A	ISO	Published	2004	Chapter 2
Renewable Energy Generation (Fuel Cell)	ISO 16110-1	It is related to hydrogen generators using fuel processing technologies Part 1 is related to overall safety Part 2 is related to test method and performance	A	ISO	Published	16110-1-1 (2007) 16110-1-2 (2010)	Chapter 2
Renewable Energy Generation (Fuel Cell)	ISO 16111	It is related to transportable gas storage devices (hydrogen absorbed in reversible metal hydrides)	A	ISO	Published	2008	Chapter 2
Renewable Energy Generation (Fuel Cell)	ISO TS 20100	It deals with the service stations of gaseous hydrogen	A	ISO	Published	2008	Chapter 2
Renewable Energy Generation (Fuel Cell)	ISO 22734-1	It deals with hydrogen generators using water electrolysis process	A	ISO	Published	2008	Chapter 2

Part 1 deals with industrial and commercial applications
Part 2 of this standard is related to the residential applications

Category	Standard	Description		Organization	Status	Reference
Renewable Energy Generation (Fuel Cell)	ISO 26142	It deals with the hydrogen detector apparatus (stationary applications)	A	ISO	Published 2010	Chapter 2
Renewable Energy Generation (Fuel Cell)	OIML R 81	It is related to dynamic measuring devices and systems for cryogenic liquids	A	International Organization of Legal Metrology (OIML)	Published 2006	Chapter 2
Renewable Energy Generation (Fuel Cell)	OIML R 139	It is related to metrological and technical requirements of compressed gaseous fuel measuring systems for vehicle	A	OIML	Published 2007	Chapter 2
Renewable Energy Generation (Geothermal)	DIN 8901	The standard is related to refrigerating systems and heat pumps (protection of soil, ground, and surface water)	A	Deutsches Institut für Normung (DIN)	Published 2002	Chapter 2
Renewable Energy Generation (Geothermal)	DVGW W 110	This standard deals with the investigations in bore holes and wells sunk to tap ground water, compilation of methods	A	Deutsche Vereinigung des Gas- und Wasserfaches (DVGW)	Published 2005	Chapter 2
Renewable Energy Generation (Geothermal)	DVGW W 115	The standard is related to well drilling, that is, boreholes for exploration, capture, and observation of groundwater	A	DVGW	Published 2008	Chapter 2

Application or service areas	Standard	Brief descriptions	Type	Issuer	Progress state	Issuing date	Location in this book
Renewable Energy Generation (Geothermal)	DVGW W 116	The standard is related to the use of mud additives in drilling fluids for drilling in groundwater	A	DVGW	Published	1998	Chapter 2
Renewable Energy Generation (Geothermal)	EN 255-3	The standard mainly deals with testing of hot water units like air conditioners, liquid chilling packages, and heat pumps with electrically driven compressors (heating mode)	A	Nederlandse Norm (NEN)	Published	2008	Chapter 2
Renewable Energy Generation (Geothermal)	EN 378	The standard is related to refrigerating systems and heat pumps. Part 1–4 deals with safety and environmental requirements	A	BSI	Published	2008	Chapter 2
Renewable Energy Generation (Geothermal)	EN 14511	The standard is mainly related to air conditioners, liquid chilling packages, and heat pumps with electrically driven compressors for space heating and cooling	A	BSI	Published	2011	Chapter 2
Renewable Energy Generation (Geothermal)	EN 15450	The standard deals with heating systems in buildings (design of heat pump heating systems)	A	BSI	Published	2007	Chapter 2
Renewable Energy Generation (Geothermal)	ISO 5149	The standard is related to mechanical refrigerating systems used for cooling and heating (safety requirements)	A	ISO	Published	1993	Chapter 2

Renewable Energy Generation (Geothermal)	ISO 5151	The standard is related to the non-ducted air conditioners and heat pumps (testing and rating for performance nationalized (e.g., in Great Britain)	A ISO	Published	2010	Chapter 2
Renewable Energy Generation (Geothermal)	ISO 13256	The standard deals with water-source heat pumps (testing and rating for performance). Particularly the rating and testing of the heat pumps that are used in Denmark and Netherland	A ISO	Published	1998	Chapter 2
Renewable Energy Generation (Geothermal)	ONORM M 7755-1	Part 1 of ONORMM7755 standard deals with general requirements for design and construction of heat pump heating systems	A Osterreichische Narm Austria (ONORM)	Published	2003	Chapter 2
Renewable Energy Generation (Geothermal)	ONORM M 7753	It is related to the heat pumps with electrically driven compressors for direct expansion, ground coupled (testing and indication of the producer)	A ONORM	Published	1995	Chapter 2
Renewable Energy Generation (Geothermal)	ONORM M 7755-2+3	It is related to the design and installation of ground source heat pump systems	A ONORM	Published	2000	Chapter 2
Renewable Energy Generation (Geothermal)	OWAV RB 207	It is related to the systems for the exploitation of geothermal heat	A Water and Waste Management Association (OWAV)	Published	2009	Chapter 2

Application or service areas	Standard	Brief descriptions	Type	Issuer	Progress state	Issuing date	Location in this book
Renewable Energy Generation (Geothermal)	VDI 2067 Blatt6	The standard is related to the economy calculation of heat consuming installations of heat pumps	A	Association of German Engineers (VDI)	Published	1989	Chapter 2
Renewable Energy Generation (Geothermal)	VDI 4650 Blatt 1	The standard deals with calculation of heat pumps (short-cut method for the calculation of the annual effort figure of heat pumps)	A	VDI	Published	2003	Chapter 2
Renewable Energy Generation (Geothermal)	VDI 4640 Blatt 1-4	The standard is related to design and installation of heat pump systems. (Thermal use of the underground heat system)	A	VDI	Published	2002	Chapter 2
Renewable Energy Generation (Hydropower)	IEC 61850-7-410	This standard is of high relevance to the Smart Grid. It specifies the additional common data class, the logical nodes, and data objects required for the use of IEC 61850 standard in a hydropower plant	A	IEC	Published	2007	Chapter 2
Renewable Energy Generation (Hydropower)	IEC-EN 61116	The standard applies to installations having outputs of less than 5 MW and turbines with diameters less than 3 m	A	IEC	Published	1995	Chapter 2
Renewable Energy Generation (Hydropower)	IEC 60041	The standard is related to field acceptance tests in order to determine the hydraulic performance of hydraulic turbines, storage pumps, and turbines	A	IEC	Published	1991	Chapter 2

					Published	Chapter
Renewable Energy Generation (Hydropower)	IEC 60193	The standard is related to hydraulic turbines, storage pumps, and pump turbines model acceptance tests	A	IEC	Published 1999	Chapter 2
Renewable Energy Generation (Hydropower)	IEC 60308	The standard specifies international code for testing of speed governing systems for hydraulic turbines	A	IEC	Published 2005	Chapter 2
Renewable Energy Generation (Hydropower)	IEC 60545	The standard is a guide for commissioning, operation, and maintenance of hydraulic turbines	A	IEC	Published 1976	Chapter 2
Renewable Energy Generation (Hydropower)	IEC 60609	Part 1 of the standard is related to cavitations pitting evaluation in hydraulic turbines, storage pumps, and pump turbines Part 2 is a guide for commissioning, operation, and maintenance of storage pumps and of pump turbines operating as pumps (evaluation in Pelton turbines)	A	IEC	Published 60609-1 (2004) 60609-2 (1997)	Chapter 2
Renewable Energy Generation (Hydropower)	IEC 60805	It provides electromechanical equipment guide for small hydroelectric installations	A	IEC	Published 1985	Chapter 2
Renewable Energy Generation (Hydropower)	IEC 60994	The standard is basically a guide for field measurement of vibrations and pulsations in hydraulic machines	A	IEC	Published 1991	Chapter 2
Renewable Energy Generation (Hydropower)	IEC 61116	The standard is a guide to specification of hydraulic turbine control systems	A	IEC	Published 1992	Chapter 2

Application or service areas	Standard	Type	Issuer	Progress state	Issuing date	Location in this book
Renewable Energy Generation (Hydropower)	IEC 61362	A	IEC	Published	1998	Chapter 2
Renewable Energy Generation (Hydropower)	IEC 61364	A	IEC	Published	1999	Chapter 2
Renewable Energy Generation (Hydropower)	IEC 61366	A	IEC	Published	1998	Chapter 2

Standard	Brief descriptions
IEC 61362	It provides guidelines for hydraulic turbine governing system
IEC 61364	It provides specification for hydraulic turbine governing system (nomenclature for hydroelectric power plant machinery)
IEC 61366	The standard is related to hydraulic turbines, storage pumps, and pump-turbines Part 1 specifies general and annexes Part 2 provides guidelines for technical specifications for Francis turbines Part 3 provides guidelines for technical specifications for Pelton turbines Part 4 provides guidelines for technical specifications for Kaplan and propeller turbines Part 5 provides guidelines for technical specifications for tubular turbines Part 6 provides guidelines for technical specifications for pump-turbines

Renewable Energy Generation (Solar energy)	IEC-EN 61427	Part 7 provides guidelines for technical specifications for storage pumps The standard basically provides information about secondary cells and batteries for solar PV energy systems. If storage is used then secondary cells is equal to the rechargeable batteries	A IEC	Published 2005	Chapter 2
Renewable Energy Generation (Solar energy)	IEC-EN 61724	The standard recommends procedures for the monitoring of energy-related PV system characteristics, and for the exchange, and analysis of monitored data. The main objective of designing this standard is to assess the overall performance of PV systems	A IEC	Published 1998	Chapter 2
Renewable Energy Generation (Solar energy)	IEC-EN 61727	The specifications in this standard apply to utility-interconnected PV power systems operating in parallel with the utility and utilizing static non-islanding inverters for the conversion of DC to AC. It lays down requirements for interconnection of PV systems to the utility distribution system	A IEC	Published 1998	Chapter 2

Application or service areas	Standard	Brief descriptions	Type	Issuer	Progress state	Issuing date	Location in this book
Renewable Energy Generation (Solar energy)	IEC/EN 61215	The standard lays down requirements for the design qualification and type approval of terrestrial PV modules suitable for long-term operation, as defined in IEC 60721-2-1. It determines the electrical and thermal characteristics of the module, like its capability of withstanding prolonged exposure in certain climates	A	IEC	Published	2005	Chapter 2
Renewable Energy Generation (Solar energy)	IEC 61646	The IEC 61646 standard lays down requirements for the design qualification and type approval of terrestrial, thin-film PV modules suitable for long-term operation as defined in IEC 60721-2-1. This standard applies to all terrestrial flat plate module materials not covered by IEC 61215. The significant technical change with respect to the previous edition deals with the pass/fail criteria	A	IEC	Published	2008	Chapter 2

Renewable Energy Generation (Solar energy)	IEC/EN 61730	The standard describes the fundamental construction requirements for PV modules in order to provide safe electrical and mechanical operation during their expected lifetime. It addresses the prevention of electrical shock, fire hazards, and personal injury due to mechanical and environmental stresses. It pertains to the particular requirements of construction and is to be used in conjunction with IEC 61215 or IEC 61646	A	IEC	Published	2007	Chapter 2
Renewable Energy Generation (Solar energy)	IEC 60891	IEC 60891 defines procedures to be followed for temperature and irradiance corrections to the measured I–V (Current–Voltage) characteristics of PV devices. It also defines the procedures used to determine factors relevant for these corrections. Requirements for I–V measurement of PV devices are laid down in IEC 60904-1	A	IEC	Published	2009	Chapter 2
Renewable Energy Generation (Solar energy)	IEC 60904-1	The standard is related to measurement of PV current–voltage characteristics	A	IEC	Published	2006	Chapter 2

Application or service areas	Standard	Brief descriptions	Type	Issuer	Progress state	Issuing date	Location in this book
Renewable Energy Generation (Solar energy)	IEC 61194	The standard defines major electrical, mechanical, and environmental parameters for the description and performance analysis of stand-alone PV systems	A	IEC	Published	1992	Chapter 2
Renewable Energy Generation (Solar energy)	IEC 61215	It is related to crystalline silicon terrestrial PV modules (mainly deals with the design qualification and type approval)	A	IEC	Published	2005	Chapter 2
Renewable Energy Generation (Solar energy)	IEC 61345	The standard determines the ability of a PV module to withstand exposure to the UV radiation from 280 to 400 nm	A	IEC	Published	1998	Chapter 2
Renewable Energy Generation (Wind energy)/Integration of Renewable Resources	IEC 61400	Part 1 is related to design requirements of wind turbines Part 2 is related to design requirements for small-scale wind turbine systems Part 11 of this standard deals with the acoustic noise measurement techniques Part 13 deals with measurement and assessment of power quality characteristics of grid connected wind turbines	A	IEC	Published	61400-1 (2005) 61400-2 (2006) 61400-11 (2006) 61400-13 (2002)	Chapter 2/ Chapter 9

Category	Standard	A	Org	Description	Status	Chapter
				Part 21 deals with the full-scale structural testing of rotor blades. Part 22 is related to wind turbine's conformity testing and certifications. Part 23 is related to wind turbine generator systems. Specifically, it deals with full scale structural testing of rotor blades	61400-21 (2008) 61400-22 (2005) 61400-23 (2002)	
Renewable Energy Generation (Solar energy)	IEC 61427	A	IEC	It provides recommendations for secondary cells and batteries for photovoltaic energy systems (PVES). It deals with general requirements and methods of test revision to include latest battery technology	Published 2005	Chapter 2
Renewable Energy Generation (Solar energy)	IEC 61646	A	IEC	The standard is related to thin-film terrestrial PV modules (mainly design qualification and type approval)	Published 2008	Chapter 2
Renewable Energy Generation (Solar energy)	IEC 61701	A	IEC	The standard is related to salt mist corrosion testing of PV modules	Published 2011	Chapter 2
Renewable Energy Generation (Solar energy)	IEC 61730-1	A	IEC	The standard defines PV module safety qualification (Part 1 requirements for construction). In this standard, some amendments have been done and published in 2010	Published 2004	Chapter 2

Application or service areas	Standard	Brief descriptions	Type	Issuer	Progress state	Issuing date	Location in this book
Renewable Energy Generation (Solar energy)	IEC 61702	The standard defines predicted short-term characteristics of direct coupled PV water pumping systems	A	IEC	Published	1995	Chapter 2
Renewable Energy Generation (Solar energy)	IEC 61829	It describes procedures for onsite measurement of crystalline silicon PV array characteristics and for extrapolating these data to standard test conditions (STC) or other selected temperatures and irradiance values	A	IEC	Published	1995	Chapter 2
Renewable Energy Generation (Solar energy)	IEC 61853	It defines PV module performance testing and energy rating Part 1 is related to irradiance and temperature performance measurements and power rating Part 2 spectral response, incidence angle, and module operating temperature measurements	A	IEC	Published	2011	Chapter 2
Renewable Energy Generation (Solar energy)	IEC/TS 62257	The standard provides recommendations for small renewable energy and hybrid systems for rural electrification Part 1 is about general introduction to rural electrification	A	IEC	Published		Chapter 2

Part 2 of this standard is from requirements to a range of electrification systems	62257-1 (2003)
Part 3 is related to project development and management	62257-2 (2004)
Part 4 is related to the system selection and design	62257-3 (2004)
Part 5 is related to protection against electrical hazards	62257-4 (2005)
Part 6 is related to acceptance, operation, maintenance, and replacement	62257-5 (2005)
Part 7 of this standard is related to generators	62257-6 (2005)
Part 7-1 is related to generators – PV arrays	62257-7 (2008)
Part 7-3 is related to generator set, that is, selection of generator sets for rural electrification systems	62257-7-1 (2010)
Part 8-1 mainly deals with the selection of batteries and battery management systems for stand-alone electrification systems. In addition, there are specific cases of automotive flooded lead-acid batteries, which is available in developing countries	62257-7-3 (2008)
Part 9-1 of IEC/TS 62257 is related to PV micropower systems	62257-8-1 (2007)

Application or service areas	Standard	Brief descriptions	Type	Issuer	Progress state	Issuing date	Location in this book
		Part 9-2 of IEC/TS 62257 is also related to PV micropower systems (microgrids)				62257-9-1 (2008)	
		Part 9-3 of IEC/TS 62257 is related to integrated system (user interface)				62257-9-2 (2006); 62257-9-3 (2006)	
		Part 9-4 of IEC/TS 62257 is related to integrated system (user installation)				62257-9-4 (2006)	
		Part 9-5 mainly deals with integrated system specifically in selection of portable PV lanterns for rural electrification projects				62257-9-5 (2007)	
		Part 9-6 is related to integrated system; specifically it deals with the election of photovoltaic individual electrification systems (PV-IES)				62257-9-6 (2008)	
		Part 12-1 deals with the selection of Self-Ballasted Compact Fluorescent Lamps (CFL) for rural electrification systems and recommendations for household lighting equipment				62257-12-1 (2007)	
Renewable Energy Generation (Solar energy)	IEC62108	It specifies characteristics, installation, and relevant information for CSP solar cells	A	IEC	Published	2007	Chapter 2

Domain	Standard	Description		Organization	Status	Year	Reference
Renewable Energy Generation (Wind energy)	AGMA 6006-A03	The standard supersedes AGMA 921-A97. It is developed for design and specification of gearboxes of wind turbines	A	American Gear Manufacturers Association (AGMA)	Published	2003	Chapter 2
Renewable Energy Generation (Wind energy)	BSI BS EN 45510-5-3	Part 5-3 of this standard is a guide for procurement of power station equipment	A	BSI	Published	1998	Chapter 2
Renewable Energy Generation (Wind energy)	BSI BS EN 50308	The standard is a guide of protective measures requirements for design, operation, and maintenance of wind turbines	A	BSI	Published	2004	Chapter 2
Renewable Energy Generation (Wind energy)	BSI PD CLC/TR 50373	The standard is related to wind turbines electromagnetic compatibility	A	BSI	Published	2004	Chapter 2
Renewable Energy Generation (Wind energy)	BSI BS EN 61400-12	Part 12 of this standard is related to generator systems of wind turbine's power performance testing	A	BSI	Published	2006	Chapter 2
Renewable Energy Generation (Wind energy)	BSI PD IEC WT 01	The standard defines rules and procedures for conformity testing and certification of wind turbines	A	IEC	Published	2001	Chapter 2
Renewable Energy Generation (Wind energy)	CSA F417-M91-CAN/CSA	The standard defines the performance and general instructions of wind energy conversion systems	A	Canadian Standards Association (CSA)	Published	1991	Chapter 2
Renewable Energy Generation (Wind energy)	DIN EN 61400-25-4	Part 25-4 of 61400 is related to communications for monitoring and controlling of wind power plants	A	DIN	Published	2009	Chapter 2

Application or service areas	Standard	Brief descriptions	Type	Issuer	Progress state	Issuing date	Location in this book
Renewable Energy Generation (Wind energy)	DS DS/EN 61400-12-1	Part 12-1 of 61400 is related to wind turbines' performance measurements of electricity production	A	Danish Standards (DS)	Published	2009	Chapter 2
Renewable Energy Generation (Wind energy)	DNV DNV-OS-J101	This is a design of off-shore wind turbines structures	A	Det-Norske-Veritas (DNV)	Published	2011	Chapter 2
Renewable Energy Generation (Wind energy)	GOST R 51237	The standard defines terms and conditions of nontraditional power engineering	A	Euro-Asian Council for Standardization, Metrology, and Certification (EASC)	Published	1998	Chapter 2
Renewable Energy Generation (Wind energy)	IEC 60050-415	Part 415 of IEC 60050 is related to wind turbine generator systems' international electrotechnical vocabulary	A	IEC	Published	1999	Chapter 2
Advanced Distribution Management	IEC 60834	*Teleprotection equipment of power systems – performance and testing* Part 1: Command systems Part 2: Analog comparison system	D	IEC	Published	1995–2003	Chapter 3
Advanced Distribution Management	IEC 60870-5	*Telecontrol equipment and systems – Part 5: Transmission protocols* Part 5-101: Transmission protocols, companion standards especially for basic telecontrol tasks	D	IEC	Published	1990–2006	Chapter 3

Advanced Distribution Management	IEC 60870-6	*Part 5-102: Companion standard for the transmission of integrated totals in electric power systems*	D	IEC	Published 1990–2006	Chapter 3
		Part 5-103: Transmission protocols, companion standard for the informative interface of protection equipment				
		Part 5-104: Transmission protocols, network access for IEC 60870-5-101 using standard transport profiles				
		Telecontrol equipment and systems – part 6: inter-control center communications				
		Part 6-1: Application context and organization of standards				
		Part 6-2: Use of basic standards (Open System Interconnection (OSI) layers 1–4)				
		Part 6-501: TASE.1 service definitions				
		Part 6-502: TASE.1 protocol definitions				
		Part 6-503: TASE.2 services and protocol				
		Part 6-504: TASE.1 user conventions				

Application or service areas	Standard	Brief descriptions	Type	Issuer	Progress state	Issuing date	Location in this book
		Part 6-601: Functional profile for providing the connection-oriented transport service in an end system connected via permanent access to a packet switched data network					
		Part 6-602: TASE transport profiles					
		Part 6-701: Functional profile for providing the TASE.1 application service in end systems					
		Part 6-702: Functional profile for providing the TASE.2 application service in end systems					
Advanced Distribution Management	IEC 61968	Common information model (CIM)/distribution management	D	IEC	Published	2004–2012	Chapter 3
		Part 1: Interface architecture and general requirements					
		Part 2: Glossary					
		Part 3: Interface for network operations					
		Part 4: Interfaces for records and asset management					
		Part 9: Interface standard for meter reading and control					

		Part 11: Common Information Model (CIM) extensions for distribution				
		Part 13: Common information model (CIM) RDF model exchange format for distribution				
Distributed Energy Resources	IEC 60255-24	*Electrical relays – part 24: Common Format for Transient Data Exchange (COMTRADE) for power systems*	D	IEC	Published 2001	Chapter 3
Distributed Energy Resources	IEC 61400-25	*Wind turbines – part 25: communications for monitoring and control of wind power plants*	D	IEC	Published 2006–2012	Chapter 3
Distributed Energy Resources	IEC 61954	*Power electronics for electrical transmission and distribution systems*	C	IEC	Published 2011	Chapter 3
Energy Management System	IEC 61970	*Common Information Model (CIM)/energy management*	D	IEC	Published 2005–2009	Chapter 3
		Part 1: Guidelines and general requirements				
		Part 2: Glossary				
		Part 3: Common Information Model (CIM)				
		Part 4: Specific Communication Service Mapping (SCSM)				
		Part 501: Specific Communication Service Mapping (SCSM) Common Information Model (CIM) XML codification for programmable reference and model data exchange				

Application or service areas	Standard	Brief descriptions	Type	Issuer	Progress state	Issuing date	Location in this book
Smart Substation Automation/Integration of Renewable Resources	IEC 61850	*Substation automation* *Part 1: Introduction and overview* *Part 2: Glossary* *Part 3: General requirements* *Part 4: System and project management – Ed.2* *Part 5: Communication requirements for functions and device models* *Part 6: Configuration language for communication in electrical substations related to IEDs – Ed.2* *Part 7: Basic communication structure for substation and feeder equipment* *Part 8: Specific Communication Service Mapping (SCSM)* *Part 9: Specific Communication Service Mapping (SCSM)* *Part 10: Conformance testing*	C	IEC	Published	2003– 2010	Chapter 3/ Chapter 9

Distributed Energy Source	IEC 61850-420	*Communication networks and systems for power utility automation-part 7-420: basic communication structure-distributed energy resources logical nodes*: this standard addresses the IEC 61850 information modeling for DER. The IEC 61850 information models for DER are to be used in the exchange of information with DER including reciprocating engines, fuel cells, microturbines, photovoltaics, combined heat and power, and energy storage, and so on	F	IEC	Published 2009	Chapter 4
Electric Storage/Distributed Energy Source/Interoperability/ Renewable Energy Generation (Wind Energy)/Integration of Renewable Resources	IEEE 1547	*Standard for interconnecting distributed resources with electric power systems*	F/D/A	Institute of Electrical and Electronic Engineers (IEEE)	Published	Chapter 4/ Chapter 8/ Chapter 2/ Chapter 9
		Part 1: Standard for conformance tests procedures for equipment interconnecting distributed resources with electric power systems		1547-1 (2005)		

Application or service areas	Standard	Brief descriptions	Type	Issuer	Progress state	Issuing date	Location in this book
		Part 2: Application guide for IEEE 1547 standard for interconnecting distributed resources with electric power systems				1547-2 (2008)	
		Part 3: Guide for monitoring, information exchange, and control of distributed resources interconnected with electric power systems				1547-3 (2007)	
		Part 4: Guide for design, operation, and integration of distributed resource island systems with electric power systems				1547-4 (2011)	
		Part 5: Draft technical guidelines for interconnection of electric power sources greater than 10 MVA to the power transmission grid				1547-5 (withdrawn)	
		Part 6: Draft recommended practice for interconnecting distributed resources with electric power systems distribution secondary networks				1547-6 (2011)	
		Part 7: Draft guide to conducting distribution impact studies for distributed resource interconnection				1547-7 (under development)	

	Part 8: Recommended practice for establishing methods and procedures that provide supplemental support for implementation strategies for expanded use of IEEE standard 1547		1547-8 (under development)		
	1547a: Standard for interconnecting distributed resources with electric power systems – amendment 1		1547-a (under development)		
	IEEE 1547 defines the interconnection of DER with electric power systems (EPSs) and provides the requirements relevant to the performance, operation, testing, safety, and maintenance of the interconnection				
Electric Storage/ Interoperability	IEEE P2030-2	*Draft gide for the interoperability of energy storage systems integrated with the electric power infrastructure:* this standard is developed to provide guidelines for discrete and hybrid storage systems that are integrated with the electric power infrastructure, including end-use applications, and loads	F/D	IEEE Published 2011	Chapter 4/ Chapter 8

Application or service areas	Standard	Type	Issuer	Brief descriptions	Progress state	Issuing date	Location in this book
Electric Storage	IEEE P2030-3	F	IEEE	*Standard for test procedures for electric energy storage equipment and systems for electric power systems applications:* this standard defines test procedures for electric energy storage equipment and systems for electric power systems applications	Published	2011	Chapter 4
Electric Storage	IEC 61850-7-410	F	IEC	*Communication networks and systems for power utility automation-part 7-410: hydroelectric power plants-communication for monitoring and control:* it defines the connection of hydroelectric power plants to the power automation. It specifies the additional common data classes, logical nodes, and data objects required for the application of IEC 61850 in a hydroelectric power plant	Published	61850-7-410 (ed 1.0 2007, ed2.0 2012)	Chapter 4
Electric Vehicle	CHAdeMO	F	CHAdeMO Association	CHAdeMO is the D.C. fast charging standard, which provides EV drivers with an opportunity to charge within 5–10 min for 40–60 km drive, and 80% charge in less than 30 min	Published	2010	Chapter 4

Electric Vehicle	IEC/ISO 15118	*Vehicle to grid communication interface* *Part 1: Definitions and use-case* *Part 2: Sequence diagrams and communication layers* *Part 3: Physical communication layers* The ISO/IEC 15118 series, which are still under development, specify the communication between EVs and EVSE	F	IEC/ISO	Under development Chapter 4
Electric Vehicle	IEC 61851	*Electric vehicle conductive charging system:* the IEC 61851 series specify the requirements for charging EVs by on-board and off-board equipment. IEC 61851 also defines the data interface for EV-EVSE communication via a control pilot wire using a pulse width modulated (PWM) signal with a variable voltage level	F	IEC	Published 61851-1 (2010) 61851-21 (2001) 61851-22 (2001) TRF 61851-1 (2012) TRF 61851 -1;22 (2012) Chapter 4

Application or service areas	Standard	Type	Issuer	Brief descriptions	Progress state	Issuing date	Location in this book
						TRF	
						61851 -22 (2012)	
						61851-31 (2010)	
						61851-32 (2010)	
Electric Vehicle	IEC 62196	F	IEC	Plugs, socket-outlets, vehicle couplers, and vehicle inlets – conductive charging of electric vehicles	Published		Chapter 4
				Part 1: General requirements		62196-1 (2011)	
				Part 2: Dimensional interchangeability requirements for A.C. pin and contact-tube accessories		62196-2 (2011)	
				Part 3: Dimensional compatibility and interchangeability requirements for D.C. and A.C./D.C. pin and tube-type contact vehicle couplers		TRF 62196 -1;2 (2012)	
				IEC 62196 standards define the requirements for plugs, socket-outlets, connectors, inlets, and cable assemblies for charging of EVs		TRF 62196-2 (2012)	
						62196-3 (under devel-op-ment)	

Electric Vehicle	SAE J1711	Recommended practice for measuring the exhaust emissions and fuel economy of hybrid electric vehicles: this standard document specifies the test procedures for HEVs for measuring and calculating the exhaust emissions and fuel economy	F	Society of Automotive Engineers (SAE) International	Published	2010	Chapter 4
Electric Vehicle	SAE J1772	SAE electric vehicle and plug in hybrid electric vehicle conductive charge coupler: the SAE J1772 standard specifies the conductive charge coupler for EV charging in North America, which has been included in the international IEC 62196-2 standard. It covers the general physical, electrical, functional, and performance requirements to facilitate conductive charging using a single phase SAE J1772 connector	F	SAE International	Published	SAE J1772 (2009, R2012)	Chapter 4
Electric Vehicle	SAE J1797	Recommended practice for packaging of electric vehicle battery modules: this document provides practice for common battery designs	F	SAE International	Published	2008	Chapter 4

Application or service areas	Standard	Brief descriptions	Type	Issuer	Progress state	Issuing date	Location in this book
Electric Vehicle	SAE J2288	*Life cycle testing of electric vehicle battery modules*: this document specifies test procedures for life cycles of EV batteries	F	SAE International	Published	2008	Chapter 4
Electric Vehicle	SAE J2289	*Electric-drive battery pack system functional guidelines*: this document specifies guidelines for the design of EV battery systems	F	SAE International	Published	2008	Chapter 4
Electric Vehicle	SAE J2293	*Energy transfer system for electric vehicles* *Part 1: Functional requirements and system architectures* *Part 2: Communication requirements and network architectures* SAE J2293 specifies the requirements for transferring electrical energy from utility to EV in North America. It defines all characteristics of the energy transfer system (ETS), which is responsible for the conversion of A.C. electrical energy into D.C. electrical energy	F	SAE International	Published	J2293-1 (2008) J2293-2 (2008)	Chapter 4

Electric Vehicle	SAE J2344	*Guidelines for electric vehicle safety*: this document provides guidelines for safety operation which should be taken into consideration when designing EVs	F	SAE International	Published	2010	Chapter 4
Electric Vehicle	SAE J2464	*Electric vehicle battery abuse testing*: this document specifies the abuse testing to characterize the response of the rechargeable energy storage system (RESS) to off-normal conditions or environments	F	SAE International	Published	2009	Chapter 4
Electric Vehicle	SAE J2758	*Determination of the maximum available power from a rechargeable energy storage system on a hybrid electric vehicle*: this document describes a test procedure for rating peak power of the rechargeable energy storage system (RESS) used in HEVs	F	SAE International	Published	2010	Chapter 4
Electric Vehicle	SAE J2380	*Vibration testing of electric vehicle batteries*: this document specifies the vibration durability testing of a single battery (test unit) consisting of either an electric vehicle battery module or an electric vehicle battery pack	F	SAE International	Published	2009	Chapter 4

Application or service areas	Standard	Brief descriptions	Type	Issuer	Progress state	Issuing date	Location in this book
Electric Vehicle	SAE J2836	Part 1: Use cases for communication between plug-in vehicles and the utility grid	F	SAE International	Published	J2836-1 (2010)	Chapter 4
		Part 2: Use cases for communication between plug-in vehicles and the supply equipment (EVSE)				J2836-2 (2011)	
		Part 3: Use cases for communication between plug-in vehicles and the utility grid for reverse power flow				J2836-3 (2013)	
		Part 4: Use cases for diagnostic communication for plug-in vehicles				J2836-4 (under development)	
		Part 5: Use cases for communication between plug-in vehicles and their customers				J2836-5 (under development)	
		Part 6: Use cases for wireless charging communication between plug-in electric vehicles and the utility grid				J2836-6 (under development)	

Electric Vehicle	SAE J2847	SAE J2836 specifies the use cases for communications between PEV and utility grid, EVSE, and utility for reverse power flow. Use cases for diagnostics between PEV and EVSE for charge or discharge sessions have been also specified	F	SAE International	Published	SAE J2847-1 (2011)	Chapter 4
		Part 1: Communication between plug-in vehicles and the utility grid					
		Part 2: Communication between plug-in vehicles and off-board D.C. chargers				SAE J2847-2 (2012)	
		Part 3: Communication between plug-in vehicles and the utility grid for reverse power flow				SAE J2847-3 (under development)	
		Part 4: Diagnostic communication for plug-in vehicle				SAE J2847-4 (under development)	
		Part 5: Communication between plug-in vehicles and their customers				SAE J2847-5 (under development)	

Application or service areas	Standard	Type	Issuer	Progress state	Issuing date	Brief descriptions	Location in this book
						SAE J2847 is critical to enable communications between PEVs and the utility grid, the EVSE, customers, and the utility for reverse power flow. Requirements for diagnostics between PEV and EVSE for charge or discharge sessions have been also specified	
Electric Vehicle	SAE J2929	F	SAE International	Published	2013	*Electric and hybrid vehicle propulsion battery system safety standard*: this document specifies safety criteria and requirements for lithium-based rechargeable battery systems	Chapter 4
Electric Vehicle	SAE J2931	F	SAE International	Under development	J2931-1 (2012) J2931-2 (under development) J2931-3 (under development)	*Part 1: Electric vehicle supply equipment communication model* *Part 2: Inband signaling communication for plug-in electric vehicles* *Part 3: PLC communication for plug-in electric vehicles*	Chapter 4

		Part 4: Broadband PLC communication for plug-in electric vehicles	J2931-4 (2012)				
		Part 5: Telematics Smart Grid communications between customers, plug-in electric vehicles (PEVs), energy service providers (ESPs), and home area networks (HANs)	J2931-5 (under development)				
		Part 6: Digital communication for wireless charging plug-in electric vehicles	J2931-6 (under development)				
		Part 7: Security for plug-in electric vehicle communications	J2931-7 (under development)				
		SAE J2931 specifies the requirements for physical layer communications using inband signaling, PLC, broadband PLC, and digital communications between PEV, EVSE, utility, AMI, and HAN					
Electric Vehicle	SAE J2936	Vehicle battery labeling guidelines: this document specifies the labeling guidelines for electrical storage devices	F	SAE International	Published	2012	Chapter 4

Application or service areas	Standard	Brief descriptions	Type	Issuer	Progress state	Issuing date	Location in this book
Electric Vehicle	SAE J2953	*Part 1: Plug-in electric vehicle (PEV) interoperability with electric vehicle supply equipment (EVSE)* *Part 2: Test procedures for the plug-in electric vehicle (PEV) interoperability with electric vehicle supply equipment (EVSE)* SAE J2953 specifies the requirements for PEV interoperability with EVSE and corresponding test procedures	F	SAE International	Under development	J2953-1 (under development) J2953-2 (under development)	Chapter 4
Electric Vehicle	SAE J537	*Storage batteries*: this document specifies the testing procedures of automotive 12 V storage batteries	F	SAE International	Published	2011	Chapter 4
Electric Vehicle	SAE J551/5	*Performance levels and methods of measurement of magnetic and electric field strength from electric vehicles, broadband, 9 kHz to 30 MHz*: this document specifies the test procedures and performance levels for the measurement of magnetic and electric field strengths over the frequency range 9 kHz to 30 MHz and conducted emissions over the frequency range of 450 kHz to 30 MHz	F	SAE International	Published	2012	Chapter 4

Advanced Metering Infrastructure	IEC 62056	Data exchange for meter reading, tariff, and load control	B	IEC	Published 62056	Chapter 5
		Part 21: Direct local data exchange			62056-21 (2002)	
		Part 31: Use of local area networks on twisted pair with carrier signaling			62056-31 (1999)	
		Part 41: Data exchange using wide area networks: public switched telephone network (PSTN) with link + protocol			62056-41 (1998)	
		Part 42: Physical layer services and procedures for connection-oriented asynchronous data exchange			62056-42 (2002)	
		Part 46: Data link layer using HDLC protocol			62056-53 (2013)	
		Part 47: COSEM transport layers for IPv4 networks				
		Part 53: COSEM application Layer			62056-61 (2013)	
		Part 61: Object identification system (OBIS)			62056-62 (2013)	
		Part 62: Interface classes			62056-76 (2013)	
		Part 76: The 3-layer, connection-oriented HDLC based communication profile			62056-83 (2013)	
		Part 83: Communication profile for PLC S-FSK neighborhood networks				

Application or service areas	Standard	Brief descriptions	Type	Issuer	Progress state	Issuing date	Location in this book
		The IEC 62056 suite of standards provides a modern meter reading protocol widely used in Europe, which is based on device language message specification (DLMS), and companion specification for energy metering (COSEM)					
Advanced Metering Infrastructure	IEC 62058	*Electricity metering equipment (A.C.) – acceptance inspection* *Part 11: General acceptance inspection methods* *Part 21: Particular requirements for electromeachnical meters for active energy* *Part 31: Particular requirements for static meters for active energy (classes 0.2, 0.5, 1, and 2)* The IEC 62058 standards specify the acceptance inspection requirements for electricity metering equipment	B	IEC	Published	2008	Chapter 5
Advanced Metering Infrastructure	IEC/TR 62059	*Electricity metering equipment – dependability* *Part 11: General concepts* *Part 21: Collection of meter dependability data from the field*	B	IEC	Published	62059-11 (2002) 62059-21 (2002)	Chapter 5

Domain	Standard	Description		Organization	Status	Reference	Chapter
Advanced Metering Infrastructure	IEC 61107	*Part 31-1: Accelerated reliability testing – elevated temperature and humidity* *Part 32-1: Durability – testing of the stability of metrological characteristics by applying elevated temperature* *Part 41: Reliability prediction* The IEC/TR 62059 standards specify the bill prediction and assessment methods for electricity metering equipment *Data exchange for meter reading, tariff, and load control – direct local data exchange*: the IEC 61107 standard specify the data exchange methods for meter reading, tariff, and load control	B	IEC	Withdrawn	62059-31-1 (2008) 62059-32-1 (2011) 62059-41 (2006) 61107 ed1.0 (1992) 61107 ed2.0 (1996)	Chapter 5
Advanced Metering Infrastructure	NEMA SG-AMI	*Requirements for smart meter upgradeability*: this document defines requirements for smart meter firmware upgradeability for AMI systems	B	National Electrical Manufactures Association (NEMA)	Published 2009		Chapter 5
Advanced Metering Infrastructure	OPEN meter Deliverables	The OPEN meter project is initiated by EU to specify a comprehensive set of open and public standards for AMI. A broad set of domains are addressed in this project, including regulatory environments, smart metering functions, communication media, protocols, and data formats	B	OPEN meter consortium	Published 2010–2011		Chapter 5

Application or service areas	Standard	Brief descriptions	Type	Issuer	Progress state	Issuing date	Location in this book
Advanced Metering Infrastructure	Utility AMI High Level Requirements	*Utility AMI high level requirements*: the high level requirements specified by the UtilityAMI are to provide AMI vendors with some general guidelines when designing or developing AMI systems or components	B	IEC	Published	2006	Chapter 5
Demand Response/Load Control	OpenADR 1.0	*OpenADR 1.0 system requirements specification*: the OpenADR 1.0 specification was accepted as part of the OASIS energy interoperation (EI) standard, which was developed to define an information and communication model to enable demand response and energy transactions	B	UCAIug OpenSG OpenADR task force	Published	2009	Chapter 5
Demand Response/Load Control	OpenADR 2.0	*OpenADR 2.0 profile specification*: OpenADR 2.0 was developed to define profiles which are specific to DR and DER applications. OpenADR 2.0 is comprised with OpenADR 2.0a, OpenADR 2.0b, and OpenADR 2.0c	B	OpenADR alliance		Open ADR 2.0a (2011)	Chapter 5

Demand Response/Load Control	OASIS EMIX 1.0	*OASIS energy market information exchange (EMIX) version 1.0:* this specification has been developed to standardize messages communicating market information including energy price, delivery time, characteristics, availability, and schedules, and so on. All of these various types of information are necessary to enable the full automation of DR decision making	B	Organization for the Advancement of Structured Information Standards (OASIS) Consortium	Published 2011	Open ADR 2.0b (2012) OpenADR 2.0c (under development)	Chapter 5
Communications	ATIS 0900105.07	*Synchronous optical network (SONET) – sub STS-1 interface rates and formats specification:* it defines the data rates and formats specifications for SUB STS-1 SONET interfaces including the definitions and content of the associated overhead channels	D	Alliance for telecommunications industry solutions (ATIS)	Published 2008		Chapter 6

Application or service areas	Standard	Brief descriptions	Type	Issuer	Progress state	Issuing date	Location in this book
Communications	CDMA2000 (TIA/EIA/ IS-2000)	*cdma2000® spread spectrum systems:* is defined by TIA and backward compatible with CDMA-one (IS-95). It is approved standard for ITU's IMT-2000 (also called 3G)	D	Telecommunications Industry Association (TIA)	Published	2000	Chapter 6
Communications	CDMA-one (TIA/EIA/ IS-95-A)	*Mobile station – base station compatibility standard for dual-mode wideband spread spectrum cellular system:* it defines compatibility standard for 800 MHz cellular mobile telecommunications systems and 1.8–2.0 GHz code division multiple access (CDMA) personal communications service (PCS) systems. It is developed by Qualcomm Inc. as IS-95	D	Telecommunications Industry Association (TIA)	Published	1995	Chapter 6
Communications	EDGE	*Enhanced data rates for GSM evolution:* it belongs to the GSM family and is backward compatible. Physical layer is maintained under the same standard as GSM (TS 45.001) by 3GPP	D	International Telecommunication Union (ITU)	Published	1999	Chapter 6

Communications	EPC Class-1 High Frequency (HF)	*EPC Class-1 HF RFID air interface protocol for communications at 13.56 MHz*: it defines physical and logical requirements for a passive-backscatter, Interrogator-talks-first (ITF), radio-frequency identification (RFID) system operating at 13.65 MHz frequency	D EPCglobal	Published	2011	Chapter 6
Communications	GMR/ETSI TS 101 376	*GEO-mobile radio interface specifications*: it defines geostationary earth orbit mobile radio interface (GMR) developed by ETSI and maintained by 3GPP, to support access to GSM/UMTS core networks	D European Telecommunications Standards Institute (ETSI)	Published	2001	Chapter 6
Communications	GSM	*Global system for mobile communications*: it defines standard developed by ETSI and maintained under 3GPP as TS 45.001 (PHY) and TS 23.002 (network architecture). It belongs to second generation cellular systems	D European Telecommunications Standards Institute (ETSI)	Published	1990	Chapter 6
Communications	HD-PLC	*High definition power line communication (HD-PLC)*: it defines PLC technology which uses a high frequency efficient wavelet-OFDM modulation method. The theoretical maximum data transmission rate is up to 210 Mbps	D HD-PLC alliance	Published	2010	Chapter 6

Application or service areas	Standard	Brief descriptions	Type	Issuer	Progress state	Issuing date	Location in this book
Communications	HomePlug Green PHY	*HomePlug Green PHY specification*: it defines a subset of HomePlug AV that is intended for use in the Smart Grid. It has peak rates of 10 Mbit/s and is designed to go into smart meters and smaller appliances such as HVAC thermostats, home appliances, and plug-in electric vehicles. It is interoperable with HomePlug AV and HomePlug AV2 devices and is IEEE 1901 standard compliant. The	D	HomePlug alliance	Published	2012	Chapter 6
Communications	HSPA	*High speed packet data access*: it defines an enhancement of W-CDMA technology. The downlink (HSDPA) enhancement is defined in Release 5 and uplink enhancement (HSUPA) is defined in Release 6. Further enhancements of HSPA technology are in Release 7 and later releases, known as HSPA+	D	3rd Generation partnership project (3GPP)	Published	2002– 2008	Chapter 6
Communications	IEEE 802.11	*Wireless local area networks (WLANs)*: it defines set of technologies developed by IEEE containing many	D	IEEE	Published	1997	Chapter 6

| Communications | IEEE 802.15.1 | Wireless medium access control (MAC) and physical layer (PHY) specifications for wireless personal area networks: it defines wireless personal area network (WPAN) technology for connecting peripherals. It is developed by IEEE and is a base for the widely used bluetooth technology, which has been further enhanced by the bluetooth special interest group | D | IEEE | Published | 2002 | Chapter 6 |
| Communications | IEEE 802.15.4 | Low-rate wireless personal area networks (LR-WPANs): it defines wireless personal area network (WPAN) technology for low-rate transmission is developed by IEEE and a base for many widely used technologies for sensor networks and M2M communication including ZigBee, and so on | D | IEEE | Published | 2003 | Chapter 6 |

amendments for improving security, quality of service, data rates, interworking, and so on. It is a base standard for the widely spread Wi-Fi technology, which is specified by the Wi-Fi alliance

Application or service areas	Standard	Brief descriptions	Type	Issuer	Progress state	Issuing date	Location in this book
Communications	IEEE 802.22	*Cognitive wireless RAN medium access control (MAC) and physical Layer (PHY) specifications:* it defines a standard for wireless regional area network using TV "white space" spectrum and using cognitive radio techniques to allow sharing of geographically unused spectrum. It aims to bring broadband access to hard-to-reach, low population density areas, typical of rural environments	D	IEEE	Published	2011	Chapter 6
Communications	IEEE 802.3ah	*Carrier sense multiple access with collision detection (CSMA/CD) access method and physical layer specifications amendment: media access control parameters, physical layers, and management parameters for subscriber access networks:* it defines physical layer specifications for ethernet links providing 1000 Mbps over PONs up to at least 10 km (1000BASE-PX10) and up to at least 20 km (1000BASE-PX20). It is also known as "ethernet in the first mile"	D	IEEE	Published	2004	Chapter 6

Communications	IEEE 802.3av	*Carrier sense multiple access with collision detection (CSMA/CD) access method and physical layer specifications – amendment 1: physical layer specifications and management parameters for 10 Gb/s passive optical networks:* it defines physical layer specifications and management parameters for 10 Gbps ethernet passive optical networks (10 GE-PONs)	D	IEEE	Published	2009	Chapter 6
Communications	ISO/IEC 18092	*Near field communication – interface and protocol (NFCIP-1):* it defines communication modes for near field communication interface and protocol (NFCIP-1) using inductive coupled devices operating at the center frequency of 13.56 MHz for interconnection of computer peripherals. It is also known as "ECMA-340"	D	ISO	Published	2004	Chapter 6
Communications	ITU-T G.651.1	*Characteristics of a 50/125 μm multimode graded index optical fiber cable for the optical access network:* it is ITU-T recommendation for a quartz multimode fiber to be used for the access network in specific	D	International Telecommunication Union (ITU)	Published	2007	Chapter 6

Application or service areas	Standard	Brief descriptions	Type	Issuer	Progress state	Issuing date	Location in this book
		environments. The recommended multimode fiber supports the cost-effective use of 1 Gbit/s Ethernet systems over link lengths up to 550 m, usually based upon the use of 850 nm transceivers					
Communications	ITU-T G.652	*Characteristics of a single-mode optical fiber and cable*: it is ITU-T recommendation describes the geometrical, mechanical, and transmission attributes of a single-mode optical fiber with wavelength around 1310 nm, but can be also used in the 1550 nm region	D	ITU	Published	2000	Chapter 6
Communications	ITU-T G.707	*Network node interface for the synchronous digital hierarchy (SDH)*: it is ITU-T recommendation for network node interface for the synchronous digital hierarchy (SDH) networks	D	ITU	Published	2007	Chapter 6
Communications	ITU-T G.783	*Characteristics of synchronous digital hierarchy (SDH) equipment functional blocks*: it defines a library of basic building blocks and set of rules for digital transmission equipment creation	D	ITU	Published	2006	Chapter 6

Communications	ITU-T G.803	Architecture of transport networks based on the synchronous digital hierarchy (SDH)	D ITU	Published 2000	Chapter 6
Communications	ITU-T G.959	Optical transport network physical layer interfaces: it is ITU-T recommendation ITU-T G.959.1 provides physical layer inter-domain interface specifications for optical networks which may employ wavelength division multiplexing (WDM)	D ITU	Published 2008	Chapter 6
Communications	ITU-T G.983.x	Broadband optical access systems based on passive optical networks (PONs): it defines a series of recommendations for broadband passive optical networks	D ITU	Published 2005	Chapter 6
Communications	ITU-T G.984.x	Gigabit-capable passive optical networks (GPONs): it defines a series of recommendations for GPON access networks	D ITU	Published 2008	Chapter 6
Communications	ITU-T G.9955	Narrow-band OFDM power line communication transceivers: it defines an ITU-T recommendation containing the physical layer specification for narrowband OFDM PLC communications via alternating current and direct current electric power lines over frequencies below 500 kHz. It addresses grid to utility meter applications, AMI, and other Smart Grid applications	D ITU	Published 2011	Chapter 6

Application or service areas	Standard	Brief descriptions	Type	Issuer	Progress state	Issuing date	Location in this book
Communications	JIS X 6319-4	*Specification of implementation for integrated circuit(s) cards – Part 4: High speed proximity cards*: it defines the physical characteristics, air interface, transmission protocols, file structure, and commands of high-speed contactless proximity integrated circuit cards. This Japanese standard Is also known as "Felica"	D	Japanese Industrial Standards (JIS)	Published	2010	Chapter 6
Communications	LTE	*Long term evolution (LTE)* is a technology developed by 3GPP specified in Release 8 and further enhanced in later releases. It is a candidate for ITU's IMT-advanced (also called 4G) as its further enhancements (LTE advanced) satisfy the requirements defined in ITU's IMT-advanced	D	3rd Generation Partnership Project (3GPP)	Published	2007	Chapter 6
Communications	NG-PON2	*NG-PON2: Next generation passive optical network 2*: it defines PON with capacity at least 40 Gbps and which can deliver services of 1 Gbps. It is proposal for future ITU standardization	D	Full service access network (FSAN)	In development	–	Chapter 6

Domain	Standard		Status	Year	Reference	Description
Communication	IEEE P1901	D IEEE	Published	2010	Chapter 6	*IEEE standard for broadband over power line networks: medium access control and physical layer specifications*: it defines a standard for high speed (> 100 Mbps at the physical layer) communication over power line. The standard uses transmission frequencies below 100 MHz. It is key standard for broadband PLC
Communications	UMTS	D 3GPP	Published	2000	Chapter 6	*Universal mobile telecommunications system*: it is defined by 3GPP in Release 99 and approved by ITU's IMT-2000 (also called 3G). Radio network technology is W-CDMA
Communications	Weightless	D Weightless SIG	In development		Chapter 6	*Weightless*: it defines an application agnostic standard for long-range machine-to-machine communication operating in TV "white space" spectrum supporting very large number of devices and long sleep cycle capability

Application or service areas	Standard	Brief descriptions	Type	Issuer	Progress state	Issuing date	Location in this book
Communications	WiMAX	*Worldwide interoperability for microwave access (WiMAX)*: it defines a technology based on IEEE 802.16 family of standards and is maintained and promoted by WiMAX forum. Mobile WiMAX is a candidate for ITU's IMT-Advanced (also called 4G) technology together with LTE	D	IEEE/WiMAX	Published	2001–2005	Chapter 6
Communication	Common Industrial Protocol (CIP)	An industrial protocol for industrial automation applications, which is supported by Open DeviceNet Vendors Association (ODVA)	D	Open DeviceNet Vendors Association (ODVA)	Published	2002	Chapter 7
Communication	CC-Link	High-speed field network able to simultaneously handle both control and information data	D	CC-Link Partner Association (CLPA)	Published	2000	Chapter 7
Data Security and Functional Safety	CC-Link Safety	Safety extension of CC-Link	E	CLPA	Published	2005	Chapter 7
Data Security and Functional Safety	CIP-Safety	Safety extension of CIP	E	Open DeviceNet Vendors Association (ODVA)	Published	2005	Chapter 7
Communication	EtherCAT	An open real-time Ethernet network originally developed by Beckhoff, which sets new standards for real-time performance and topology flexibility	D	Beckhoff	Published	2003	Chapter 7

Category	Standard	Description	Type / Organization	Status	Year	Reference
Data Security and Functional Safety	IEC 61508	Functional safety of electrical/electronic/programmable electronic safety-related systems	E IEC	Published	2010	Chapter 7
Data Security and Functional Safety	IEC 62351	Standard which handles the security of TC 57 series of protocols including IEC 60870-5 series, IEC 60870-6 series, IEC 61850 series, IEC 61970 series, and IEC 61968 series	E IE	Published	2007–2010	Chapter 7
Communication	ISA100.11a	Wireless systems for industrial automation: process control and related applications	D International Society of Automation (ISA)	Published	2009	Chapter 7
Communication	Powerlink	A deterministic real-time protocol for standard Ethernet	D Ethernet Powerlink Standardization Group (EPSG)	Published	2001	Chapter 7
Data Security and Functional Safety	Powerlink safety	Safety extension of Powerlink	E EPSG	Published	2006	Chapter 7
Communication	PROFIBUS	A standard for field bus communication in automation technology	D Twenty-one companies and institutes devised a master project plan called "field bus"	Published	1987	Chapter 7
Communication	PROFINET	The open industrial Ethernet standard of PROFIBUS and PROFINET International (PI) for automation	D PNO (PROFIBUS National Organization)	Published	2001	Chapter 7

Application or service areas	Standard	Brief descriptions	Type	Issuer	Progress state	Issuing date	Location in this book
Data Security and Functional Safety	PROFIsafe	Safety extension of PROFIBUS and PROFINET	E	Siemens	Published	1999	Chapter 7
Data Security and Functional Safety	TwinSAFE	Safety extension of EtherCAT	E	Beckhoff	Published	2007	Chapter 7
Communication	WirelessHART	A wireless sensor networking technology based on the highway addressable remote transducer protocol (HART)	D	Highway Addressable Remote Transducer (HART) Communication Foundation (HCF)	Published	2007	Chapter 7
Interoperability	ANSI/ASHRAE 135-2008/ISO 16484-5 BACnet	BACnet a home and building automation standard, which defines information model and messages for building system communications at the consumer end. It integrates a range of networking technologies to achieve scalability from very small systems to multi-building operations that span wide geographic areas using IP protocols	D	ANSI/ASHRAE/ISO	Published	2008	Chapter 8

Interoperability	ANSI/CEA 709 and CEA 852.1	ANSI/CEA 709 and ANSI/CEA 852.1: Local Control Network protocol suite is a general purpose LAN protocol. It is used for variety of applications such as electric meters, street lighting, and in home building automation	D	American National Standards Institute (ANSI)/ Consumer Electronics Association (CEA)	Published	2010	Chapter 8
Interoperability	ANSI/CEA 709.1-B	Part 1 of the standard is about a specific physical layer protocol Part 2 is an explicit physical layer protocol designed for use with ANSI/CEA 709.1-B-2002 Part 3 is also is an explicit physical layer protocol designed for use with ANSI/CEA 709.1-B-2002 Part 4 is a protocol offers a way by using a user datagram protocol (UDP) to tunnel local operating network messages through an IP, thus making a way to produce larger internetworks	D	ANSI/CEA	Published	709.1-B (2002) 709.2-AR (2006) 709.3 R (2004) 709.4 (1999)	Chapter 8
Interoperability	IEC 60870-6/ TASE.2	IEC 60870-6/TASE.2 is an open, and mature standard that is widely implemented with compliance testing. This is part of the IEC 60870 Suite included in PAP14	D	IEC	Published	2005	Chapter 8

Application or service areas	Standard	Brief descriptions	Type	Issuer	Progress state	Issuing date	Location in this book
Interoperability	IEC 61968/61970 Suites	IEC 61968/61970 are open standards that are widely implemented and maintained by an SDO with support from a users group. They are part of PAPs relating to integration with IEC 61850 and multispeak	D	IEC	Published	–	Chapter 8
Interoperability	IEEE C37.118	IEEE C37.118 is an open standard, widely developed, and maintained by an SDO. It includes some requirements for communications and measurement. Currently, it is being updated by IEEE power system relaying committee (PSRC) relaying communications subcommittee working group H11	D	IEEE	Published	2005	Chapter 8
Interoperability	IEEE 1686-2007	The IEEE 1686-2007 is an IEEE standard for substation intelligent electronic devices (IEDs) cyber security capabilities	D	IEEE	Published	2007	Chapter 8

| Interoperability | ISO 19136 (Open Geospatial Consortium Geography Markup Language (GML)) | GML is an open standard, which is in compliance with ISO 19118 for the transport and storage of geographic information modeled according to the conceptual modeling framework used in the ISO 19100 series of International Standards. GML is widely used along with supporting open source software. It is also used in disaster management, home/building, and equipment location information bases | D | Open geospatial consortium (OGC) | Published | 2007 | Chapter 8 |
| Interoperability | NIST Special Publication (SP) 800-53, NIST SP 800-82 | This family of standards is open source, which is developed by NIST. SP800-53 describes security measured required for government standards of United States. SP800-82 is in the completion process, which specifies security specially for industrial control systems, such as the electric grid | D | National Institute of Standards and Technology (NIST) | Published | 2005 | Chapter 8 |

Application or service areas	Standard	Brief descriptions	Type	Issuer	Progress state	Issuing date	Location in this book
Integration of Renewable Resources	UL 1741	*Standard for safety inverters, converters, controllers, and interconnection system equipment for use with distributed energy resources:* the UL 1741 standard defines requirements for inverters, converters, charge controllers, and interconnection system equipment (ISE) of distributed energy resources (DERs). UL 1741 is intended to supplement and be used in conjunction with IEEE 1547	F/D/A	Underwriters laboratories (UL)	Published 1999		Chapter 9

PEM, proton exchange membrane; PV, photovoltaic; UV, ultra violet; PV-IES, photovoltaic individual electrification systems; CSP, concentrated solar power; TASE, telecontrol application service element; XML, extensible markup language; EV, electric vehicle; HEV, hybrid electric vehicle; PLC, power line communication; AMI, advanced metering infrastructure; IP, internet protocol; S-FSK, spread-frequency shift keying; EU, European Union; DR, demand response; STS-1, synchronous transport signal 1; GSM, global system for mobile communication; EPC, evolved packet core; UMTS, universal mobile telecommunications system; OFDM, orthogonal frequency division multiplexing; HVAC, heating ventilation and air Conditioning; W-CDMA, wideband code division multiple access; SIG, special interest group; HCF, HART communication foundation; ASHRAE, American society of heating, refrigeration and air-conditioning engineers; LAN, local area network; SDO, Standards Developing Organizations; PAP, priority action plan.

Index

Smart Grid Standards: Specifications, Requirements, and Technologies, First Edition. Takuro Sato,
Daniel M. Kammen, Bin Duan, Martin Macuha, Zhenyu Zhou, Jun Wu, Muhammad Tariq and Solomon Abebe Asfaw.
© 2015 John Wiley & Sons, Ltd. Published 2015 by John Wiley & Sons, Ltd.

Printed and bound by CPI Group (UK) Ltd, Croydon, CR0 4YY